The Holistic Inspirations of Physics

The Holistic Inspirations of Physics

THE UNDERGROUND HISTORY OF ELECTROMAGNETIC THEORY

VAL DUSEK

Rutgers University Press

New Brunswick, New Jersey, and London

Library of Congress Cataloging-in-Publication Data

Dusek, Val, 1941–
    The holistic inspiration of physics : the underground history of
electromagnetic theory / Val Dusek.
        p.   cm.
    Includes bibliographical references and index.
    ISBN 0–8135–2634–5 (alk. paper). — ISBN 0–8135–2635–3 (pbl. : alk.
paper)
    1. Electromagnetic theory—History.   I. Title.
QC670.D88   1999
530.14'1—dc21   99–14095
                                                                      CIP

British Cataloging-in-Publication data for this book is available from the British Library

Manufactured in the United States of America

For my daughter, Athena, in memory,
January 31, 1998

CONTENTS

PREFACE AND ACKNOWLEDGMENTS

This work is largely based on the research of specialized scholars in Chinese, Renaissance, and German studies, and in the history of early modern science. I believe it is valuable as a comparison and synthesis of diverse strands of work previously available only in technical journals and in specialist works that have not, for the most part, been philosophically or sociologically analyzed.

The historical research of the 1960s broadened our conception of the nature and roots of science and stimulated this project. This included works of Joseph Needham on Chinese science, of Frances Yates on Renaissance occult and hermetic science, and of Pearce Williams on the influence of German Romantic philosophy and science on the origins of field theory. My survey of Chinese science and society is indebted not only to Joseph Needham's staggeringly large achievement, but also to Benjamin Schwartz, A. C. Graham, and Chad Hansen, who each synthesized the last few decades of research on ancient Chinese philosophy, and to Livia Kohn and Nathan Sivin, who have made results of their original research on Chinese religion and science accessible to nonsinologists. Chance contacts with D. P. Walker and Charles Webster led me to study their accounts of early modern occultism and science. While reading Pearce Williams's biography of Michael Faraday, I was struck by the fact that the Chinese, hermetic, and Romantic alternative sciences all made important contributions to the prehistory of field theory, and that the common features of these approaches contributed to an important part of contemporary physics.

Also in the background of this work are varied studies of the social foundations of science by such Marxist writers as Boris Hessen, Edgar Zilsel, and Franz Borkenau, as well as work by Robert K. Merton, Christopher Hill, and Charles Webster on the relationship of science to the Puritan Revolution in seventeenth-century England. On the notion of the "underground," my reading of Friedrich Heer's *Intellectual History of Europe* long ago was key. My teacher, Robert Palter, provided me with an inspiring example of a meticulous and thorough philosopher and historian of science, though he will undoubtedly disagree with many of the theses here.

In the late 1970s and early 1980s the pathbreaking work of a number of feminist scholars reoriented our conception of science and its social context in relation to gender. The work of Evelyn Fox Keller on the role of gender metaphors for the relation of the scientist to nature and of Carolyn Merchant on the gendered transformation of the concept of nature in the early modern period were particularly illuminating to me. Keller's treatment of alternative approaches to science was also suggestive in that it placed the value of proliferation and diversity in science within a social and gender context.

Much of my adult education occurred at the Boston Colloquium for the Philosophy of Science. Here, scientists and humanists from all over the world have presented work that dealt with science in a broad social and political context, as well as with philosophy and science outside the mainstream— including Spinoza, Goethe, Hegel, and Marx in relation to natural science. Robert S. Cohen and the late Marx Wartofsky supported a "multicultural" approach to science, long before that term existed, and sponsored investigations of the social context of science combined with a commitment to scientific objectivity.

To my spouse, Christy Hammer, I owe appreciation for the encouragement, confidence, and inspiration finally to put this work on paper, as well as for helpful criticism and suggestions. She encouraged me to dictate a first draft of this work to her, despite her own full schedule of work and dissertation. I thank Franklin Rosemont Jr., for reading and gently criticizing the sections on China. My colleagues Bill de Vries and Bill Woodward read drafts of the manuscript and made many helpful suggestions. I gained insights from Hisa Kuriyama and Jan Golinski, historians of science, on traditional Chinese and late-eighteenth-century science, respectively. I am grateful for the support given by the University of New Hampshire Center for the Humanities in the form of a Senior Fellowship, which allowed me to finish this work. Outside readers' comments were exceedingly helpful, and I thank Rutgers University Press editor Doreen Valentine, who saw potential in the manuscript and maintained great enthusiasm and support. Diane Grobman did an extraordinary job of checking names and improving style. I wish to thank Nancy Hammer for preparing the name index. Finally, I owe thanks to my parents, who supported and financed my education and library, and who always encouraged me to pursue subjects and a career of my own choice, even if my interests strayed far from the family vocation of the visual arts. This work arose in part, I believe, from the conflict between, and search for synthesis of, the aesthetic and scientific interest in nature.

The Holistic Inspirations of Physics

CHAPTER 1

Introduction: Historical Undergrounds, Holistic Theories, and Classical Physics

*A*lternative, holistic worldviews played a role in the development, understanding, and application of electromagnetic field theory. These holistic approaches include the only three traditions radically different from the mainstream mechanistic image of science: medieval Chinese, Western Renaissance occultist, and German Romantic. I will situate these "alternative" approaches to science in their social context and background, and will trace their connection with components of "accepted" physical science in relation to a number of social movements and philosophical theories, uncovering parallels between modern physics and holistic thinking.

While several popular books have claimed parallels between modern physics and Eastern philosophy, none deals with the actual, historical influences of both traditional Chinese thought and nonmechanistic, holistic Western thought on the philosophies of the scientists who developed field theory. Likewise, a number of works have emphasized the nonmechanistic and nonreductionistic interpretations of twentieth-century quantum physics and chaos theory, but none has shown the extent to which classical field theory is a product of organic and holistic philosophies and frameworks. Although historians of science have done detailed work on each of the individual traditions and periods treated here, there is no comprehensive work that emphasizes the structural similarities of the forms of thought of these traditions, noting how each presented alternative views of nature to the mainstream, Western tradition of atomistic, mechanistic science. There has not been a synoptic work linking these three traditions in terms of their contributions, both in theory and experiment, to the present science and technology of a central part of classical and modern physics—electromagnetic field theory—within a philosophical analysis and defense of the notions of holism and organic unity in science.

The Chinese, Renaissance hermetic, and Romantic traditions do not deal with atoms or atomistic analysis, but with the relation of the whole to its parts, giving priority to the whole and claiming that the parts reflect or mirror the whole, rather than that the whole is constructed from its parts. They all share this holistic, monistic (undivided whole) account of reality. These similarities can be appreciated by examining the role of analogy (relationships of similarity) and polarity (opposing poles of forces, or opposing qualities in general) in the conceptual structure of the three alternative sciences. The social background of holistic ideology is defined in terms of nonindividualistic community and is based in peasant and communal societies. The intellectual contribution of these social groups is marginalized as "primitive" in the thought dominant among the urban and elite. The mediation of the ideas of peasant and communal societies into the more elite and formalistic (and individualistic) society, especially in recent European history, is often downplayed or ignored.

Unlike traditional Marxist history of science, this work focuses on the contribution of peasant thought to ways of conceptualizing, rather than on artisan and bourgeois thought. A leading Marxist historian identified the "mechanicks" with the mechanical worldview.[1] Marx made the famous claim that René Descartes "saw with the eyes of the manufacturing period."[2] Marx and classical Marxists, in contrast to some later Maoist and third world Marxist thinkers, tended to portray peasants as reactionary compared to the urban working class, consistent with Marx's references to "the idiocy of rural life" and to rural France as a great sack of potatoes.[3] My own claim is that, even if the mechanical devices and activities of artisans encouraged a mechanistic view, the peasant sources and background of urban artisans could have contributed to an holistic viewpoint.

Feminist work on the history of science emphasizes the male nature of science, the exclusion of women from the scientific professions, and the biographies of the heroic few who managed to succeed as scientists, mostly during the last century or so. The present work looks at earlier science and does not focus on women working in the science or protoscience of that period, but examines those ideologies and worldviews which were, at least on the theoretical level, more friendly to the "female" principle (such as Taoism, Renaissance occultism, and Romanticism) and the extent to which women's culture helped supply (often without attribution) the philosophical background ideas that were absorbed into the mainstream science of field theory.

The Chinese contributions included invention of the compass and discovery of the earth's magnetic field and of magnetic declination. The Chinese knew about variations in the earth's magnetic field before Europeans even knew about the compass. The contributions of Renaissance European hermetic and occultist science include support for Copernicus's sun-centered universe, and Tycho Brahe's rejection of the idea of crystalline spheres holding the planets. Neo-Pythagorean concepts of the harmony of the spheres and mathematical mysticism contributed to Johannes Kepler's first attempt at a truly physical explanation (going beyond a purely mathematical description) of the motion of the planets, based on a theory of magnetism and on remnants of astrology. Alchemical ideas of active forces and "occult" influences contributed to Isaac Newton's theory of gravitation force as action at a distance, rather than as a result of purely mechanical collisions and contact action. German Romantic science of the early nineteenth century contributed to Hans Christian Ørsted's discovery of electromagnetic induction (the interaction between electricity and magnetism) and to the formulation of the electromagnetic field concept by Michael Faraday. Both were influenced by German Romantic *Naturphilosophie* (nature philosophy), with its conception of the unity of the forces of nature.

The Holistic versus the Atomistic Conception of the Objects of Science

Scientific theories not only explain and predict particular observations or experimental results but also involve images of reality. Theories may assume that the world is made of separate parts or is a single continuous stuff. Such views are sometimes called metaphysical assumptions or presuppositions. All that is meant by *metaphysical* here is the description of reality, of what really exists. Different scientific theories can have different metaphysical presuppositions.

The popular image of modern science is atomic. The discovery of the atomic building blocks of matter is believed to provide the firm foundation to our understanding of the physical world and forms the basis of reductionistic thinking in the West. For instance, logicians speak of "atomic" and "molecular" sentences where they mean simple and compound sentences. The atomistic beads on a string, or what Ernst Mayr calls "beanbag genetics" model, of the human genome, has dominated genetic analysis.

The usual story of the rise of modern science from the sixteenth to eighteenth centuries is of the rise of the atomistic, mechanical model of nature.[4] This model is supplemented by a model of causal interaction in terms of pushes or pulls that can take the form of physical collisions (contact actions) in the simplest cases, or with addition of exertion of additional forces (action at a distance) in more developed versions. The rise of the mechanistic model in the theory of nature has been paralleled and reinforced by the mechanization of technology. In the twentieth century, the atomic model has undergone major supplementation and transformation. Fields were added to the particle model, and particles and fields have become aspects of a perplexing physical reality. The straightforward mechanical causality that was part of the earlier atomic model was replaced by indeterminacy and statistical causality in quantum theory.

Nevertheless, the strength of the atomic mechanical image of science remains great. Many physicists working in atomic, nuclear, and particle physics discuss and think of the physical realm in terms of particles or atomlike entities, despite the notion that fields are equally real and important. The very term *elementary particle physics*, for a theory that holds fields and particles as interchangeable and complementary, shows the deep cultural bias toward atomistic as opposed to field thinking even in the specialty that has most developed field theory. Popularizations of quarks as the fundamental particulate building blocks of matter reflect the ease with which the idea of quarks is assimilated into the traditional notion of solid, minute atomic particles. In fact, quarks, like other particles in modern physics, have both particulate and wave or field aspects.

The popularity of Richard Feynman's particle formulation of quantum electrodynamics (and his revealing talk of "partons" instead of "quarks")[5] and the relative unpopularity of Julian Schwinger's equivalent field formulation, reflect not merely personality differences and differences in concreteness of exposition by these two physicists, but also the bias toward particle thinking in the West. Schwinger formulated his field ideas in extremely compact and abstract mathematical expressions, while the gregarious and particle-oriented Feynman had an unmatched talent for physical intuition, striking pictures, and informality of presentation.[6] A more difficult entanglement of the personal, political, and theoretical is the earlier field-oriented work of Pascual Jordan. Jordan focused on a field conception of quantum mechanics and approached matter in terms of quantitized matter waves, claiming to have "derived the very existence of particles." This claim was left out of later textbooks, "perhaps as 'too philosophical,'"[7] yet his brilliant work as an early forerunner of quantum field theory is insufficiently credited, due in part to his unimpressive oral presentations, but due more to his biological and parapsychological speculations, which sometimes verged on the crackpot, and due also to his vociferous support of the Nazis.[8]

The power of the atomic model in science owes as much to nonscientific social background as it does to actual discoveries directly supporting the notion.[9] Concepts of organic unity of physical entities (or of the universe) in major, alternative traditions were important in the development of science. In physics, the idea of forces and fields has depended on such philosophical views of the world.

Numerous books appeared in the 1970s and 1980s emphasizing comparisons between modern subatomic physics and various Asian philosophies, starting with (and imitating) Fritjof Capra's *Tao of Physics*.[10] Some eminent physicist-popularizers with more philistine arrogance than intellectual care have, with ridicule directed at the more superficial analogies, dismissed all such analogies.[11] What has been overlooked in both sympathetic comparisons of modern physics with Eastern thought and angry rebuttals of them, however, is the fact that in earlier Western science there were philosophical traditions quite different from the dominant one of atomic mechanism.

In the very beginning of Western philosophy, the pre-Socratic philosopher-scientists of ancient Greece presented a number of different philosophical alternatives to atomism. The atomic theory of Democritus was heralded as the most significant until twentieth-century physics forced recognition that the other alternative models of nature produced by the earliest Greeks have relevance to contemporary science. These alternative models include Anaxagoras's image of the world as continuous and undivided, Parmenides's view of the world as an unchanging unity (the One), Heraclitus's view of the world as pure process, and Pythagoras's world as made of abstract mathematical structures. These ancient views have all returned as relevant to modern physics. Indeed, two founders of quantum mechanics turned to the study of the earliest Greek philosophers in search of suggestions for understanding the perplexing new physics.[12]

The historical image of the rise of modern science as the rise of atomic mechanism has neglected equally important Western and Eastern scientific traditions of organic holism. Many Renaissance scientists who first investigated nature relatively independent of theological constraints saw the world as an organic unity held together by occult influences. Although this image of nature was soon replaced by that of atomic mechanism, the forerunners of the idea of the physical field drew on ideas of the organic unity of nature.

During the Romantic era there was a revival of concepts of holism and of nature as an organic unity. The Romantics saw the unity and holism of living organisms as the proper model for physics—just the opposite of the Enlightenment, mechanist ideal of Thomas Hobbes, Julien Offroy de La Mettrie, and others to explain (and explain away) biological organization in terms of mechanistic physics. Romantic ideas of physical unity were important heuristics for the discovery of electromagnetic induction and the development of the concept of the electromagnetic field.[13] Romantic ideas of nature as an organic unity were again submerged with the development of mathematical field theory in the late nineteenth century; however, new versions of holism arose with the development of quantum mechanics and subatomic physics in the twentieth century.[14]

Beyond parallels and analogies between Eastern philosophy and modern physics, there are a number of actual early results of Eastern ways of thought that contributed to the development of the modern theory of electromagnetism in Western science. For example, the discovery of the compass and of the distribution of magnetism upon the globe occurred far earlier in China than in the West. The notion of Earth's magnetic field probably could not have been discovered on the basis of the mechanistic atomism of the West alone. German philosopher and mathematician Gottfried Wilhelm Leibniz became very interested in Chinese language, philosophy, and the book of divination known as the *I Ching*. Leibniz was the source of many of the ideas that would develop into field theory a century after his death.

The Social Background of Holistic Thinking: The Underground of History, Including Peasants, and Women

The atomic, mechanistic worldview is strongly reinforced by the existence of individualistic and competitive societies. From Abdera in ancient Greece to Britain and France in the seventeenth and eighteenth centuries to most of the developed world today, the rise of the atomic, mechanistic conception of science has accompanied the rise of individualistic, competitive societies.[15]

Holistic, nonatomic models of reality originated in movements of those outside of or excluded from the modern, individualistic society, which is dedicated and indebted to the domination and exploitation of nature. Traditional, nonindustrial, peasant societies and the cultures of women worldwide have been the source of holistic and organic views that periodically have invaded the dominant individualistic society and worldview. Two major periods when the views of the excluded and oppressed outsiders entered into the literate discourse of the social establishments were the Renaissance and the French Revolution. During both of these periods the dominant structures of society broke down, the relations of the social classes became more fluid, and the holistic thought of women and peasants was mediated by déclassé intellectuals.

Although there are some significant exceptions, the scientists and philosophers who contributed to the holistic tradition were mostly upper-class males. However, these writers were themselves receptive to the thought of women and peasants, both refracting and refining this thought through the forms of the high culture of their day while expressing resistance to it. During the last three decades a great outpouring of work on the history of popular culture, peasant and working-class communities, and women has occurred. Many historians, literary theorists, and others turned their focus to those who were "peoples without history,"[16] or "hidden from history,"[17] or wrote "history from the bottom up."[18]

Meanwhile, the works of the Continental theorists (such as the French *Annales* school; the deconstructionists following Jacques Derrida; cultural-studies theorists and anthropologists following Mikhail Bakhtin;[19] and various individual writers on peasant and popular thought, such as Carlo Ginzburg,[20] Antonio Gramsci,[21] and Hans Mayer[22]) were translated and absorbed into the American scholarly canon. The *Annales* discuss history of the *longue durée*, the "long period," emphasizing economic and agricultural eras, and "the history of mentalities," often stressing peasant thought.[23] Bakhtin reveals the presence of peasant and popular forms of expression in literature. Gramsci, an Italian Marxist, discusses the Marxist social movement in relation to such earlier movements as the Reformation and highlights the role of "organic intellectuals." Gramsci's term *hegemony*, used with respect to the dominant ruling-class belief structure, became popular in postmodern circles, even if his writings as a whole are not studied so widely and seriously as the frequency of the word *hegemony* in literary discourse would suggest.[24] Many postmodernists and practitioners of cultural studies discuss the role of the "Other," of Hegelian origin and used by Derrida, but probably borrowed in its modern form from Simone de Beauvoir's *Second Sex*[25] and used by many writers subsequently to denote women, gays, people of color, and the colonized.

One scholar who earlier dealt with themes of illiterate peasant and lower-class cultures in their relation to "higher" intellectual culture was Friedrich Heer, an Austrian Roman Catholic professor of the history of ideas. Heer, with his notion of the "underground" of history, gave an account of the mediation between the thought of excluded groups (such as peasants, the lower classes, and women) and the high culture of a period, sometimes subverting and always energizing the latter, but often simultaneously being partially co-opted by it. Heer's use of the term *underground* has a surprisingly contemporary ring, probably stemming both from Dostoyevsky's "underground man" and the European Resistance movements during World War II.

Heer's major works appeared decades before the widespread interest in the thought of subordinate groups;[26] however, his views have had surprisingly little impact on historians and other writers in the English-speaking world. He gives a fascinating account of the absorption and transformation of lower-class, popular culture by bohemian intellectual movements and aristocratic society throughout the course of European history. His discussion of the interaction of lower-class with elite culture is much broader in scope than even that of Bakhtin, Ginzburg, or Lucien Febvre,[27] who, despite their broad knowledge, limit themselves to the use of peasant culture and expressions by a single writer (Rabelais in the case of Bakhtin and Febvre) or a single movement (witch cults in the case of Ginzburg). Heer, in contrast, deals with the entire scope of European intellectual history from 200 C.E. to the present.

In *The Medieval World*, Heer included a chapter on outsiders, Jews, and women, emphasizing themes that would become fashionable only decades later. Although he is strongest on the medieval period, he has also written on the nineteenth and twentieth centuries,[28] on youth movements, and on the comparison of religious and political terror. He wrote about the importance of Hildegard of Bingen, an extraordinary medieval writer of hymns, visions, medical tracts, and historical speculation three decades before the feminist movement revived interest in her work in the 1980s.[29]

By the underground Heer means the currents of the popular culture that are not under the hegemony of the elite and high culture. The underground includes not only the archaic, peasant, nonlettered culture, but also the culture of the bohemians, those intellectuals who are not part of the institutions of the official intellectual culture and who are in contact with the lower-class elements. This lumpen intelligentsia acts as a mediator between the lower-class culture and the literate intellectual culture.[30] What Heer shows over and over again, sometimes inadvertently, is how a brittle ruling-class culture throughout the feudal and capitalist eras could maintain and revitalize itself only by co-opting the energies of the popular culture and bending them to the purpose of the rulers. Heer fears the total eruption of popular culture through the facade of upper-class culture. He speaks of the tremendous resentment of the masses, and sees Nazi and Stalinist totalitarianism as the result of the genuine surfacing of popular energies. Although he is politically on the side of the liberal wing of the aristocracy, his fascination with popular and underground culture gets the best of him, and the significance of his history far outruns his ideological conclusions.

During the feudal period, and during the capitalist period before the nineteenth century in Britain and America, genuine folk and proletarian culture was rarely documented. When it was, it was generally written by men of the intermediate stratum who had contact with both the lower classes and the literary culture of the upper classes. When Marx spoke of the leap from prehistory to history, his remarks could be applied to the vast majority of humanity, who are still in prehistory, even in the simple sense that their lives are not recorded in writing. Their words die with them. The underground culture, insofar as it was recorded, was mediated by various artisans and bohemians. While occasionally these were males of peasant and working-class origin, usually they were former members of the aristocracy or the intelligentsia who, because of economic crisis, came into close contact with the lower classes.

Throughout the late Middle Ages and early modern period these economi-
cally unfortunate intellectuals mixed with the oppressed classes and produced
eschatological, utopian, and revolutionary theories, which were not generally the
direct recording of the ideas, myths, and programs for action of the oppressed, of
peasants and proletarians. Instead, the genesis of underground culture was the result of the mediating individual who had somehow received an education in the
high culture of the day as well as been exposed to the moods, situations, and problems of the oppressed. In one sense, the underground culture was a reflection of
lower-class culture, but primarily in terms of mood and attitude. The particular
theoretical contents of the countercultural system were not, however, a simple borrowing of peasant myths or proletarian politics. The countertheories were an expression of problems and prereflective orientations as modified and restructured
in terms of concepts and ideas borrowed from the literate ruling culture.

Common features pervade Heer's underground culture. Since Heer is fearful of and not sympathetic to these features, he has not intended to produce an
idealized or admirable image of the counterculture. The common elements that
Heer finds, despite his antagonistic attitude, are therefore all the more significant.
One of these features is the eschatological consciousness.

For the purpose of social analysis, Heer's treatment of the underground is
not sufficiently specific on the differences between historical periods and social
strata. With different historical periods—the decline of the Roman Empire, the later
Middle Ages, the Renaissance, the early-nineteenth-century Romantic movement,
and the fin-de-siècle movements of nihilism and anarchism—both the nature of
the lower classes (whether peasants, slaves, or salaried workers), and the specific
form of the bohemian mediating intelligentsia varied. Heer tends to emphasize the
similarities in term of eschatological transformation, biological this-worldliness,
and antiauthoritarian brotherhood, but does not emphasize the differences of the
groups involved thoughout history. What Heer does show, however, is the tremendous innovation produced by the displaced intelligentsia forced into solidarity with
the oppressed productive classes.

The era of the late Roman Empire and early Christianity was one such time
of innovation, a period of social disruption in which dispossessed and discouraged
intellectuals were thrown into contact with slaves and proletarians in various mystery cults, Christian sects, and heresies. It was a period of occultism and massive
superstition, but also a time of great ethical and religious innovation.

The late Middle Ages, Renaissance, and Reformation also saw tremendous
cultural innovation. These were, like the late Roman Empire, eras of occultism
and superstition, but also one of ethical and religious innovation. Antonio Gramsci
was interested in the differences between Renaissance humanism, as a movement
of elite intellectuals, and the Reformation, involving peasant religious and political movements as well as changes in the high culture.

The period from the early nineteenth to the early twentieth century is another
era of fusion between the lower-class popular culture and the intellectual culture.
The Romantic movement around 1800 and the bohemian artistic movements of the
late nineteenth and early twentieth centuries marked periods when mediating

déclassé intellectuals assumed successively greater importance and the underground began to burst through into both the high culture and dominant political movements.

Certainly Heer's underground notions are applicable to the counterculture or bohemian movements, including the Beatniks of the 1950s, hippies of the 1960s, and the New Age culture of the late twentieth century. Zbigniew Brzezinski may not have been far off the mark to call the student revolt of the 1960s a kind of "peasant revolt" of humanist lumpen intelligensia made obsolete by the electronic society. In the case of the 1960s counterculture, the bohemian stratum of students became numerous enough to form a minority political movement of their own in the New Left. In the case of the New Age culture, the numbers and ferment are less obvious on the public scene, but the political significance in relation to ecological issues and gender equality may be far more significant than is now realized.

Economic dislocations forced together intellectuals with the illiterate yet skilled craftspeople in giving birth to modern science in the Renaissance.[31] The Renaissance man, the artist-craftsman exemplified by Leonardo da Vinci and Leon Battista Alberti, is a product of this fusion. Renaissance figures such as Paracelsus and Bernard Palissy, who contributed to modern medicine and geology, respectively, because of their exposure to the knowledge of peasants and artisans resemble wandering bohemians in their uprooted lifestyle on the edge of society.

Critics of the counterculture have rightly pointed to the regression to the occult as a mark against it; yet the two greatest past ages of cultural transition—the late Roman Empire (with the rise of Christianity) and the Renaissance—were also high waters of the occult, astrology, and various superstitions. The intellectual-artisan culture of the Renaissance that gave birth to modern science was steeped in alchemy, astrology, and magic. In some respects the popularity of the occult among the cultured classes during the Renaissance was greater than it was in the Dark Ages, and was surpassed only by that in the late Roman Empire during the transition to Christianity and feudalism. If Kepler's astronomy and Paracelsus's chemistry and medicine emerged from their occultist preoccupations, similar phenomena may be possible with our own counterculture. Ecological consciousness may supply the cultural milieu that could give rise to new scientific concepts. Although these outlooks in the late twentieth century often are related to religious and political movements, one can trace the impact of similar, earlier views on the origins of field theory, an important part of classical and modern physical theory.

Chinese society has maintained and managed to incorporate various thought forms of agrarian and prestate society at the highest level of cultural elaboration and state organization. The centrality of the family in social structure and organization, the role of large public-works projects in irrigation and flood control, the lack of independence of cities, and the weakness of religious movements independent of the central government led to a society in which high cultural sophistication developed without the formalistic and atomistic-individualistic tendencies found in the West.

At the same time, Taoism within Chinese society served as a religious and cultural outlet for various antisocial and rebellious tendencies as well as for the preservation of peasant ceremonies and traditions in juxtaposition to the state religion centering around the emperor and Confucian ritual. Although the revolutionary nature of early Taoism was soon co-opted, it remained a milieu in which antisocial undercurrents and somewhat Romantic (in Western terms) appreciation of nature was cultivated. Earlier philosophical texts, such as those of Chuang Tzu, became central to philosophical Taoism among Chinese literati, as Taoism became the philosophy of choice for Chinese hermits, dropouts, and retirees who returned to nature. Early Taoist religion also supplied roles with relatively more status for women (except at the very highest level—head of the religion), and preserved sexual ceremonies presumably resembling much earlier Chinese peasant religion. It also developed a physiological, sexual alchemy that emphasized female orgasm, though for male benefit, which contrasts most strongly with Western, Judeo-Christian attitudes toward female sexuality.

Renaissance medical reformer and philosopher Paracelsus incorporated local peasant lore into his views.[32] He recommended that his readers learn from Tatars and from peasants, female healers, and wisewomen.[33] Magician Henry Cornelius Agrippa similarly contrasted the incorrect book learning of the medical doctor with the practical experience of the peasant woman healer who administers the surest remedy free of charge to everyone,[34] argued for the superiority of women, and was the first to tie the issue of male and female merits to the oppression of women.[35] Although most alchemists were male, they presented a view of nature in which the relations between male scientist and female nature were egalitarian, in sharp contrast to the model of male exploitation of female nature prominent in the writings of Francis Bacon and of the empiricist and mechanistic philosophers of the Royal Society.

The hermetic philosophers and natural magicians of the Renaissance were often associated with the feared and persecuted world of witchcraft. For instance, Agrippa was widely believed to have practiced black magic. Kepler (who was not a hermeticist, but shared the Neoplatonic worldview) had playfully portrayed his mother as a witch in a science fiction work, and had to defend her when she was jailed on charges of witchcraft. Defenders of mechanism opposed the hermeticist intellectuals as well as the revolts of the peasants, which were often couched in heretical religious and occultist forms. It is not so paradoxical that many partisans of empirical research and mechanistic science encouraged persecution of witches as genuinely in league with the Devil. In contrast, Johann Weyer, a student of Agrippa, was one of the earliest to deny the reality of witchcraft, instead treating it as a form of mental illness. This shows the closeness of some of the occultist and natural-magic doctrines to those associated with female peasant healers and practitioners of pre-Christian pagan traditions who were persecuted in the witch mania.

Romantic science and Romantic philosophy of nature had fewer direct ties with peasant culture than did either early Chinese Taoism or Renaissance occultism. However, Romanticism in its early phases was energized, and often instigated, by the events of the French Revolution, with its upsurge of ordinary people against the aristocracy and clergy, even if many Romantics were conservatives or became such after disillusionment with the Terror, the excesses of the Revolution, and the military dictatorship of Napoleon. Certainly, such Romantics as William Blake and Percy Bysshe Shelley allied themselves with the radical forces. Others, among them Wordsworth and Schelling, initially supported the French Revolution but later became disillusioned.

The Romantics revived the doctrines and views of Renaissance occultist and hermeticist figures, including Giordano Bruno and Paracelsus, whose doctrines and personalities were ridiculed and discredited during the height of the Enlightenment. The Romantic rejection of industrial society and its "dark Satanic mills" was accompanied by an aesthetic appreciation of rural life and of wild nature, untamed by civilization. Similarly, the Romantics revived appreciation of primitive mythology and the epics and sagas of various early European pagan societies. They collected and imitated in their own works the folk poetry and songs of peasants. Also out of some of the societies supportive of the radical phase of the French Revolution arose the movements that became utopian socialism, communism, and Marxism, which in turn, through purposive contact with the working classes, absorbed or celebrated aspects of indigenous working-class culture as a contrast to the high culture of the ruling powers.

One also finds among the Romantics male appreciation of the Eternal Feminine as well as powerful and important women such as Mary Shelley (author of *Frankenstein*), Caroline Schlegel-Schelling (successively married to two leading German Romantic philosophers), and Madame de Staël (novelist, political liberal, and interpreter of German Romantic philosophy to the French). The early Romantic movement was centered in salons in Jena, Berlin, and Weimar, organized and led by women who were outsiders in their society by virtue of being Jews as well as women. Among the Fourierian and Saint-Simonian movements in France were conceptions of women's liberation and even, among the followers of Comte de Saint-Simon, the search for a female deity in Egypt.[36]

Thus, in each of the three major movements connected to the prehistory of field theory one finds percolation among a marginal intelligentsia of precivilized and peasant thought, including a more positive attitude toward women than was found in the high culture of the period or in competing intellectual movements.

Polarity and Analogy: "Primitive" Modes of Thought in Modern Science and Philosophy

A feature of the worldviews and philosophical backgrounds of the theories that led to the conception of the electromagnetic and other fields of force is their emphasis on both holism and polarities within the whole and analogies treating the particular entities as images or models of the whole.[37]

Traditional Chinese thought is famously structured in terms of the yin/yang polarity. Taoism and Neo-Confucianism deal with analogies between the natural and the social world as well as between individual objects and the cosmos as a whole. Renaissance thought, particularly in its occultist and hermetic strains, emphasizes the microcosm/macrocosm analogy, where the human being is the little world that reflects the features of the larger world and is expressed in terms of polarities and oppositions. Post-Renaissance founders of early modern science and philosophy, such as René Descartes and Thomas Hobbes, rejected the microcosm/macrocosm analogy and replaced it with the mechanistic world of physical atoms, or indivisible spirit-atoms, or souls, and replaced polarities with formal, mathematical laws of motion. Thinking in terms of polarity and analogy was revived in the Romantic movement in the early nineteenth century by Friedrich Schelling and *Naturphilosophie.*

One dominant twentieth-century account holds these modes of thought to be "primitive" and claims that they are replaced by a formal, logical, atomistic, and mechanistic mode of thought in the modern era. Many noncivilized or prestate peoples think in terms of polarity and analogy. French anthropologist Lucien Lévy-Bruhl contrasted what was called "primitive" thought with that of "civilized" peoples,[38] the former characterized as "prelogical." Lévy-Bruhl came to have doubts about his own thesis, but views like his dominated the conception of the rise of logic and the transition "from religion to philosophy"[39] and supported general doctrines, during the previous two centuries, of unilinear progress and the superiority of the civilized West to the uncivilized primitives of the third world colonized by the West.

Since the mid-twentieth-century breakup of the European colonial empires of Great Britain and France this view has been widely criticized. Claude Lévi-Strauss and other structuralists found logic and algebraic patterns in primitive thought,[40] as well as primitive modes of thought widespread in civilized Western thought about virtually all areas except the technical, scientific, and academic.[41] Our everyday inferences about probabilities are seen by psychologists and risk-benefit analysts to follow patterns far from those of scientific and mathematical probability theory.[42] Deconstructionists have claimed to discover primitive, semantic polarities guiding the allegedly purely formal-logical arguments of classical and analytically oriented philosophers and theorists.

Despite the critique of the thesis of the transition from primitive, prelogical thought to civilized, logical thought, many anthropologists make an exception for the formal sciences. Marshall Sahlins and others point to the similarity of our thought about families, food, and pets to primitive systems of polarities and analogies, but often exclude formal scientific thought, in a way similar to Karl Mannheim's exclusion of mathematical and scientific thought from his analysis of social influences upon thought.[43] More recent sociology of scientific knowledge and anthropology of science do not distinguish science from other social belief systems in the sense of attributing to it an authoritative truth status that puts it beyond the sort of social analysis one would apply to other belief systems.[44]

The contributions of traditional Chinese thought, Renaissance occultist cosmologies, and Romantic philosophy of nature to the rise of concepts of electromagnetic forces and fields exemplify how even in the physical sciences, thinking in terms of polarity and analogy has played a role. Even James Clerk Maxwell, who transformed the qualitative field theory of Michael Faraday into the mathematical formation of modern electromagnetic theory, which appeared to leave behind the realm of holistic thinking for formalism, was as a philosopher a theorist of analogy.

Despite the prevalence of the positivistic and hypothetico-deductive models of scientific theory, there is a minority tradition in the philosophy of science that has emphasized the role of models and analogies.[45] According to the hypothetico-deductive model, a theory is to be understood in terms of the formal-logical or mathematical syntax of the laws and experimental conditions from which are formally deduced the descriptions or predictions of particular events. Here, analogies or models are merely of psychological or pedagogical value. They are useful for teaching the novice or popularizing the theory, but are not of *logical* significance. They may be relevant to "the context of discovery" but not to the "context of justification." That is, they may help us in a psychological way to discover our theories, but they have no role in the proof or disproof, or in the rational acceptance or rejection of theories. Here, models play a role in the cognitive structure of the theory itself. Indeed, models may play a role in considerations of the "logic of discovery" (when considered as the evaluation of worthiness of entertainment of theories) and of what Larry Laudan calls the "context of pursuit" of theories. Models also have a place in the logical structure of explanation and testing in science in the account of the dynamics of theories, in which the strategies and rules that guide the paths of elaboration and fine-tuning of theories—what might be called the "context of deployment," as in Lakatos's "research programmes"[46] (or Laudan's "research traditions"[47])—are opposed to the consideration of theories as static structures.

The "Inverse Whig Theory of History"

The traditional view in the history of science in E. J. Dijksterhuis's *Mechanization of the World Picture* is that science culminates in atomism and mechanism. This is the general story of science given in survey texts today and almost all scholarly work up until the 1960s. However, the holistic and field-theoretical tradition in science coexisted with the mechanistic tradition even in the latter's heyday.

The "Whig theory of history" in science holds that the history of science culminates in contemporary results of science. Past theories are evaluated on the basis of their degree of similarity to presently accepted theories. Past, false theories are dismissed as ignorant superstition. Since the 1960s most historians and philosophers have claimed to reject the Whig theory. There has been a rehabilitation of rejected theories presented in their historical context. Histories of these rejected theories often bracket considerations of truth and present an extremely relativistic view of science. This study treats theories in terms of the positive contribution of viewpoints presently considered well confirmed. In this respect, my argument resembles the much despised Whig theory of history except that the present results of science (in which the story of science culminates) are rather different than those which are emphasized in the story of the mechanization of the world picture.

An example of anti-Whiggism concerns Renaissance occultism. Many who debate the relevance of occultism for later, so-called genuine science have accepted a distinction between forward-looking and backward-looking historians of science. The forward-looking historians gauge past science by its correctness when compared with our contemporary scientific knowledge, and select figures and incidents in the history of science insofar as they contribute to contemporary science. Backward-looking historians emphasize the sources, origins, and backgrounds of early science. They emphasize the point of view of the earlier scientists, and judge them in terms of the reigning criteria of the scientists' times, rather than the criteria of contemporary scientific knowledge.

Frances Yates, a historian of hermeticism, makes this contrast in a plea for the value of her own history of hermeticism. She claims that a history of science that is "read backwards"[48] is needed to supplement the usual sort of history which is "solely forward looking." Ironically, partisans of the view that hermeticism (or, more generally, occultism) is important for understanding the history of science accept this dichotomy and align themselves with the backward-looking historians. Social historian of science Arnold Thackray, in criticizing Mary Hesse's questioning of the importance of hermeticism for science, requests that Hesse "allow less traditional and less determinedly 'forward-looking' historians unhindered license peacefully to pursue their own hermetic hares."[49]

The defense of the relevance of occultism for understanding early modern science need not be purely backward-looking and antiquarian. Our very interest in the history of science or the philosophical study of past scientific ideas has built into it an implicit reference to ideas and activities that we consider to be significant for contemporary science.[50] We cannot pretend to do history of science totally unstructured by our current ideas of science. Dismissing all discussion of past thinkers in relation to our present conceptions as Whig history of science is inadequate. We can recognize elements in past inquiries into nature that contributed to or were absorbed into modern science without falling into naive conceptions of progress or debunking of past ideas that are different from our present ones. True, past Whig history often made crude distinctions between science and superstition in cases in which the superstitions were really protoscientific ideas or even scientific ideas that later writers did not recognize because of an excessively narrow image of science, often based on crude popularizations. Presentist history also is guilty of claiming past thinkers had correct ideas on the basis of superficial resemblances between past and present ideas. The search for precursors, both by Whiggish historians debunking past superstition and contrasting it with modern truth, and by medievalists who wish to give prestige to older thinkers by drawing superficial comparisons with modern ideas, can result in misleading evaluations. Avoiding crude accusations of stupidity against great thinkers of the past need not force us to claim, self-deceptively, that we are evaluating the ideas of past thinkers solely in their own terms. A half century of discussions of the philosophy of history and of hermeneutics shows us that both naive presentism (viewing past theories in terms of their correspondence or resemblance to present theories taken as "the truth") and naive claims to discuss past thinkers while being uninfluenced by present values are equally mistaken. Ironically, many of the antiquarians who discuss past science and occultism in a solely backward-looking mode do so because they themselves have fallen victim to some of the naively oversimplified, stereotypical, enlightenment conceptions of modern science that they accuse Whig historians of holding. They defend their own inquiries into occultism on purely antiquarian grounds, because they think it totally unlike modern science. (Yates initially claimed of her *Bruno* that "with the history of genuine science . . . this book has absolutely nothing to do. That story belongs to the history of science proper.")[51] Were these antiquarians to look more carefully into the role of fields and forces in the models of contemporary science and into the role of metaphysics in contemporary scientific inquiry, they would feel less obligated to investigate their own subjects in a solely backward-looking manner.

The Whig theory of history is rightly criticized for presentism; however, it also incorporated a set of beliefs about the nature of contemporary science as atomistic and mechanistic, purely empirical and eschewing wild speculation. These theses have been undermined by decades of philosophers' criticisms of naive versions of empiricism, operationalism, and inductive method, and by historical accounts of the role of metaphysical, religious, and social presuppositions in the history of the development of scientific theories. Theories now considered false are not for that reason unscientific (as has been emphasized by Karl Popper with his falsifiability criterion for science). Rather, they were testable theories that were tested and found in error. Many theories now considered to be false greatly contributed to the development of science, either as a source or object of criticism of opposing theories or as the source of ideas incorporated into opposing programs.

Theories that are highly speculative and not tightly tied to empirical data have also been recognized as scientific, indeed sometimes more scientific than those which do not go beyond the available data. Philosophers and historians of science have documented how theoretical assumptions often structure perception and description of the data at least as much as data issue in or guide the development of theory. Many philosophers, historians, and sociologists of science have analyzed how paradigms, metaphors, and metaphysical presuppositions have guided theory formation and selection. Even an account of science with an eye toward its outcome in contemporary science need not and should not be an account that culminates in a supposedly nonphilosophical and purely empirical science, any more than in a supposedly atomistic and mechanistic science.

A history of science that claims not to judge past theories in terms of their relation to modern science has the problem of what to include within the category of "science." That is, the very term *science* in its present usage (not merely *scientia* or *Wissenschaft* or knowledge in general) is a recent one. Terms such as *scientist* (as opposed to *natural philosopher*) and *physicist* are products of the early nineteenth century, introduced by French utopian socialist Saint-Simon and British polymath William Whewell. When discussing the science of earlier periods or of other cultures, one is guided by this more modern usage.[52] *Science*, once the term came into use, was construed to include the early modern science of Galileo, Newton, and others whose work was seen as in the tradition that culminated in nineteenth-century science. Greek science was also usually included, although it did not have the controlled experiment or the modern treatment of observations (indeed there is debate concerning whether Greek science involved experiments at all). Because the Middle Ages were stereotyped as the dark ages of religious superstition and opposition to science, this period was excluded from the eighteenth- and nineteenth-century accounts of science. It sometimes still is, as when Imre Lakatos, in the 1970s, refers to "(alleged) medieval science."[53] Only with the work of physicist Pierre Duhem in the early twentieth century (which was not fully published and did not gain a major scholarly following in the English-speaking world until the mid-twentieth century) did it become generally acceptable to discuss medieval science. Only because of the heroic work of Joseph Needham did traditional Chinese science come to be recognized as brilliant and important. Today it is still contested terrain to discuss traditional African science.

This volume emphasizes the importance of non-Western science as well as of medieval science and argues against the dismissal of such traditions as not being science. Western science is indebted to Chinese, Indian, and Arabic science at least as much as to Greek. The West has much to learn from traditional, Native American, and sub-Saharan African knowledge of agriculture, medicines, crafts, and survival techniques, as well as religious and philosophical attitudes toward nature. Nevertheless, when we identify this as "science" as opposed to "thought about nature" or "literature about nature," we are signaling that the selection of material is guided by concern with the scope and limits of modern science. Theories and histories of science that arrogantly and provincially reject past scientific theories or sciences of traditional and non-Western societies are now rightly rejected themselves. At the same time, twentieth-century Western historians, sociologists, and other theorists of thought about nature in the past, or in non-Western societies, cannot pretend to have totally suspended all interests and orientations that arise from contemporary Western science and society. The very interest in alternative approaches to science in other cultures is structured by a critical interest in modern science.

An Introductory Survey of Holism

Holism is the idea that the whole is more than the sum of its parts, that the whole determines the character and dispositions of its parts. The concept is an old one, but the term originated with Jan Smuts,[54] who used it to mean that the biological organism is not fully accounted for in terms of its parts. Holism and *organicism* are basically the same doctrine, though organicism is applied primarily to biology.

During the Romantic period in the early nineteenth century, organic wholes were contrasted with mechanisms and synthesis was contrasted with analysis. Most recently, the term *holism* has appeared popularly in "holistic medicine" and in other approaches associated with New Age thought. Much of the U.S. public today, if it had heard of holism at all, associates it with holistic medicine. Although the organic metaphor and many of the inspirations for holism in the last two centuries have come from biology or from the use of biological imagery, the notion of holism has applicability to other realms.

The notion of organic unity was applied to works of art (indeed, Immanuel Kant treated works of art and biological organisms in the same text, his Third Critique, or *The Critique of Judgement*).[55] Nineteenth-century social theorists used the notion of organic wholes and the contrast between organic and mechanical in their treatment of societies. Holistic notions are used in the treatment of biography and personality in both history and psychology. In psychology, in fact, the gestalt movement of the early and mid-twentieth century treated perceptual fields and the relationship of ground to figure in terms of wholes.

In the twentieth century, the interpretation of measurement in subatomic physics, as developed by Niels Bohr, involves a holism of object and measuring apparatus in which the measuring apparatus and the measured subatomic particle cannot be successfully separated by analysis during the act of measurement or from the final result of the measurement. An alternative approach to subatomic physics, that of David Bohm and John S. Bell, eliminates the subjectivism of the influence of the observer on the object in measurement, but holism reappears in the nonlocal nature of the interactions and the relation of the subatomic particle to the rest of the universe, which is described as a monistic unity. Other forms of holism have appeared in twentieth-century physics in discussions of field theory, particularly of nonlinear fields in Einstein's general relativity theory and of the universe as a whole in relativistic cosmology.

More recently, interest in nonlinear systems in classical mechanics and in modern physics (including chaos theory) has stimulated holistic interpretations of phenomena previously described in an atomistic fashion. It is perhaps significant that Mitchell Feigenbaum, the discoverer of an important constant in chaos theory, was an admirer of Goethe and German Romantic thought.[56]

Atomism: Its History and Impact

To understand holism, it helps to characterize *atomism*, the major approach opposed to holism. Physical atomism in the West began with the theories of Leucippus and Democritus in the fifth century B.C.E., who argued, on purely speculative, nonexperimental grounds that matter is not infinitely divisible. They concluded that the universe is composed of a multitude of separate, indivisible entities that they called *atoms* based on the Greek word for "undivided." All the objects in the universe are combinations and configurations of atoms in empty space.

Atomism has been associated historically with social individualism. Metaphysical individualism—the assumption that the ultimate realities are particular individuals—was never a major movement in China, and flourished only in the short-lived Mohist school (fifth to third centuries B.C.E.). Mohists are variously described as slaves, craftspeople, and military engineers—definitely not aristocrats. Their individualistic approach, concentrating on physics, logic, love of neighbor, and utilitarian ethics, stands in stark contrast with other Chinese schools. Unlike Taoism and Confucianism, Mohism did not last into the empire. Atomism was introduced into China from India much later via Buddhist systems and was never influential.

Although atomism was nonexistent in ancient China, and Mohist particularism was short-lived, atomistic thought in the West persisted longer in the ancient world and was revived in the Renaissance. In ancient Greece the first atomist of whom we have extensive fragments is Democritus of Abdera, which produced two major philosophers: Democritus the atomist and Protagoras the relativist. Democritus claimed that all there is "in reality are atoms and the void."[57] Protagoras is famous for his statement that "man is the measure of all things."[58] Democritus and Protagoras are said to be the philosophers of democracy.[59] Given that their philosophies are most similar to those of modern democratic individualism and modern atomistic materialism, it is important to know about Abdera, a city in northern Greece that was famed for its coinage but not for art or culture. Abdera was a highly money oriented society, center of the grain trade, and the greatest producer of coinage in the ancient Greek world. This is consistent with the association of atomism, democracy, and relativism with market society. Atomism was continued in the ancient world by Epicurus (341–270 B.C.E.) and the poet Lucretius (ca. 95–55 B.C.E.), but Epicurus and his followers used atomism for constructing an ethics, not for its scientific value.

Ancient Greece produced the idea of physical atomism and developed the formal proof and logical analysis that laid the foundation of the atomistic form of early modern science. Besides the Greek physical atomists, the Megaric and Stoic logicians laid the foundation for an atomic approach to logic, although this was not further developed or generally recognized (with the important exception of Leibniz) until the nineteenth century.

With the revival of individualism in the sixteenth century C.E. there was a revival of the ancient Greek atomic theory of Leucippus and Democritus. Just as ancient Greece was notable for competition of all sorts, early modern Europe (particularly England with its "possessive individualism") shared an individualist and competitive social orientation. (In significant contrast to Greece, ancient China lacked much individual competition in sports.)[60] The similarities of Hellenic Greece and early modern Europe in terms of individualism, money, the market, and competition show a commonality of the social basis of atomism.[61] Atomist thought was preserved in the primarily ethical writings of the Epicurean school of the later phases of Greek thought and the Roman Empire, culminating in Lucretius's poem *De Rerum Natura* (On the Nature of Things). The rediscovery of a single copy of Lucretius's poem in the fifteenth century led to the widespread reception of atomistic ideas in the seventeenth century.

This revival led to a fantastically successful approach to physics and chemistry. Figures including Isaac Newton in physics and Robert Boyle in chemistry made the philosophy of physical atomism central to science in the late 1600s. Newton had conjectured that "God in the Beginning formed Matter in solid, massy, hard, impenetrable moveable Particles."[62] The success of atomic thought in physical science led to imitation in other fields. Indeed, the trend was to replace continuous or organic approaches with discrete or atomistic methods in all areas of knowledge. Atomic methodology blossomed in introspective psychology and logical analysis of language.

The atomistic, empiricist approach in psychology was developed in the eighteenth and early nineteenth centuries by such figures as Hartley and James Mill, father of liberal economist and political philosopher John Stuart Mill.[63] More recent psychologies have perpetuated atomistic notions with respect to behavior and cognition. The terminology of *molecular* and *molar* was carried over into behavioral psychology by E. C. Tolman.[64] Evolutionary psychologists (such as Richard Dawkins) discuss the "meme" as a conceptual or cultural unit parallel to the gene in biology. Factory time-motion studies introduced the "therblig" as a unit of human motion. In analytical philosophy of action the notion of "basic actions" is used by Arthur Danto as a kind of atomic unit.[65]

Logical atomism as an interpretation of symbolic logic starts with atomic sentences, which cannot be broken down into smaller sentences as part of the meaning of the larger sentence. Molecular, or compound, sentences are constructed as logical combinations of the atomic sentences by means of connectives such as *and* and *or*. Logical atomism can be associated with a metaphysics that claims that the world is an aggregate of atomic facts corresponding to the atomic propositions. Ludwig Wittgenstein in *Tractatus*[66] and Bertrand Russell in "The Philosophy of Logical Atomism"[67] presented versions of such a view of the nature of reality corresponding to the structure of elementary logic of sentences. Russell's (mis-)interpretation of Wittgenstein (which was angrily rejected by Wittgenstein himself) gave birth to logical atomism.

The recent explosion of computer technology and its economic importance has further legitimized atomistic thinking in the form of bits of information, as pixels (those minimal dots which compose pictures on computer terminal screens) and as discrete methods of problem solution approximating or replacing continuous methods. The terminology of *bits* of information has become ubiquitous. New forms of psychological and neurophysiological atomism have been revived because of the appeal of the computer model of mind and the digital model of thought and nature. This tendency of thought is reinforced by everyday technological changes, such as the digital watch succeeding the continuous circular dial and the replacement of analog recording with digital recording.

A Brief History of Holism

Characterizations of holistic thought were not explicitly made until after the triumph of atomism in the eighteenth century. Most thought in precivilized societies and early civilizations was implicitly holistic in that it made numerous global associations among different realms of experience—social and natural and psychological.

Plato and Aristotle rejected Democritean atomism. (One, probably apocryphal, anecdote claims that Plato bought up all available copies of the works of Democritus and burned them.) For Plato, the cosmos was an organism. Aristotle treated biological organisms as purposive unities. Holism, monism, and continuism were propounded by various Greek thinkers. An organic or biological model of the universe in which physical processes were portrayed as goal directed was prefigured by pre-Socratic philosopher Empedocles (who in other respects had a protochemical model of the cosmos) and developed into a total system by Aristotle. Greek thought also contributed the monistic theory of Being as undivided and unchanging, originating in the writings of pre-Socratic Parmenides and developed in late antiquity by Plotinus and Neoplatonists. Anaxagoras developed the theory of the universe as a continuum, no part of which is divided from any other part.

Medieval thought continued holism in many areas, especially those which incorporated aspects of the indigenous tribal and pagan European myths. Medieval thought also incorporated the holistic aspects of the systems of the Neoplatonists. In the high Middle Ages Aristotle's theories were revived and combined with Neoplatonism and primitive indigenous pagan elements into a theory of the universe. Simultaneously, however, a more analytic, logical approach was introduced with the revival of Aristotelian logic in Scholasticism. In later medieval philosophical nominalism, strongly atomistic approaches developed from Scholasticism, although the organic aspects of Aristotelian natural philosophy prevented the development of a thoroughgoing atomism among all but a few of the Scholastics.[68] The Church and intellectual classes adhered to a logical and formal theology and legal system superimposed upon the more traditional, pagan beliefs of the illiterate peasants.

Early Renaissance hermeticist, occultist, naturalistic, and Neoplatonic movements, were strongly holistic. Many Renaissance philosophies were more holistic than late medieval Scholasticism, as many emphasized astrology and magical doctrines of powers and correspondences, especially since Scholastic theology was no longer at the center of philosophy.[69] Renaissance thought emphasized the correspondence between the macrocosm (the universe) and the microcosm (the individual human or other individual organism) and made the holism of indigenous paganism explicit and systematic. By the late 1600s, however, Renaissance occultism which lingered on in Kepler's works and in Newton's secret writings, was rejected by the scientific community and formal and atomistic thought became dominant in the self-image of physics.

Explicitly holistic approaches were revived in the Romantic era of the late 1700s and early 1800s. Goethe and Blake denounced Newton's analysis of light and colors. Schelling and Coleridge emphasized the organic unity of nature. Nature philosophy flourished briefly in Germany, even in the physical sciences, and involved a speculative approach to physics based on notions of organic unity and polarity. Philosophically, it was similar in certain respects both to Eastern thought and to Renaissance hermeticism. It played an important role in the development of electromagnetic field theory, but was generally ridiculed and rejected by mid-nineteenth-century materialism and early-twentieth-century logical positivism. Chemist Justus von Liebig called it a "black death" of natural science, and it was considered the epitome of irresponsible speculation and metaphysics.[70] Many of the mid-nineteenth century materialists in physics were more influenced by nature philosophy than they cared to admit publicly.

While Darwinism extended this approach into the biological world, especially by eliminating purpose and divine intervention from evolution, in fact the young Darwin was far more sympathetic to organic thought than was recognized either by his contemporaries or by many of our own.[71] The same is true of Karl Marx, whose early writings contain Romantic themes (especially via incorporation of the creative, constructive activity of the self in the philosophy of Johann Gottlieb Fichte), but whose later works largely reflect prevalent notions of mechanistic science and scientific law, after the failure of the revolutions of 1848. Marx's followers in the late-nineteenth-century social democratic movement made Marxism even more mechanistic. In physics the work of Ritter, Ørsted, and Faraday was inspired to varying degrees by German Romantic holism, and nature philosophy was incorporated into a mechanistic and/or antispeculative science by being reinterpreted in terms of Newtonian action at a distance and other more purely mechanical models. Faraday, Marx, and Darwin had all absorbed elements of early-nineteenth-century Romanticism but were understood via late-nineteenth-century thinkers in a mechanistic and often positivist manner.

In the last quarter of the nineteenth century and the early twentieth century, holistic notions made a partial comeback in some areas. In German philosophy of culture and history, historicism and theories of empathy and of expression dominated. For example, Wilhelm Dilthey had the notion of organically unitary worldviews and of artistic styles. There was a neo-Hegelian revival in both Britain and Germany. Theorists of history and sociology, such as Dilthey and Max Weber, supported various doctrines of *Verstehen* ("understanding") and aesthetic theorists, such as Theodor Lipps, had theories of empathy (*Einfühlung*) or empathetic understanding. In biology, Hans Driesch's vitalism and philosophy of organism were influential. Driesch moved from a mechanistic understanding of embryology through "developmental mechanics" to an appeal to *entelechy*, a term borrowed from Aristotle, to characterize the vital organizing principle of development. As biologist Richard Lewontin quipped in reviewing this tale, just as there are no atheists in foxholes, "there are no atheists in the embryology lab." The early twentieth century saw a number of antimechanistic biologists, including John Scott Haldane,[72] Ludwig von Bertalanffy (later proponent of so-called general systems theory),[73] Ralph Lilly (a follower of A. N. Whitehead's philosophy of organism),[74] E. S. Russell,[75] and Jan Smuts.

Since World War I, many of the major proponents of an antireductionist or emergentist biology, even if they did not always call themselves holists, have been self-proclaimed Marxists. These include biochemical embryologist Joseph Needham, J.B.S. Haldane (son of holist J. S. Haldane and contributor to population genetics and biochemical reaction theory), and John Desmond Bernal ("the sage" who worked in x-ray crystallography of biological molecules and on the origin of life).[76] In the aftermath of the Vietnam War, another coterie of Marxist-influenced, antireductionist biologists appeared at Harvard, including the brilliant evolutionists Richard Levins, Richard Lewontin,[77] and Stephen Jay Gould. Other mid-twentieth-century holists, who were not Marxists, were influenced by Marxism. C. H. Waddington, an embryologist, was an associate of Needham and other

Cambridge Marxist scientists.[78] François Jacob and Jacques Monod (despite the latter's later violent rejection of Marxism) were affected by currents of Marxism in the French Resistance. Their early work on genetic control and feedback (the "operon") complicates the linear logic of the "central dogma" of DNA.[79]

Most significantly, modern physics reintroduced various sorts of holism. It was explicitly appealed to in Niels Bohr's Copenhagen interpretation of quantum mechanics, which is still the official interpretation of subatomic physics. Later opponents of Bohr, such as David Bohm and John S. Bell (of Bell's theorem), rejected Bohr's holistic analysis of the act of measurement or observation but replaced it with an equally holistic account of the physical reality itself. In the holistic realist interpretation of quantum mechanics, the universe itself is an inseparable whole in which all elements of reality are interconnected.

Meanwhile, in Einstein's general relativity theory, or theory of gravitation, implicitly holistic conceptions appeared even if physicists and philosophers did not discuss them as such. The general relativistic field is a nonlinear field, governed by nonlinear differential equations in which the whole is not equal to the sum of its parts.[80] In consideration of cosmology and the origins of the universe, the initial conditions of the universe as a whole may not be fully distinguishable from the laws of nature. In addition, topological or "global" features of space-time are emphasized in the explosion of work on general relativity since the 1960s.

Holistic elements remained implicit even in Newtonian mechanics, although they were not widely recognized until the late twentieth century, with the popularization of chaos theory. Certain holistic and nonreductionistic themes have reappeared in nonlinear phenomena in mechanics and other branches of science. Topological and global approaches to classical physics, such as the qualitative theory of differential equations, were begun before the turn of the twentieth century by Henri Poincaré,[81] but the philosophical impact was overlooked in the excitement of the development of quantum mechanics and relativity theory. During the late twentieth century, Poincaré's ideas were greatly elaborated, axiomatized, and made accessible via computer graphics, and with them holistic themes have reemerged.

Continuism and Holism

Holistic thought has often emphasized continuity (analog) as opposed to discreteness (digital); however, continuity as opposed to discreteness alone is insufficient to constitute holism. A holism where the entire universe is the only whole is also continuism. (Much of traditional Chinese thought about the world is both holist and continuist in this manner.) But a holism with separate wholes, such as organisms or societies, with the whole controlling the parts or possessing nonreducible properties absent from the parts, can have a discreteness among distinct wholes. For instance, organismic biology treating animal and plant organisms as wholes, but not treating populations, species, or the entire biosphere as individuals, is holistic with respect to organisms (claiming they cannot be reduced to their atoms), but is atomistic with respect to the relations of individual organisms in populations.[82]

Anaxagoras, a pre-Socratic philosopher, was the founder of continuity as the fundamental principle in Western tradition. Anaxagoras said that "things are not cut off from one another with a hatchet."[83] In Chinese philosophy (and perhaps even in ordinary Chinese language) continuity takes precedence over individual objects. The continuous "All" is primary. Individual objects are cut out of the continuum from a variety of perspectives; that is, individuals are not the fundamental entities out of which the rest is made, as in atomism, but individuals are artifacts of our purposes of classification.

Continuisms are also found in the philosophies of Leibniz in the seventeenth century, in continuum physics of the eighteenth and nineteenth centuries, and in the philosophy of Charles S. Peirce in the late nineteenth century. These are not holisms. Leibniz's philosophy contains elements of both logical atomism and physical holism. Continuism—the belief that the fundamental stuff is continuous rather than atomic or discrete—does not necessarily assert that the whole of the continuum exerts influence on its parts. It does not claim that the whole has properties that are not accountable by the properties of the parts, or that the parts are coordinated into some pattern of the whole.

Eighteenth-century European physicists working on continuum mechanics, following Descartes[84] and Leibniz more than Newton in their natural philosophy, were for the most part neither holists nor organicists. Similarly, the logic of the Peirce made the mathematics of the continuum central,[85] and he incorporated into his own philosophy the notion of continuity both from nineteenth-century evolutionary theory and from Hegelian philosophy. Because of this, Peirce developed a logic different in significant ways from that of a logical atomist such as Bertrand Russell. However, Peirce's logic was not organicist, as were the "logics" developed by some Romantics and some philosophers who followed Hegel. Continuism is not organicism, in the sense that the elements of the continuum need not be structured into some sort of organic whole.

Peirce's logic, based on continuity, has some similarities to Taoist ideas. G. Spencer Brown, whose graphs in *Laws of Form*[86] strongly resemble Peirce's existential graphs, begins his book with a quote from Lao Tzu. Brown based his logic on the notion of drawing a distinction within the continuum. As in Peirce's existential graphs, one begins not with a set of atomic sentences or objects but with an undifferentiated continuum, then separates out regions. The earliest, best-known, and most simple, logical graphs (Euler diagrams) were devised by Leonard Euler, who perhaps contributed more than any other physicist to classical continuum mechanics.

Continuism is usually antiatomistic, although some early forms of continuism supposed that the continuum is made out of an aggregate of separate physical points. Descartes and Galileo each argued for a continuum made out of a sum of distinct points. *Dynamism*, or the theory of point atoms as centers of force, can have affinities with this position. Similar theories were developed in the eighteenth century by Boscovich and Kant as precursors to field theory.

Monism and Holism

Besides continuism, the other position that needs to be distinguished from holism is monism. Some authors confusingly equate the two.[87] The term *monism* is often also used to denote the doctrine that there is numerically only one entity in the world, called *substantival* monism. A second kind of monism, called *attributive monism,* holds that there is only one *kind* of thing in the world. Examples of attributive monism are materialism (everything is material), and idealism or panpsychism (everything is mental). An opposing doctrine to attributive monism is *dualism*, which holds that there is both mind and matter in the world (à la Descartes).

The monism of relevance to holism is substantival monism. Indeed, if God has a place in this philosophy, God and the universe are one (pantheism). A religious, substantival monism must be pantheism (God is the universe). The opposing doctrine to substantival monism is called *pluralism*, meaning that there are numerically many entities in the world. A materialism that claims that there are many atoms, all material, is an attributive monism. However, it is also a pluralism in the second sense of opposition to substantival monism.

Monism emphasizes the oneness of reality in a stronger sense than does holism. Holism claims that there are organizing structures that unify groups of entities. The characteristics of the individual entities may be modified by the holistic structure or unity of which they are a part; however, they can maintain some of their status as individual entities. In monism, the unity is greater and the reality of the apparently separate individuals in the universe is denied. One can conceive of monism as the most extreme pole of holism—at the opposite end of the scale from atomistic individualism.

There are basically two ways in which the individuality of separate objects can be denied in monism. One can either claim that the appearance of separate objects is an illusion or one can treat the separate objects as modifications or modes of the whole. The independent status of the plurality of entities is denied in both versions of monism, but in the second there is some feature of the One, or Whole, that answers to each apparently separate individual. The individuality is an illusion; however, the characteristic observed is not wholly illusory. The first version of monism in which individuality and separate individuals are wholly illusory was manifest in Parmenides and in some of the monistic systems of Indian mysticism. The late-nineteenth and early-twentieth-century British Neo-Hegelians or absolute idealists (such as Francis Herbert Bradley), also adhered to this version of monism.[88]

Parmenides started with the view that we cannot truly say that which is not. Thought must correspond to reality. There is no nonbeing, and nothing does not exist. Nothing cannot be, as Parmenides claims.[89] From this, Parmenides infers that there is no empty space or change from something to what it is not. Also, by inference there is no coming to be or passing away. Things are indivisible, insofar as things cannot be divided into spatial parts. Reality is undifferentiated. It cannot be qualitatively distinguished into parts. There is only the One. In a poem Parmenides calls this the "Way of Truth."

Parmenides's philosophy contrasts with the philosophy of radical change, process, and flux of Heraclitus. Parmenides denies change and claims that change is an illusion. His disciple Zeno developed arguments to prove that the concept of motion is self-contradictory. It is possible that Hippasus of Metapontum, like Zeno, was proving the irrationality of plurality and change, not like Zeno in order to reject it, but following Heraclitus in order to embrace it.[90]

Parmenides's philosophy also contrasts with that of the atomists Leucippus and Democritus, who assert the reality of empty space and a plurality of ultimate individuals—atoms. One contested interpretation of the atomists is that they develop their philosophy in reaction to the dilemmas propounded by Parmenides. They made empty space into a kind of positive replacement for nonbeing or nothingness and retained the unitary indivisibility of Parmenides's One in each of their myriad atoms. In this spirit, philosopher of science Rom Harré has called ultimate point atoms "Parmenidean individuals" despite the fact that there is a vast plurality of them.[91]

Early-twentieth-century philosopher Emile Meyerson[92] claimed that the drive of classical physical science is Parmenidean. The laws of conservation of matter and of energy reduce what appears to be a plurality of different things and processes to an unchanging unity and totality. Extreme determinism seems to remove the novelty of change by showing that future events are in some sense already contained in the present and past. Meyerson claims that the Parmenidean drive of classical physics is countered by the reality of the directionality of time and is constantly undermined by the empirical drive for new discoveries to be incorporated into the unitary synthesis of deterministic laws of motion and universal laws of conservation.

Some interpreters of Einstein's general relativity theory have claimed an affinity to the Parmenidean worldview.[93] Logician Kurt Gödel (who for leisure activity would help his friend Einstein solve field equations) held the view that relativity theory supports Parmenides, who denies the reality of time.[94] When set theorist and science popularizer Rudy Rucker asked Gödel what causes the illusion of the passage of time, Gödel questioned whether "there is a perceived passage of time at all." He claimed, "The illusion of the passage of time arises from the confusing of the *given* with the *real*. Passage of time arises because we think of occupying different realities. In fact, we occupy only different givens. There is only one reality."[95]

Einstein intimately married space and time into a unitary space-time. In special relativity, the rapid motion of an object relative to an external observer leads to the measurement of the length of the moving object by the observer to be contracted or shorted in the direction of motion. Also, the lapse of time of processes in an object moving rapidly relative to an external observer leads those processes as measured by the observer to slow down. Einstein's unification of space and time into a four-dimensional space-time, combined with the determinism of Einstein's theory, allows for an interpretation of physical processes as simply predetermined paths within four-dimensional space-time. It is then claimed that the flow of time or temporal becoming is simply a subjective illusion on the part of the observer.

As relativity theorist Hermann Weyl describes it, consciousness is like a worm crawling along the predetermined world-line, successively illuminating different small regions of the preexisting reality.[96]

I recall a speaker using the image of a dark room in which a flashlight successively illuminates small regions of a complex pattern on a carpet in order to describe the status of events in general relativity. This interpretation of general relativity theory is criticized by some philosophers as a confusion in which the treatment of time as a dimension of space in a graph or four-dimensional model makes time itself a dimension of space. This is no different from the representation of time as a dimension or axis on a graph in classical science. Many attributes are represented as axes on graphs without being identified with dimensions of a physical space. Given the distinguishability of spacelike and timelike separation in relativity theory and the difference between time and space, the four-dimensional world is now more properly spoken of as a 3+1 dimensional world. Nevertheless, the denial of the flow of time or temporal becoming a part of the world picture of physics is strongly argued for by many physicists and philosophers of science, and the Parmenidean interpretation of relativity theory is a common one.

In summary, holism has been contrasted with atomism and compared to its closer relatives, continuism and substantival monism. On the one hand, holism can be seen to possess much more central organization than continuism. On the other hand, the elements of components in systems described in terms of organic holism have more reality and more independent status than the illusory status of differentiated appearances in a strict substantival monism. We have seen that although the term *holism* has relatively recent origins and was initially related to biological theory, it has powerful applicability to physics.

PART I

The Contributions of Chinese Science and Technology

CHAPTER 2

Discovery of the Compass and Magnetic Declination

$\mathcal{T}$he modern theory of electromagnetism arose on the basis of knowledge of magnetism and the compass. The compass arose in China many centuries before it arrived in the West. Chinese thought allowed not only discovery and use of the compass, but acceptance of the Earth's magnetic field and its variations before Europeans had even begun to use the compass as a purely practical aid in navigation.

In 1620 Francis Bacon claimed that the modern period in Europe was superior to the age of the ancient Greeks and Romans because of the technological superiority of the moderns. Bacon emphasized three inventions central to the rise of the modern age: gunpowder, the compass, and printing. "It is well to observe the force and virtue and consequences of discoveries; and these are to be seen nowhere more conspicuously than in those three which were unknown to the ancients . . . ; namely printing, gunpowder and the magnet. For these three have changed the whole face and state of things throughout the world."[1]

Ironically the compass and gunpowder were products not of modern Europe, but of ancient or medieval China, and were brought to the West in the late Middle Ages and Renaissance. Printing, although independently invented (though perhaps reconstructed from reports of printing in Asia) likewise arose centuries earlier in China than in Europe. The examples chosen in Bacon's claim of European superiority over the Greeks rested largely on Europeans' ability to borrow Chinese technology, not on their own inventions.

Until the mid–twentieth century virtually no one in the West believed in a significant Chinese science. Texts in the history of science and technology searched for the reasons why the Chinese made no significant technological advances.[2] Westerners saw that China was politically disintegrating and militarily defeated by Western powers. The Chinese opium addiction (forced on China by the British in the Opium War in the 1830s) was seen as evidence of the inherent degeneracy of the Chinese. The admiration the eighteenth-century Enlightenment felt for the Chinese was replaced by disdain.

The Travels of Chinese Science

Despite modern Western ignorance of the contributions of Chinese science and technology, a great deal of Chinese technology and scientific knowledge had traveled to the West along with trade goods in ancient and medieval times. The trade was usually indirect: from China to Central Asia or India to Persia or Arabia to Europe. Thus many Chinese products and ideas were said to have come from the Arabs or Persians, not from the Chinese. One major route of trade was the famous Old Silk Road.

There were more travelers from the West to China than vice versa, but a few Chinese travelers got as far as the Middle East. Chang-Ch'ien journeyed to Central Asia to make treaties against the Huns with the tribe of the Yueh-Chih. Chang-Ch'ien was captured by the Huns, detained, but eventually released. His journey may have paved the way for the Old Silk Road.[3] Another Chinese traveled west as ambassador to the Byzantine court in Constantinople, but he was discouraged from proceeding by Persian sailors defending their trade monopoly. One Chinese scientist studied with Arab scientist al-Rāzi for a year and copied the medical works of the Greco-Roman physician Galen, but he may not have made it back to China.[4]

Along with the trade in silk and other goods, numerous vegetables traveled from Europe to be grown in China, and numerous fruits and garden flowers were carried from China to be cultivated in Europe.[5] Along with goods a number of technological devices and knowledge of the possibility of constructing other devices diffused between China and Europe. Jugglers and magicians brought knowledge of the mechanics involved in stage devices to China from the Middle East, but the flow of technology from China to Europe was far more important than that in the opposite direction.

The inventions that diffused from Asia to Europe determined much of European history. Buddhist missionaries from India who journeyed to China developed a rudimentary toe-loop stirrup, which developed into the modern stirrup in China and was carried west by nomads. This led to mounted warfare and gave birth to the feudal system in Europe with Charles Martel's distribution of lands for the support of mounted knights in 732 C.E.[6]

The technology of gunpowder, also a Chinese invention, has been claimed to have led to the demise of feudalism. The armored knight was relatively impervious to sword or ax attacks by foot soldiers but could be brought down by bullets. Similarly the medieval castle had to change its form with the advent of the cannonball. Thus two Chinese inventions, respectively, initiated and destroyed the feudal knight and the whole system of social arrangements and morality of chivalry that went with him.

Printing with movable type, another Chinese invention, revolutionized the Western intellectual world and made possible widespread literacy and the expansion of linear ways of thinking as well as contributing to the religious Reformation by making the Bible widely available.[7]

Ironically, the ship's rudder, or steering device, and the mariner's magnetic compass, both Chinese inventions, made possible the European Age of Exploration.[8] Along with the Chinese invention of gunpowder, these inventions supplied the means of transport and the weapons that made possible the exploration and conquest of the non-European world, including China, which had given birth to the inventions and which was left in the defeated state that made nineteenth- and early-twentieth-century Westerners deny the existence of any significant Chinese science and technology.

Joseph Needham: The Foremost Western Rediscoverer of Traditional Chinese Science

The awareness among Western historians of the riches of traditional Chinese science began to dawn only in the 1960s with the work of British biochemical embryologist Joseph Needham (1900–1995). Needham's multivolume work, *Science and Civilization in China*, the later volumes written with the help of Chinese and other collaborators, belongs to the most monumental works of historical scholarship.

Needham, a Cambridge academic and a practicing Christian, was exposed with a jolt to Marxist social history and philosophy of science with the unannounced appearance of a delegation of leading Soviet scientists at a London conference in 1931. He later recalled that he heard the "trumpet blast" of a new approach to the history of science.[9] In an article written for his own *Festschrift* under the pseudonym Henry Holorenshaw (who knows Needham "better than most"),[10] he describes his initiation into the study of Chinese science. During the 1930s he befriended several Chinese students and became fascinated with Chinese culture. These Chinese students asked him why modern science arose only in Europe.[11] Needham had never himself asked this question and could not answer them. His fifty-year quest for an answer became one of the great scholarly adventures of the century. During World War II he traveled to China with a delegation from the Royal Society and then stayed on, helping to set up the Sino-British Science Cooperation Office. During this period Needham learned about traditional Chinese science.[12] He notes that it was his excited realization that the Chinese had discovered magnetic declination four centuries before the West that stimulated him to write a history of Chinese science.

Needham's eclectic religious and philosophical orientations were ideal in aiding him to appreciate Chinese thought. Although a Christian, Needham called himself an "honorary Taoist" and claimed that his Christian background helped him understand traditional China with its centrality of religion. His sympathy with Marxism allowed him to remain in friendly contact with mainland China after it became communist and was entirely isolated from the United States during the early cold war and the Korean War. Needham was influenced by the "philosophy of organism" of mathematician and logician Alfred North Whitehead. He saw in it affinities both to his work in organismic biology and to the Marxist dialectic of nature and nonmechanistic materialism as well as to Chinese thought about nature and humanity.[13] Needham was also familiar with and somewhat sympathetic to some of the mechanistic tendencies in seventeenth-century British thought, such as Cambridge Platonism. Thus, he was philosophically very well prepared to appreciate Chinese "organic naturalism."

Needham's rediscovery of the riches of traditional Chinese science and technology not only proved Western opinion about the absence or lowly state of traditional Chinese science to be in error, it also contributed to world civilization by showing how inventions and discoveries were transferred from East to West in early times.[14]

In the 1950s, the leading American historian of science, Charles Coulston Gillispie, reviewed and dismissed an early volume of Needham's work on the grounds that Needham was a Marxist and was on friendly terms with people in Communist China and hence could not be trusted to be a reliable source. Gillispie warned the reader about Needham's masterpiece: "The whole discussion is extremely interesting—if it is true. . . . This is not simply a history of Chinese science. It is a Marxist history of Chinese science. . . . The whole treatment is shot through with Marxism." He hoped that "in the future volumes Professor Needham will spare us his philosophy. For how useful . . . would be . . . a straightforward account which could be trusted by readers." Although Marxists have written such purely factual books, "the present volume is not one of them."[15]

Gillispie's influential history of science, *The Edge of Objectivity*, presents the rise of science as a constant struggle against the excesses of mystical insight and romantic, organismic ideas. The present work can be considered in many respects as a mirror image of Gillispie's. What Gillispie claims to be the main enemy of science in fact contributed to major areas of science but was suppressed in the modern self-image of science. The holistic ideas that Gillispie denounces are portrayed by him as legacies of less civilized eras and live on among the lower classes constantly threatening to erupt and destroy the fragile edifices of science in the name of a more integrated, less alienating view of the world. Gillispie's scientists, self-restrained in their skeptical distance from emotionally satisfying but speculatively dangerous doctrines, maintain the atomistic, mechanistic view of science against the superstitious, romantic, and ideological as epitomized by Lamarck and the French revolutionaries. In some ways Gillispie's whole interpretation of the history of science may be seen as an overreaction to the crackpot Lamarckian biology of the politically ruthless agronomist Trofim Lysenko, which dominated biology under Stalin in the Soviet Union.[16] Now that this period of political distortion of science of the former Soviet union has come to an end, perhaps a less partisan evaluation of the organic peasant and romantic doctrines in earlier science will be possible.

As the magisterial volumes of Needham's work appeared, and as U.S. isolation from mainland China diminished, such attitudes began to erode. Gillispie, however, in *The Edge of Objectivity* (1960), could still assert that China and India did not create science. He also presents his fears that when non-Western countries attain the atom bomb, they will not be so restrained as the Christian West has been in its use (forgetting Hiroshima and Nagasaki). He asks, "And what will the day hold when China wields the bomb? And Egypt? Will Aurora light a rosy-fingered dawn out of the East? Or Nemesis?" Needham comments, "This is certainly very near the edge."[17]

By the end of the cold war, Needham's claims for the Chinese origin of a number of major inventions, whether appropriated by the West (like gunpowder) or rediscovered centuries later by the West (sex hormone therapy and biological pest control), were no longer so controversial within the history of science. Some scientists, however, still react against the conception of the origins of science as multicultural. A review of a popular book on scientific method, by a nuclear physicist and science educator, states that the author "nails his thesis against the doors of what he perceives as the current orthodoxies of New Age romanticism, political correctness and multiculturalism, reiterating his view that the core of scientific thinking was a uniquely Western discovery."[18] This author says that, despite sending vessels to Africa, "the Chinese, seeing everything from their own perspective, learned nothing about the outside world."[19] Probably the author is unaware that the most complete map (covering Europe and Africa in detail) known prior to Columbus comes from Confucian Korea.[20] Students of Chinese knowledge of geography and of eighth-century T'ang dynasty cultural synthesis will be surprised at the claim that the Chinese learned nothing about the outside world.[21] The author emphasizes China's "ethnocentrism," citing Daniel Boorstin's account of invention, which ignores non-Westerners,[22] and notes that the Chinese "didn't develop objectivity,"[23] apparently unlike himself.

By the 1980s Needham's basic claim that advances such as printing, the compass, and gunpowder were Chinese, not Western, inventions had entered popular education along with widespread claims of the medical efficacy of acupuncture (though the history of other Chinese biotechnologies and agricultural technologies has not). Third world scholars have taken up projects similar to Needham's, concerning the traditional sciences of other non-Western societies; nevertheless, genuinely comparative work has not taken a significant place in the academic history of science. One of Needham's later collaborators claims that this is because behind the cultural relativism that appears to dominate history and the social sciences, there lurks a tacit assumption of Western superiority.[24] This claim may seem harsh, but it is supported by the leading survey of science from the Middle Ages to the early modern period, which not only limits itself to Western science, but insufficiently covers the well-known Arab contribution to medieval European science. It was precisely in the early Middle Ages, or Dark Ages, while Western civilization was at its nadir, that Baghdad, China, and Mayan Central America were at their zeniths.[25] Contemporary historians of Western science no longer claim inferiority of non-Western cultures, but still hold to a doctrine of "separate but equal" with emphasis on the separate, and with little realized commitment to the claim of "equal."[26] One of the reasons that Needham could give the respect that he did to Chinese science was his belief in "ecumenical science" and its representation of reality. He seems old fashioned to postmodernists, but he gives more genuine respect to Chinese culture than do many supposed relativists. Following Needham's lead, the present work is also traditional with respect to its identification of a world science to which China mightily contributed, and attempts the comparative approach.

The Chinese Discovery of the Compass and Magnetic Declination

Chinese correlative thinking played an important role in the invention of the compass. The magnetic stone, called the lodestone, was known in both China and the West early in the first millennium B.C.E., but the discovery of the directive power of the lodestone was first made and applied in China. Around 1190 C.E., knowledge suddenly appeared in Europe of both the directive or direction-seeking quality of the lodestone and the ability to magnetize pieces of iron by touching them to the lodestone.[27]

There are mentions in literature of the needle compass a century earlier in China, and evidence of its use several centuries before that. Furthermore, the Chinese knew of magnetic declination—the discrepancy between magnetic north (or south), which is the direction in which the compass points, and geographical north and south. (Chinese compasses and other directional devices point south rather than north, as they do in the West.) *Brush Talks from Dream Brook* (or *Dream Pool Essays*, as Needham calls it) by Shen Kua (1031–1095), written around 1088, mentions not only of the magnetic compass but magnetic declination as well.[28] Declination was discovered in the West only in the mid-1400s, and the discovery is often mistakenly attributed to Columbus in 1492.

Needham suggests that the origin of the magnetically directed spoon, the precursor of the compass, was associated with an early form of Chinese chess.[29] Here the importance of a cosmology in which the unity of the larger universe with earthly affairs is emphasized played an important role in the Chinese discovery of the compass. The early form of chess was not military but "Image Chess." It symbolized the conflicts of the yin and yang forces of the universe.[30] The chess pieces were evidently magnetic and the board was oriented to the Pole Star so that the game positions could reflect cosmic arrangements. Needham suggests that the so-called Image Chess was later simplified and militarized, probably in India, and Persian military chess led to the chess with which Westerners are familiar.

Astronomical and astrological imagery continued to be associated with chess down to the modern period despite the loss of astronomical symbolism of the playing pieces. Early precursors of this cosmic chess probably involved the initial throwing of the figures, or pieces, similar to dice games or the *I Ching*. Native Americans had similar games in which human and animal figures, which had marks of the four cosmic directions, were thrown during play.[31] In the earlier forms of Chinese games conclusions were drawn from the angles in which the figures rested.

Several early references to chess in China also involved references to magnetism. There are mentions of chessmen hitting one another presumably by magnetic attraction. There is also reference to the lodestone lifting chessmen, and this is associated with iron needles, perhaps precursors to compass needles. Chess pieces were moved on a silver plate in one Chinese account. This may possibly be related to Saint Augustine's report of iron pieces moved on a silver plate. In the first century C.E., there is report of a lodestone dipper associated with chess, similar to the magnetic spoons (spoon-shaped magnetite) that were used for orientation on diviner's boards.

In the work *Discourses Weighed in the Balance* (*Lun Heng*)[32] by Wang Ch'ung (27–97 C.E.), naturalistic skeptic and critic of the pseudosciences, there is a passage on the south-pointing spoon on the diviner's board. The spoon represented the constellation of the Big Dipper, and the seasons of the year were correlated with the positions of the Dipper.

The diviner's board (*Shih*), called the "cosmic board," "cosmographic model,"[33] or "cosmograph,"[34] was an astrological device, representing the circular heavens and the square Earth by a circle within a square. The board could not represent the three-dimensional arrangement of the heavens above the earth, nor could it (for practical purposes having to do with fitting the markings of various positions in the heavens on the board), represent, even in two-dimensions, the accurate portrayal of the physical covering of the Earth by the heavens. Also it did not measure accurately distances on the edge of the circle, but represented only qualitatively a sequence of twelve equal divisions. Thus the diviner's board was misleading as a scale model of the heavens and was soon replaced for scientific purposes with a three-dimensional one. Nevertheless the board found its way into a range of Chinese ceremonial and meditational practices far beyond its immediate use as an astrological diviner's aid.[35] Some early Chinese mirrors, among them those called "TLV mirrors" because of their pattern using motifs resembling our letters *T, L,* and *V,* replicate in decorative form the cosmograph.[36] It has been suggested that the magical power they were believed to have stems in large part from this relation to the diviner's board. The constellation corresponding to our Big Dipper played a large, lasting role in Chinese religious cosmology. For instance, Taoist meditation techniques, which in part involved becoming one with the heavens or the stars, centered on the North Pole.[37] There is also the Taoist practice of "pacing the Big Dipper," that is, walking in the limping steps of Yü in the pattern of the Northern Bushel, the constellation that we call Ursa Major, or the Big Dipper.[38] By pacing off the pattern of the constellation on the ground one can participate in the pattern of the heavens, although not so fully as those immortals who literally "pace the void" by traveling about among the stars.[39]

The earliest compasses had a fish-shaped piece of wood that carried the lodestone needle or a fish-shaped leaf of iron floating in water.[40] One of the earliest Western names for the compass was the Latin and Italian *calamita,* which means "tadpole." The Chinese character for tadpole contains that for spoon or ladle, and perhaps that is a source for this Western terminology for compass.[41] Later, dry compasses were developed, originally in the shape of a turtle containing a lodestone suspended on a pivot. Needles were initially used to sharpen or make more precise the orientation of the fish or turtle.

One of the diviner's boards from Korea is decorated in the same colors as the siting compass. Similarly, the compasses carry inscriptions on their back similar to some of the cosmographs.[42] The original use of the compass in China was for siting, or orientation, of tombs and buildings (*geomancy*). The surveyor's compass that first appeared in the West was south pointing and was likely the geomantic compass that had arrived by land via Central Asia and Russia. Siting harmonized the location and orientation of structures with the local patterns of cosmic breath. Just as the winds that appear in the name for siting include the cosmic winds of *ch'i*, so the waters of siting include invisible, underground streams that supposedly circulate and deposit minerals. The idea of these underground veins of earth is mentioned as early as the fourth century B.C.E. The *Kuan Tzu* (a large, miscellaneous work on philosophical and military topics, possibly by the Chi-Hsia Academy around 319–280 B.C.E.)[43] also speaks of passages within the body of the Earth which are like sinews and veins.[44]

Geomancy continues in China to the present. It has had a great influence on the aesthetics of the placement of tombs, homes, and cities. Traditional Chinese landscape paintings are strongly influenced by the idea of siting in terms of the location of humans and their buildings in relationship to the landscape.[45] In Hong Kong in the 1990s, geomancy has become an important activity for the orientation and arrangement of businesses and homes.[46] According to principles of siting, a building is supposed to be backed by tall rocks or cliffs and should face south on low land and water.[47] Defenders of siting note that this orientation is ideal for solar heating. Though the government of mainland China had earlier suppressed siting,[48] it has returned to some extent in mainland China in recent years. One of Hong Kong's leading geomancers has a short television program in mainland China.

Although siting is likely to be dismissed as a silly superstition by those in the West with a scientific outlook, one can find in the siting compass evidence of Chinese recognition of magnetic declination. That is, one can find the divergence of the magnetic north from geographical north. A good example of the somewhat racist dismissal of the strange markings on the siting compass is that of the famous student of Chinese religion, J.J.M. de Groot. He claims that the traditional compass face "fully shows the dense cloud of ignorance which hovers over the Chinese people: it exhibits in all its nakedness the low condition of their native culture illustrating the fact that natural philosophy in that part of the globe is a huge mountain of learning without a single trace of knowledge in it."[49] De Groot concluded that the Chinese did not know about magnetic declination, based on questioning of peasants and popular geomancers. Needham compares his technique and conclusion to investigating British nuclear physics by quizzing folk dancers.[50] The traditional siting compass, of the sort that is still used by geomancers in China, has some eighteen circles of markings and symbols around the compass face. These include traditional directions and astronomical constellations, symbols from the *I Ching*, and various calendar units. "Fossilized," as Joseph Needham calls it, in these compass rings are the records of the shift in the direction of magnetic north over the centuries in the Middle Ages.[51]

Before examining the features of the traditional compass markings that show evidence of magnetic declination, it is perhaps worth noting that some other rings of the traditional compass may also show a shift in the nature of the Chinese compass. The first and fifth circles around the compass dial show trigrams from the *I Ching*. In the first they are in the symbolic form of broken and solid lines, and in the fifth they are in the form of the ideographs of the Chinese names for the trigrams. These two circles differ in the association of the trigrams with the directions. The first has the trigrams of the active, male principle associated with the south and of the passive, female principle associated with the north. The other ring associates these trigrams with the northwest and the southwest. These correspond to two different historical associations of the trigrams with directions. The first is called the "prior to Heaven order" and the second is the "posterior to Heaven order." Both of these orders are discussed in the appendixes to the *I Ching*. Needham suggests that in the third century C.E. the shift from the old divining board and spoon to the needle in the compass influenced the change of the association of the directions with the trigrams.[52]

The contemporary Chinese siting compass shows evidence for changes in magnetic declination over history. Comparing the fifth circle with the twelfth and seventeenth circle, one finds shifts in the direction of south (the direction Chinese compasses pointed). In the twelfth circle, the direction south is shifted 7.5 degrees east, while in the seventeenth circle south is shifted 7.5 degrees west. These two circles are traditionally associated with famous early siters. Master Ch'iu between 713 and 741 C.E. introduced the *cheng chen*, or "correct needle system." Around 889 C.E. Master Yang introduced the *fengshen*, or "seam needle" system. In this second system, the declination had shifted 7.5 degrees to the west. In the twelfth century, Master Lai introduced the *chung chen* system, or central needle system, which was 7.5 degrees eastward of the astronomical north-south line.[53]

These systems correctly reflected the degree of magnetic declination at the time they were introduced. There was a shift from declination west to declination east of the North Pole in the latter two dates. Shen Kua had explicitly written about the slight declination of the needle as a scientist in 1088 C.E. However, the siting compass implicitly records in "fossil" form the practical awareness of the declination in earlier centuries. Nineteenth-century sinologist Alexander Wylie claimed to have found an eighth-century reference to declination in a Buddhist work, but no one since has been able to find his reference.[54] Another early reference is in the *Nine-Heaven Mysterious Girl Blue Bag Sea Angle Manual*. (The *blue bag* is a Taoist term for the universe; the *sea angle* suggests navigation.)[55] The earliest datable reference is in a poem by the founder of the Fusion School of Siting, quoted by Chinese encyclopedias: "It rides upon all three," making reference to the "three Souths," that is, the three systems of siting orientation found in the traditional compass rings.[56] These references to declination are over two centuries before the recognition of declination in Europe and before even the simple description of the magnetic compass appeared in Europe. It was only in the sixteenth century that the fact that the declination shifted with time was explicitly recognized and recorded.

There is also some possible indirect evidence of the development of awareness of magnetic declination in China in terms of the orientation lines of cities. It has been noted by observers that several of the old cities seem to have two differ-ent lines of orientation. For instance, in Nanjing the walls are oriented 13 degrees east. In K'ai-feng the walls are oriented 11.5 degrees east while the palace is ori-ented directly north and south.[57] With further archaeological investigation in terms of dating inner cities versus walls of some of the old Chinese cities, it may be possible to date the introduction of declination by the medieval Chinese siters.

Siting predates the compass, but of the two major schools, the one school of siting that used the compass arose in the maritime region of Fukien. There was less navigational use of the compass in China than there was in the West since much navigation was done in China's rivers and canals. The siting compass was then adapted for navigation. Possible evidence is that the ceremony for a ship's Taoist temple makes extensive reference to the compass and to famous Taoist siters, both real and mythical.[58] Taoist priests sailed on ships using the compass, and they may have kept the working of the compass secret from foreigners. This may account for the fact that early medieval Arab travelers do not report the compass, which perpetuated the misconception held by Westerners that the Chinese mariner's compass did not exist during the time of the Arab travelers.

The mariner's compass was used with great accuracy for ocean navigation in Southeast Asia by the Chinese. It may be that there were two routes of the com-pass to the West. The inland astronomer's compass pointed south as late as 1670 while the mariner's compasses pointed north. Hence, the astronomer's compass in Europe may have derived from the Chinese siting compass rather than from the maritime compass. The more probable route of the compass to Europe was the inland route. We do not find references to the compass among the Arabs before references appear in Europe. The Persian and Turkish names for compass mean "south-pointer," again suggesting Chinese origin.[59]

The relatively early date of the Chinese discovery of the compass had long been doubted because of the confusion caused by Chinese references to the "south-pointing chariot," or cart. Legends place this device at a very early date, which led some writers to claim the invention of the compass as early as 1000 B.C.E. These extreme claims to antiquity made Westerners totally skeptical of all Chinese claims to priority with respect to the compass and magnetism. For instance, Thomas Young, a British physicist (1773–1829) ridiculed this claim: "The Chinese are said to have been acquainted with the use of the compass above 3000 years ago; but in such accounts it is impossible to ascertain how much the spirit of national vanity may have induced the historian to falsify his dates."[60] However, the south-pointing chariot that led some authors to defend Chinese priority in the invention of the compass was in fact a mechanical self-regulating device that kept a pointer ori-ented toward the south as the vehicle turned.

Further confusion concerning the appearance of the compass in the West after China is caused by references to an orienting stone used by Norsemen before 1100. This were references to the double-refracting Icelandic spar crystal, which was used to locate the sun below the horizon on cloudy days.[61] The calcite mineral allowed them to find the polarization angle of light.

The sudden appearance of the mariner's compass in the West without intermediary development or geographically intermediary transmission by the Arabs suggests that the compass was brought by the Mongols. William Gilbert was perhaps close to the truth when he suggested, in *On the Magnet* in 1600, that it was Marco Polo who had brought the compass to Europe. It was not Polo himself but other earlier travelers from the Mongol Empire who most probably brought the compass. Petrus Peregrinus (Peter the Stranger, or Peter the Wayfarer) wrote a treatise on the magnet that is extraordinarily advanced and lacks predecessors. This sudden appearance of a finished treatise suggests Eastern influences on the Western theory of the compass.

CHAPTER 3

Major Movements of Chinese Thought

―――❧❦❧―――

$\mathcal{T}$he discovery of the compass was enabled and fostered by the outlook and understanding of the world prevalent in ancient and medieval China. Taoism, correlative cosmology, and the Neo-Confucian synthesis contributed to an understanding of the cosmos as a continuous whole in process whose parts spontaneously correlate. Although China did not develop lawful experimental science similar to that in the West, it did develop organic approaches to the world that aided an understanding of magnetism as well as of biological and social phenomena.

An Overview of Major Chinese Philosophical Systems

Needham emphasized the centrality of Taoism for Chinese science, focusing on the Taoist high valuation of nature. He was also sympathetic to Taoism's association with popular peasant rebellions and secret societies that had millenarian visions of future utopias. In this respect, science is linked to social movements that advocate popular influence and social betterment. Needham's own eclectic mixture of Marxism and Christianity allowed him to become a self-described "honorary Taoist," fusing a Romantic and religious appreciation of nature with an international social gospel of human betterment. The contrast of Taoism with Confucianism strengthens the argument for the importance of Taoism for Chinese science, as Confucius, like Socrates, lacked interest in the physical world. Confucius never spoke about peculiar natural phenomena or the ways of heaven. Most Confucian writers were almost entirely focused on human society rather than on nature.

The contribution of Taoism to science has been questioned, although the objections can be countered.[1] A popular Taoism absorbed many of the local spirit cults and shamanism. For the educated sophisticates there were the Taoist philosophers. Although the primacy of the aesthetic approach to nature in philosophy and the spiritual approach to nature in Taoist religion are claimed to be inconsistent with a scientific interest, this focus on nature and its creative powers might encourage positive attitudes toward the investigation of nature. The spiritualism of popular Taoism need not be an obstacle to science. There is historical evidence that in the West as well some millenarian, utopian movements with a concern for social betterment were supportive of science. Another objection to the view that Taoism was the major philosophical tendency in China encouraging science is the fact that the positive Taoist attitude toward nature did not include explicit interest in the detailed empirical investigation of nature. Although Needham interpreted passages in Chuang Tzu and Lao Tzu as encouraging the investigation of nature and taking a positive attitude toward empirical research, some have taken issue with this.[2] Even accepting this criticism, one can still claim that the Taoists' strong aesthetic and religious appreciation of the most varied phenomena of nature could be encouraging to science.

The Romantic movement in the West is often incorrectly portrayed as having a purely aesthetic attitude toward nature and being completely antiscientific. Romantic poets denounced the mechanistic approach to nature of Newton and his followers, but Goethe was concerned with investigation of phenomena in biology, geology, and psychology; Coleridge pursued interests in chemistry; and Novalis studied geology and wrote about mathematics. Romanticism's aesthetic and mystical approach to nature did encourage a number of positive scientific researches in physics, including the discovery of ultraviolet radiation, the investigation of electricity, and the formation of the electromagnetic field concept. Likewise, Needham's view of Taoism as a major positive contributor to science is defensible, despite the complications of the multiplicity of Taoist tendencies and Taoism's primarily aesthetic and mystical attitudes toward nature.

The Mohists (followers of Mo Tzu) certainly developed ideas that contributed to scientific activity. Their treatment of geometry in terms of minimal units (finite but pointlike entities), their interest in physical optics and mechanical military devices, and their development of logic and demonstration contributed to a science more like that of the West than any other Chinese school.[3] The Mohist movement and its ideas died out in the second century B.C.E., however, due to the decline of power by independent craftsmen with the consolidation of the empire.

Taoism is usually contrasted with Confucianism with respect to positive attitudes toward nature. Confucianism was the major official religion of the state bureaucracy, and bureaucrats had to concern themselves with such technological matters as military strategy and technology, agriculture, and economic production. Shen Kua, one of the most creative and prolific scientists of traditional China, spent his career as a government bureaucrat and advisor.[4] This involved overseeing of irrigation and land reclamation, defense of the borders of the empire, and keeping astronomical records in order to report peculiar phenomena to the emperor as omens.

Large government resources permitted technical projects on a scale that would not have been possible for individual amateur investigators, including the huge hydraulic irrigation projects and detailed astronomical and other records of natural phenomena for prognostication and omens. At the same time, the bureaucracy undermined original and creative scientific thought in various ways. The demands of political administration took away time and energy from systematic and extensive scientific investigation. The use of astronomy to determine the fate of the ruling dynasty contributed to the recording of novel phenomena in the heavens such as sunspots, new stars, and comets recognized as genuine celestial objects. At the same time, however, the use of these novel phenomena as political omens encouraged falsification and misdescription of events (such as eclipses) in order to encourage or criticize imperial policy. The extreme emphasis on literary and historical scholarship in the qualification tests for the bureaucracy discouraged the sort of matter-of-fact, unadorned prose, and literal, morally neutral description of natural phenomena that would have been valuable for the development of science.[5]

Ironically, Confucianism, which in many ways was not encouraging to science, was strongly antispiritualistic. Confucius dismissed or evaded questions about the afterlife and immortality. One might think that Confucian humanism with its nonspiritualistic tone would be encouraging to science, but in fact it contains a number of aspects that were less encouraging than the more spiritualistic views of the popular Taoists and the Mohists, who even believed in ghosts. This perhaps is similar to the role of religious movements and religious viewpoints in the early phases of the rise of Western science. In the late Middle Ages, theological issues about the power and omniscience of God encouraged speculation about different possible arrangements for the universe that God might have created, and undergirded the notion of the thoroughgoing rationality of the universe. Spiritualistic and theistic beliefs are more conducive to the initial stages of the development of science than antispiritual beliefs, despite the fact that the further development of science undermines and often destroys such beliefs.

There were tendencies within Confucianism that might have led to positive attitudes toward science, or at least toward technology. One was the combination of Confucianism and the correlative cosmology, but the main use of this was to justify and organize the social order, as did Tung Chung-shu, who made Confucianism the official doctrine of the emperor. The anecdote that Tung was so studious that he did not look into his garden for three years (meant to show his scholarly dedication) suggests that he had no genuine interest in nature itself but was concerned with fusing the traditional doctrines concerning nature with Confucianism for political and social reasons.[6]

In the doctrines of Hsün Tzu (the last major original Confucian thinker of the period of the warring states and of highly creative philosophy) one finds the possibility of the justification of interest in the technological exploits of nature. Hsün Tzu took over from Confucius the humanistic dismissal of concern with omens and prognostications of the heavens.[7] Many Confucians, before and after Hsün Tzu, held doctrines of the will of heaven and of fate. Hsün Tzu rejected these and claimed that the way of heaven was simply a sequence of events, that "the sage does not seek to know heaven."[8] Thus, Hsün Tzu's goal was not explanation but the search for patterns in phenomena, presumably patterns like those found in astronomy and in the weather, and which could be put to practical use. He neither wrote about technology nor had positive attitudes toward technological innovation. Nevertheless, a philosophy like his could, in principle, support technological development. "Instead of magnifying heaven and contemplating it, why not seek to domesticate and curb it? . . . Why not curb its decree and put it to use? . . . Why not unleash capabilities and transform them?"[9] Hsün Tzu was not influential when Confucianism was revived by the Neo-Confucians and was based on Mencius's rival theory of human nature. Still, within a Confucian bureaucracy concerned with matters of military, agricultural, and economic science, a philosophy similar to that of Hsün Tzu might have developed spontaneously out of the combination of Confucianism, practical politics, and technological demands.

Another trend in philosophy that gave some support to science was that of the legalists. Among the legalists were two students of Hsün Tzu who developed earlier ideas present among political advisors. These political philosophers rejected tradition in favor of universal, public principles of law. They were cynical about traditional rituals and treated them simply as customs. The legalists advocated quantitative methods and evaluation of laws in terms of objective consequences for behavior rather than good intentions or moral principles. The legalists applied their scientific and quantitative approach to the social realm but not to the natural realm, however, they developed a theory of the stages of society in terms of population growth resembling that of Thomas Malthus over two thousand years later.[10] The legalists' attempts to engineer social behavior looks even more strangely modern. They proposed to mold behavior solely by reward and punishment. Their teacher, Hsün Tzu, had described human action in terms of "arousal" and "response," similar to the twentieth-century behaviorists' concepts of stimulus and response.[11] Han Fei, the major legalist theorist, was far more behavioristic than Hsün Tzu and considered all human behavior as responses molded by rewards and punishments. He claimed that consistent enforcement of the law would be sufficient to mold society into any desired form, resembling twentieth-century American behavioral political scientists.[12] Legalists also advocated collection of quantitative information on population, land, and economic output. Legalists were developing something similar to modern behavioral, quantitative social science in a society that did not produce a quantitative physical science for two thousand years.[13] This shows that the conception of a quantitative social science does not have to be borrowed from a previously existing quantitative physical science, as it was in the West.

Just as the development of quantitative methods in the social realm did not automatically lead to the same in the natural sciences, so the development of skeptical attitudes in China did not lead to the replacement of religious viewpoints by scientific viewpoints. Chinese philosopher Wang Ch'ung of the later Han era (first century C.E.) developed a skeptical approach to many of the superstitious beliefs of his time concerning spirits, divination, and occultist forecasting.[14] Because of his skepticism, many modern writers have seen him as expressing a scientific attitude, but he did not really develop an alternative theory to the correlative cosmology that existed in his time. In fact, in some passages he seems to accept the general principles of that worldview and criticizes it only piecemeal.

When skepticism arose as a widespread Chinese philosophical attitude in the 1500s and 1600s (the same time it revived in the West), it did not lead to the triumph of empirical science as in the West. Rather, it led to the loss of belief in the traditional cosmological systems as well as in any other systems and therefore the unwillingness to replace them with new theories, including Western ones that were equally doubted.[15] Insofar as empirical methods did arise in the Ching era after 1600, they were applied to intellectual history and literary criticism. The so-called School of Evidential Research of the Ching era focused its energies on the analysis of language for the purpose of authentication and dating portions of classical texts. This did lead to a great deal of sophisticated literary scholarship; however, these inductive and quantitative methods were not used in the empirical sciences.[16]

Thus the naturalistic views of the Confucians, the quantitative methods of the legalists, the inductive methods of the literary School of Evidential Research, and the debunking of occultism by the skeptics probably all contributed less to physical science than did the aesthetic and mystical attitudes toward nature of the Taoists.

The Structure of Correlative Cosmology

Correlative cosmology is the Chinese system of correlations between several sets of qualities and aspects of the universe. A series of correlations are set up between the four directions, the four seasons, the five elements, the five planets, the five tastes, the five colors, and so on.

The earliest correlations were between the four directions and the four seasons. When the five "elements" (better called "phases"), earth, fire, water, metal, and wood, were introduced, the four directions were augmented by the "center" to make five, and the four seasons were awkwardly augmented by an intermediate period.

The attributes, states, or things that were ordered in correlative cosmology were also classified in terms of yin and yang in opposing pairs, for example, summer versus winter and fire versus water. Yin and yang originally meant simply shade versus light. Later, however, the meaning of yin and yang became generalized as referring to passive and active forces that became famous in Chinese thought.

The five phases and the two forces, yin and yang, are similar in some respects to the four elements and the principles of love and strife in the Greek pre-Socratic philosophy of Empedocles, whose four elements were taken over by Aristotle. The Chinese elements, however, differ from the Greek elements of earth, air, fire, and water (which continued to influence Western thought through the seventeenth century) in that the Chinese elements are better described as the five phases, or the five processes.

The Chinese phases are not used in a reductionistic manner as are the Greek elements.[17] That is, the Chinese phases are not the ultimate roots of things, as in Empedocles, but rather are phases of transformation of the the the all-pervasive *ch'i*.

They are "not five chemically or physically distinct substances, but five aspects of a process, the activity that drives it, or the substance that changes character in it."[18] For instance, colors or tastes or seasons are not made out of the Chinese phases.

The term "five phases"[19] emphasizes the fact that the five elements are transformed into one another, unlike the Greek elements. In fact, the transformation of one phase into another is an important part of the system of correlative cosmology. There is a conquest cycle that is based on the observations that water quenches fire, soil dams water, wood digs soil, metal chops wood, fire melts metal. There is also a generation cycle in which wood burns into fire, fire becomes ash (soil), soil produces metals, metals liquefy, and water allows wood to grow. The interrelations of these cycles can be further elaborated by noting that each phase generates the immediate predecessor of the phase it conquers.

The correlative cosmology includes not only cosmic processes but human beings and social institutions. Chinese medicine makes use of the correlations of the seasons with bodily organs and activities. This is similar to Greek, Roman, and medieval medicines' association of the seasons with the humors: blood, yellow bile, black bile, and phlegm. The Chinese system, however, is notably different from the West's in its emphasis on correlations with dynasties and behaviors of the emperor. For instance, the Yellow Emperor was naturally associated with yellow and soil, while the First Emperor was associated with black and water, as designated by the conquest cycle of his replacing the Chou dynasty, associated with red and fire.

In the *Lu Spring and Autumn* there is an account of the changes of clothing and behavior of a single emperor during the year, as well as changes of location in the palace. Earlier there were accounts of movement of the sovereign through different regions of the realm during different seasons.

The phases were associated with numbers, such as one for water, two for fire, three for wood, four for metal, five for soil, six for water, seven for fire, eight for wood, nine for metal, including a second cycle. The association of the numerical five with soil, and pairs of numbers of the two cycles, such as two and seven with fire, and the association of soil with the center and the four other phases with the four directions or cardinal points allows, with further manipulation, the construction of a magic square in which the numbers add up to fifteen in each direction. It is not known whether the magic-square principle was discovered by considerations of the five phases and their conquest cycle, or whether the numbers assigned to the five phases were derived from a magic square already discovered. Whatever the origin, the Chinese discovered magic squares earlier than any other culture and developed them into successive complexity.[20]

The *I Ching* or *Book of Changes*

The document and system that is by far the best known in the West and extremely important in Chinese correlative cosmology is the *I Ching*; or, *Book of Changes*, based on a system of divination. Because it was not destroyed in the legendary burning of the books by the First Emperor, and because it was made part of the Five Classics, later Confucians had to incorporate it into their philosophy.

The *I Ching*; *or*, *Book of Changes*, is based on a system of patterns—the hexagrams—which are combinations of broken and unbroken lines, presumably representing results of a divination process. This process has traditionally involved use of long and short sticks made from stalks of milfoil. The original divination device may have been different from the casting of the milfoil sticks, and in the contemporary West people use coins. The broken and unbroken lines are combined into eight possible combinations of three—the trigrams (which appear on the flag of the Republic of Korea). The trigrams are combined in pairs into the sixty-four hexagrams. The trigrams and hexagrams are then interpreted. The system of broken and unbroken lines was interpreted by Leibniz as a binary arithmetic (arithmetic using only 0 and 1), which today is used in the computer. Also, the sixty-four combinations formed by the hexagrams correspond to the DNA code.

The system of interpretation of the formal patterns of the hexagrams is complex both conceptually and historically. Legend claims that the *I Ching* was compiled around 1000 B.C.E. The book is based on peasant omens, current around the seventh or eighth century B.C.E. and compiled around the third century B.C.E.[21] Elaborate appendixes were added sometime between 100 B.C.E. and 100 C.E. A "summary meaning" and an "attached verbalization" is also attributed to each pattern. The verbalizations had begun to apply yin/yang cosmology. The Great Appendix, which was added by a process of accumulation in the first two centuries B.C.E., develops a more comprehensive metaphysical interpretation involving correlative cosmology and shows an interest in technological inventions.

The *I Ching* divination involves coordination or association of simultaneous events; however, the relationship to the future is probabilistic. There is an element of contingency and indeterminacy in the casting of the sticks. The forecasts or omens are not rigoristically deterministic and predictive. This structure shows some resemblance to quantum physics with its randomness and indeterminacy of statistical prediction combined with strange correlations of simultaneous events. Oddly, this resemblance is not noted by popular writers on physics and Eastern mysticism.

Commentators on the *I Ching* have related the system of hexagrams and omens to various features of the natural world. Needham notes that the possibilities for generating classifications are greater in the *I Ching* than in the five-phase cosmology. Astronomical characteristics, such as the motions of the Sun and Moon, were worked into the system after 200 C.E. Chinese alchemists, particularly concerned with the manipulation of time, in turn connected the astronomical correlations of the *I Ching* with the times of heating or adding reagents to chemical reactions. Times of successful heating (or so-called fire times) for chemistry were determined by the *I Ching* method. Biological observations were also fitted into the *I Ching* system; for instance, in the book *The Beetle and the Sea*, which accounted for the redness of blood by the traditional association of the color red with hexagram 29.[22]

There is some difficulty in assessing the extent to which the *I Ching* can be considered a positive guiding force in Chinese science. The use of the work in divination and its popularity, meant that much of its use was for nonscientific life orientation. Insofar as Confucian scholars commented on the book as a classic, however, and alchemists, physicians, and others made use of it, it did play a role in the conceptual framework of Chinese science.[23] The increased use of the *I Ching* coincided with periods of relative decline of the observational spirit of Chinese science and philosophy. The number of hexagrams combined with the ambiguity of omens and the variety of the interpretations could be used to incorporate just about anything. Nineteenth-century traditional Chinese scholars claimed that all of Western science was already included and understood in the *I Ching*. These extravagant claims, as well as the uncritical popular use of the *I Ching* in Western occultism, have led to a discrediting of the work for many in both East and West.

In spite of this, the *I Ching* did play a role in the use of correlative cosmology in science and may also have played a role in technological innovation.[24] Oddly, Needham, who is so willing to give Chinese protoscience every benefit of the doubt in comparing it to modern science, denies a positive role for the *I Ching* in technology. He says it would have been better had the *I Ching* been thrown into the sea.[25] A. C. Graham, in contrast, suggests that the openness to contingency and the clearing of the mind involved in the divination method can contribute to the freeing of the mind from preconceptions.[26] The Great Appendix attributes all major inventions to the *I Ching*, an exaggeration emphasizing interest in technology.

Needham suggests that the *I Ching* functions as a kind of "vast filing system," and suggests that it appealed to the Chinese because of its "administrative approach" to the natural world. He suggests that "associative organismic thinking" was "a mirror image of the Chinese bureaucratic society,"[27] and conjectures, "Taoists [were] in a highly organized society where bureaucracy was dominant, the atomists in a confederation of separate city states and individual merchant adventurers."[28]

As a scientist, Needham does not treat the modern, reborn organismic worldview of parts of twentieth-century physics and biology as a mirror of our bureaucratic society. Rather, he attributes our organismic science to "the fact that science today has to deal with realms of the universe that the Newtonian system did not envision."[29] However, the parallel argument is that the twentieth-century bureaucracy in the West is connected with the worldview that gave rise to quantum mechanics, general relativity theory, biological organismic theory, and more obviously, cybernetics and general systems theory.

The atomistic individualist worldview, present in parts of ancient Greece, practically disappeared during the Middle Ages, but revived in the seventeenth century along with political and economic individualism. Atomism completely dominated Western science until the mid–nineteenth century at least, and dominates popular images of science to the present, including those held by many scientists themselves. The rise of field theories in the nineteenth century was followed by the New Physics of the twentieth century, in which fields came to achieve equal status with particles. Organismic biology emphasized the important role of structures of matter with different levels of organization in biological organisms. More recently, chaos theory has been described as a nonreductionist theory of complex phenomena. All of these nonindividualistic theories have arisen during a period when economic individualism is in conflict with and modified by the rise of large governmental and corporate bureaucracies, in some ways similar to those of traditional China. Because of the similarity of the modern West and traditional China, China scholar John Schrecker has made the claim, which seems outrageous to many, that traditional China better fit Weber's conception of "modern" society than does early modern Europe. One can see his point, however. Early modern visitors to China were shocked by the Chinese refusal to participate in duels; Chinese bureaucrats would walk away from an insult.[30] Many other features of traditional Chinese bureaucracy are more like twentieth-century Western society than early modern Europe.

Historical Origins of Chinese Correlative Cosmology

Binary oppositions are universal in so-called primitive thought (and prevalent in contemporary Western thought as well). Analogues to the lists of contrasting qualities, similar to those classified under yin and yang in China, can be found in all cultures. Claude Lévi-Strauss has emphasized this universality of binary thinking and claimed that primitive thought is not different from civilized thought in its structure. The Pythagorean had lists of opposites, such as male versus female, light versus dark, odd versus even, and square versus oblong,[31] similar to the Chinese classifications, and also including a sexual polarity as well as an implicit value judgment in each pair.

Psychological tests of word associations often take responses in terms of such binary opposites as normal and other responses as deviant, revealing psychological quirks or perversions. Such polarities are present in the structure of ordinary language. Modern Western philosophy, until recently, emphasized logical argumentation and tended to dismiss polarities as primitive. As Graham points out, however, recently both British and American ordinary language analysis, based on the work of Wittgenstein, and French deconstruction, based on the work of Derrida, have brought back consideration of the primitive polarities underlying modern Western philosophy.[32] Gilbert Ryle,[33] criticizing the mind/body dichotomy, and philosophers of science who, following Thomas Kuhn,[34] examine opposing paradigms in science (such as field versus particle), have also highlighted these background polarities. Derrida has examined the value judgments implicit in the dichotomies that unconsciously guide modern philosophy, such as male versus female, speech versus writing, and same versus other. He has attempted to expose and undermine them by raising the value of the allegedly subordinate member of each pair.

Chinese correlative cosmology may be an extremely primitive one, but it is not found in written records until relatively late. It is not found in the earliest records on oracle bones (cracked bones for divination on which are found inscriptions), or in *The Book of Poetry*, or in *The Book of Documents*. The most relevant text in the historical records dates from the third century B.C.E., not from archaic times.[35] Benjamin Schwartz notes that since Chinese thought contains little in the way of myth, it cannot be subsumed under primitive, mythic thought. Certainly isolated pieces of the correlative cosmology are found earlier than its full-blown appearance—yin and yang for example. *The Spring and Autumn Annals* contain yin and yang as part of the six *ch'i*. In addition, there are references to the five phases in the *Annals*. It should be noted that the *Tso Chuan* tradition of interpretation of the *Annals* began somewhere between the late fifth century and the early second century. Like the "Ten Wings" added to the *I Ching* in the third century B.C.E., it is not as old as was traditionally claimed. Taoism contains much play with yin and yang but does not contain the correlative cosmology.

The *Historical Records* locates the beginning of a full-fledged correlative cosmology with Tsou Yen. He was a native of far northeastern China. His area of origin and the shamanistic characteristics of some of the other diviners from the northeast associated with him suggest possible shamanistic bases for his thought. Needham calls him a thorough student of nature, classifying China's geography, fauna, and flora, while Sivin says that the inventory is more one of population, geography, and natural resources.[36] Tsou then extrapolates beyond China to the rest of the world.

Needham says that Tsou Yen presented ideas that "carried with them an element of political dynamite which the feudal rulers were not slow to realize."[37] He goes even further to say that "if Tsou Yen had had the 'know-how' of the atom bomb in his possession, he could hardly have faced the rulers of the states with a steadier eye,"[38] meaning that the theory of the sequence of conquests of the five phases in its day was as powerful a weapon in its own way as the atom bomb is today. The First Emperor of China by defeating the house of Chou had left a vacuum of legitimation. Despite the multiplicity of warring states, Chou had retained the aura of the spiritual center of authority, even if no longer ruler of all China. When Chou was overthrown there was no center for spiritual legitimacy. The theory of the sequence of conquests of one by another of the five phases gave legitimization—mandate of heaven—to the new emperor, who could claim that his phase or element had conquered that of the previous rulers.[39] The First Emperor was to rule by virtue of the power of water, and was to wear black and associate himself with all other attributes correlated with water, since water followed fire (the symbol of the previous rulers) in the cycle.

The First Emperor rejected the traditional philosophical schools, such as Confucianism, and is supposed to have slaughtered the scholars and burned the philosophy books but kept practical works on agriculture and on divination. Tsou Yen was as much a diviner as a philosopher.[40] There are some resemblances between Tsou Yen and Pythagoras in terms of the combination of protoscientific, natural philosophy with conceptions of political organization and actual political activity.

Pythagoras is best remembered today for his mathematical discoveries and his doctrine of the reality of numbers ("things are numbers"); however, he also started a religious and political movement that was guided by aphorisms concerning taboos and proper behavior. He presented a table of opposing qualities, similar to that in Chinese correlative cosmology. Stories about Pythagoras suggest shamanlike activity. He was seen in two cities at the same time and bit the head off a poisonous snake. The Pythagorean movement came to was dominant in several

Italian cities that issued coinage with Pythagoras's image. Finally, however, Pythagorean rule was overthrown and Pythagoreans were burned in their meeting-house. Pythagorean rule, with its conception of an elite, its relatively high status and theoretical equality of women, and its combination of mathematical knowledge and political doctrine, was clearly a model for Plato's *Republic*.[41] The importance of Pythagorean ideas in the work of Plato, although recognized with respect to the Pythagorean belief in the reality of numbers and several other matters, is underemphasized in other areas—for instance, the Pythagorean treatment and role of women—and the existence of Pythagorean women philosophers[42] was a clear precursor and likely model for Plato's treatment of women guardians in *The Republic*. Also, Plato's later doctrines, presented orally, seem to have returned to Pythagorean number mysticism;[43] however, Pythagoras and Plato were unsuccessful in converting the Greek states to rule by their doctrines. In contrast, Tsou Yen successfully implanted his doctrines in Chinese civilization, lasting for over two thousand years.

In reaction to the political success of Tsou Yen, the philosophers began to incorporate cosmology into their systems of thought. Confucius had rejected the study of nature, but later Confucians incorporated the five phases into their theories.

Correlative Cosmology, the Discovery of the Compass, and Other Technologies

The discovery of the orientation of the magnetic stone, or lodestone, to the North Pole in the West occurred some fifteen hundred years later than in China. The mariner's compass was introduced in China at least a century before it was in the West as well. In addition, the Chinese discovery of magnetic declination, or the discrepancy between magnetic north or south and true north or south, was not made in the West until four centuries later. Magnetic induction—the process of magnetizing iron by remanence, heating it to the Curie point and then cooling it— was used some six hundred years earlier in China than in the West. Chinese correlative cosmology was a factor in the Chinese ability to make these discoveries precociously.

Magnetism was understandably mysterious. Einstein recalls how, as a child, he was fascinated by a compass his uncle gave him. In the West, mechanical causal thinking made it a struggle to accept and understand the workings of the magnet. In the Chinese correlative cosmology, all that had to be accepted was the coordination of the magnetized iron with the North and South Poles. A model of how the North Pole pulled or turned the compass needle was not necessary. In correlative cosmology, one simply discussed the orientation, or aligning, of the compass needle with an important polarity of the universe. The structure of the universe as a whole is reflected in the smaller scale operations of the compass.

A. C. Graham has claimed that causal thinking is at the basis of Chinese technology. Although Graham grants the role of inspiration by correlative cosmology in the exploration of magnetism, he writes, "The invention and development of the compass, however, in having to survive the practical testing of effects, brings us down to the realm of causal thinking, irrelevant to proto-scientific cosmology.

Causal thinking would have to be the main factor in the extraordinary fertility of invention in China. . . . The organizing of proto-sciences by correlation within a society's patterning of concepts has nothing to do with the extent to which the concepts are being used for causal explanation and practical invention."[44] Graham simply assumes that all technological thinking is causal thinking, confusing this with the valid point that primitive societies have workable technologies despite their mythic worldviews and explanations. Deploying and accepting the behavior of magnetic stones and iron itself, however, involved thinking of a different sort than that of efficient causality or mechanical causality prevalent in the West.

Correlative cosmology was influential in the development of the magnet and was also integral to a number of other Chinese observations and innovations. In geology, the seismograph was developed by the Chinese 1500 years before it was in the West. Prospecting for minerals in terms of plants that grow in the soil (geobotanical prospecting) was developed over two millennia before it was used in the West. Nonchemical, biological pest control, through introduction of insects to eat other (pest) insects, which is only now becoming prominent in the West, was used in China since the third century C.E.[45] In fact, it was introduced into the United States by USDA workers who visited China in the early twentieth century.

Additionally, the Chinese developed many innovations in the area of biochemical medicine. For instance, the science of endocrinology developed in China in the second century B.C.E. Several of these discoveries were made only in the twentieth century in the West, among them medical uses of the thyroid hormones, dating from the seventh century C.E. This was not done in the West until the 1890s. The Chinese also developed hormone therapy, using extracts from the placenta or extracting sex hormones from urine.[46] Inoculation against smallpox was developed some eight hundred years earlier in China than in the West.[47] With regard to geological innovations, the Chinese search for correlations that did not demand an understanding of physical causal connections was helpful. In the case of the magnet and biochemical innovations, the affinity of Chinese correlative cosmology for thinking in terms of fields and complex organismic systems of interaction undoubtedly helped the actual technological and medical manipulations.

Correlative Cosmology in Relation to Early-Modern and Twentieth-Century Western Theories of Nature

Whitehead's philosophy can be understood as a twentieth century, temporalized version of the earlier, more static philosophy of Leibniz, who thought his own philosophy similar to that of the Chinese Neo-Confucians and was fascinated by what he saw as the binary arithmetic of the *I Ching*. Needham was helped in understanding and sympathizing with traditional Chinese science by his interest in Whitehead's philosophy of organism, itself constructed both from the concepts of Einstein's special relativity theory and from ideas from organismic biology and philosophers of process and evolution.

The story of the understanding of the electromagnetic field itself leads to Einstein's special relativity theory. Needham suggests that Chinese cosmology resembles the organismic conception of reality, which is not simply the sum of externally related individuals or atoms, but the universe as a whole or larger units within it having an organization that shows itself in the orientation and behavior of their parts. This organismic conception also can fit well with the older doctrine of macrocosm and microcosm. The whole (macrocosm) is reflected in each part (microcosm). For instance, just as each cell contains in its DNA the genetic information of the entire individual organism, so the individual human reflects the structure of the universe as a whole.

Perhaps Needham goes too far in saying that the Chinese, in effect, attempted to leap from a prescientific worldview to a post-Einstein science without going through the intermediate Galilean phase, but the affinity of the traditional Chinese worldview for field theory is rightly emphasized by his claim. Needham writes, "There are two ways of advancing from primitive truth. One was the way taken by some of the Greeks . . . that one ended up with a mechanical explanation of the universe. . . . The other is to systematize the universe of things and events into structural patterns which conditioned all mutual influences of its different parts. On the Greek worldview, if a particle of matter occupied a particular place . . . it was because another particle had pushed it there. On the other view, the particle's behavior was governed by the fact that it was taking its place in a 'field of force.' . . . Causation here is not 'responsive' but environmental."[48]

The organismic aspect of Chinese cosmology puts it in strong contrast with a mechanical worldview. The coordination of the parts to the organization of the system as a whole is what is explanatory, rather than the physical influence of the parts on one another. As a trained biochemist and proponent of twentieth-century organismic biology, Needham was aided by his own studies in biochemical embryology in grasping the spirit of Chinese science.

One of the metaphors Needham uses is that of a "harmony of wills": "rather like dancers in a country dance, none of whom are bound by law or pushed by others, but who cooperate voluntarily."[49] Needham also speaks of "cybernetic control."[50] Another image from biology he uses is that of the "endocrine orchestra." He says that "the universe is a vast organism, with now one component, now another, taking the lead at any one time, with all the parts cooperating in a mutual service which is perfect freedom."[51]

Similarities and Differences of Correlative Cosmology and Medieval and Renaissance Western Thinking

In explicating Chinese correlative cosmology, Needham uses modern models of physical field and biological organism. Schwartz suggests that there are conflicts between some of Needham's models. In the usual model of organism, the whole is considered to control the parts. The whole exists at a higher level than the parts. Often, organismic theories are hierarchical. Schwartz points out that Needham's reference to cybernetic control infers a kind of higher level guiding principle.[52] Needham, however, wishes to contrast Chinese with Western thought in terms of the lack of such guiding principles.[53]

Schwartz says that "the difference [between Chinese and Western organismic thought], if any, is a matter of nuances rather than absolutes."[54] Needham is correct insofar as he emphasizes the spontaneous coordination aspect and is perhaps misleading insofar as he draws on Western hierarchical control models. This is where his Taoism and Marxist materialism come to his aid. Form emerges from matter, it is not imposed upon it externally. In the medieval West, Platonic and Aristotelian models of hierarchical control produced a different organismic cosmology than that of the Chinese.

Needham is correct in saying that the use of purposive souls, entelechies, animas, archaei "was exactly the path Chinese thought had *not* taken."[55] Interestingly, Giordano Bruno, a Renaissance philosopher, explicitly rejected the Aristotelian model of form imposed upon matter. In addition, he rejected as sexist the Aristotelian metaphor of form as male and matter as female.[56]

A.C. Graham questions the contrast between Chinese and prescientific Western cosmology.[57] There is a significant difference between ancient and medieval Western cosmology and traditional Chinese cosmology in their relation to science. Even if Greek cosmology rested on early theories of polarity and analogy, a mathematical, geometrical science of astronomy and of physical statics arose that is clearly far different from Chinese science. Greek Platonic philosophy with its conception of nonphysical mathematical form, Aristotelian philosophy with its conception of substance, and Greek philosophy in general with its notion of purpose clearly head in different directions from Chinese philosophy. Neither abstract form nor permanent substance nor structures of purposes play a central role in Chinese philosophy. Certainly, the Hippocratic and Aristotelian doctrines of the four elements, four humors, and so on, and later connections made to the planets have resemblances to the Chinese theory of the five phases. Indeed, the Greeks and medievals had a fifth element—the quintessence—of which the heavenly bodies were made.

In the early Middle Ages, in the West, the doctrine of the elements and humors and various correspondences dominated the worldview. Nonetheless, systematic science was not well developed. In the later Middle Ages, as systematic science and rigorous metaphysics developed, it was Aristotelian theory that gave the main impetus to scientific theorizing. Thus, although significant aspects of the system of correlations were retained, the most developed scientific and philosophical articulation was based on Greek philosophy even if considerably modified.

Substance, purpose, efficient causality, and hierarchy emanating from a creator God dominated Western systematic thinking. The notion of creation from nothing, creation ex nihilo, is a notion unique to the fusion of Greek and Hebrew thought. This led, on the one hand, to a conception of a totally logical universe in which there was no surd (incomprehensible preexisting matter), since all had been created by an all-knowing God, which led to the drive for a totally logical or mathematical understanding of everything. On the other hand, the conception of creation ex nihilo with an emphasis on the inscrutability of God's design, led to a purely empirical approach to the universe, in which the rationality of things could be discovered only by piecemeal observation. Both Western rationalism and empiricism can be traced to this doctrine of creation from nothing.

The relationship of Chinese correlative cosmology to Western Renaissance hermetic philosophy is a close one. Needham has written that the sixteenth- and seventeenth-century natural philosophers, such as Bruno and Fludd, would not have found the Chinese cosmology so alien to themselves as it is to us. Graham has used the thought of Johannes Kepler as a point of departure for comparing traditional Chinese and Renaissance protoscience. Graham, however, errs in considering the comparison with Kepler or Bruno and Chinese thought primarily in terms of "cosmos-building of a proto-scientific sort," and claims that "the only technology proper to correlative cosmos-building itself is magic as in Renaissance Europe."[58] With respect to the compass, both Chinese correlative cosmology and Renaissance hermetic cosmology contributed positively to genuine science and technology. Kepler theorized that it was magnetic force, rather than what we would call gravitational force, that bound the planets to the Sun. Renaissance occultism also guided the development of the conception of force in Newton's theorizing about gravitation and contributed a primitive precursor to the field concept.

Graham also significantly notes that German *Naturphilosophie* (the philosophy of nature) is a later version of this thought in the nineteenth century.[59] This, too, made significant contributions to Ørsted's theory of electromagnetic induction in the world and to Faraday's concept of the electromagnetic field. Oddly, Graham illustrates the contributions of German *Naturphilosophie* to science only by an example of stereotypical, racist thinking tied to the four elements. Lorenz Oken, a German biologist, correlated black, white, yellow, and red "races" to the four elements: earth, fire, air, and water. He seems not to recognize that along with such occasional, speculative, racist fantasies there were also valuable conjectures concerning the physical and biological world in German Romantic science. German Romantic nature philosophy, which has some interesting relationships to earlier forms of correlational thinking and which itself led to the universe of field theory, structurally resembles certain aspects of traditional Chinese thought. Historians after Needham may claim that Needham's comparisons of Chinese thought with modern physics and Whitehead's cosmology are ahistorical, but historical sequence can be traced from the contributions of correlative cosmology both in its Chinese form and in its Western Renaissance and Romantic forms to that of modern physics. Just as two of the founders of quantum mechanics (Erwin Schrödinger and Werner Heisenberg) returned to the sources of Greek pre-Socratic thought, so we may return to the Chinese cosmology in its role in the discovery of the Earth's magnetic polarity and the compass to find part of the origins of our own thought.

Unity or Diversity of the Tao? Taoism, Chinese Science, and Political Movements

Taoism is widely understood to be a single, unitary philosophy and movement, but several recent scholars have emphasized the diversity of philosophies, religious movements, and political tendencies called Taoist. There has been an overreaction to popular, broad-brush characterizations of Taoism when revisionist scholars deny that there are any unifying features. The redating of Lao Tzu to centuries after Confucius, and even after his supposed follower, Chuang Tzu, has challenged popular stereotypes, both Chinese and American, concerning the course of Taoism. The new studies of Taoism as a religion have led some to deny connections among the ancient Taoist philosophers, movements of folk religion (some of them revolutionary), and the later established Taoist religion politically subordinated to the official Confucian doctrines of the bureaucrats. One could similarly deny the existence of Platonism by distinguishing between Plato (including various phases), Neoplatonism, Platonism in mathematics, and various artistic uses of Platonism. The history of Taoism is complex and tangled, but certain core religious, philosophical, and political values can be found throughout its various phases.

Needham's Taoism: Scientific and Democratic

Needham's claim that Taoism was the Chinese philosophical and religious movement most supportive of science has been influential.[60] Needham favored Taoism because of its acceptance of nature, its emphasis on transformation and "process philosophy," its association with antiestablishment and individual rebellion, its early connections with peasant revolts, and its concern for female sexual satisfaction. Needham characterizes Taoism as "strongly magical, scientific, democratic, and politically revolutionary."[61] Taoism as naturalistic and individualistic is contrasted to Confucianism as state oriented and bureaucratically conservative. Taoism has shown a positive image of and role for women, it was born of revolutionary political movements, and has been the main popular religion of the Chinese people.

Revisionist Views: Denial of a Unified Notion of Taoism

Nathan Sivin has criticized this strong identification of science with Taoism and doubted the unity of a variety of what are—according to him—very different movements under the name of Taoism.[62] Michel Strickmann, one of the most acerbic modern opponents of the blanket term Taoist, demands that the term be restricted to those who have allegiance and can trace their roots to Chang Tao-ling, one of the founders of the organized Taoist religion. He condescendingly notes that the vulgar Americans, who have numerous translations of Lao Tzu and Chuang Tzu (but, until recently, none of the major Taoist religious scriptures), will of course find this view incomprehensible.[63] Strickmann wishes *tao chia* (Taoist philosophy) to revert to a mere library classification.[64]

A stronger case can be made than is allowed by the revisionists for the connection of Taoism with science, as well as with radical political movements, and as having a more egalitarian role for women than other Chinese movements. In order to do so, however, it is necessary to look at the numerous social strands and tendencies that have been unified under the name of Taoism.

Taoist Philosophy versus Taoist Religion

The simplest traditional distinction is between Taoist philosophy (*tao chia*) and Taoist religion (*tao chiao*). The former consists of the works of Lao Tzu and Chuang Tzu of the fourth and third centuries B.C.E. and those writers who commented on them, utilized their imagery, or attempted to imitate their style. Taoist religion became institutionalized in the late second century C.E. and consists of organized religious hierarchies led by the Celestial Master. Taoist religion incorporates a great deal of popular religion and is usually much less sophisticated than the writings of the early Taoist philosophers. Some Western writers have wished to separate totally so-called Taoist philosophy from Taoist religion.[65] Japanese scholars of Taoism have also claimed that Taoist philosophy and religion were in the beginning separate, not aspects of a single phenomenon.[66]

It cannot be denied, however, that the *Tao Te Ching*, attributed to Lao Tzu, is a work of philosophical Taoism that became a central scripture of religious Taoists.[67] Indeed, the members of the early Yellow Turban movement memorized Lao Tzu as part of their religious activities.[68] One of the early commentaries on Lao Tzu, the *Lao-tzu Hsiang-erh Chu,* is plausibly attributed to Chang Lu, one of the early Celestial Masters.[69] Strickmann undermines his own dismissal of the philosophical classics from Taoism by granting that over half the canon is composed of commentaries on the major philosophical texts.[70]

Lao Tzu: Legend and Text

The traditional view of the life and times of Lao Tzu placed him around 550 B.C.E. as a teacher of rites to Confucius. This "Lao Tan" is the only figure in Confucius's *Analects* to assume sufficient superiority to Confucius to call him by his first name. According to legend, this Lao Tzu became discouraged with developments in China and traveled west.[71] According to later legends, he became Buddha.

It seems likely that the original "Lao Tan" was simply a figure in Confucian literature. Later Taoists, such as Chuang Tzu, used him as a figure to criticize Confucius because of Lao Tzu's superior role as teacher of Confucius. Even in the writings of Chuang Tzu, Lao Tzu is not yet a major philosophical sage. Chuang Tzu has numerous people berating Confucius in the various stories. The "mad man of Chu" receives more attention as a critic of Confucius than does Lao Tzu. The *Chuang Tzu* may indeed have initiated the tradition of making Lao Tzu an ancient Taoist sage. Obviously the later story about Lao Tzu being Buddha was developed in the early centuries C.E. to counter the influx and popularity of Buddhism.

Recent scholarship suggests that the *Tao Te Ching* of Lao Tzu was actually composed some three hundred years later than the traditional date, around 250 B.C.E. The book is marvelously suggestive, mysterious, and enigmatic, which accounts for its seven hundred commentaries in traditional China and over a hundred different English translations. The work was probably composed from traditional sayings attributed to the old man, Lao Tzu, and may include ancient shamanistic songs of earlier, popular religion. The person who wrote or compiled the work was a genius who produced a unified, profound, and poetic work that rapidly made a huge impression. Part of its power is its reversal of the customary beliefs of the reader. "The aphorisms of *lao-tzu* hit the reader as successive blows from opposite sides."[72] The doctrines of nonaction and the emphasis on the reversal of the valuation of the traditional elements of the polarities of Chinese thought (such as the feminine being valued above the masculine, or the dark as valued above the light, or the passive as valued over the active) all had a tremendous impact.

Chuang Tzu: At Play with Philosophy and Language

Chuang Tzu's book, one of the masterpieces of world philosophy, is a tremendously rich work emphasizing the acceptance of the natural processes and of natural transformations, including that of life into death. The *Chuang Tzu* presents philosophically profound ideas in the form of stylistically brilliant fables and extremely funny jokes. It can be read either as amusing literature or as a profound absorption of the logical paradoxes of the Chinese sophists for the purpose of emphasizing the limitations of any set of linguistic categories in some areas of application.

Chuang Tzu, like Shakespeare, can be appreciated and enjoyed on many levels. The deceptively simple stories, fantasies, and jokes assimilate and criticize most of the philosophical currents of his time. Like Plato, the "dramatist of the life of reason,"[73] Chuang Tzu presents sophisticated philosophical arguments in the form of powerful and moving dramatic exchanges. Chuang Tzu, like Plato, combines deep philosophical insight with a seductive literary style. This may seem an odd comparison, since Chuang Tzu's positions are opposite to Plato's. Plato initiated the Western search for eternal absolutes, unity, and objectivity through mathematical precision. Chuang Tzu argues for the primacy of process and vagueness, the inadequacy of reason, and the naturalness of opposing perspectives that cannot be transcended with some overarching point of view. Plato devalues the body and the world of process, while Chuang Tzu values immanence and change. Plato attempts to establish a rigid value hierarchy, while Chuang Tzu advises "seeing all things as equal." Like Plato, however, who incorporated the Greek logical paradoxes of Zeno and the linguistic analysis of Cratylus into dramatic literary dialogues, Chuang Tzu turns the Chinese logical paradoxes and linguistic contradictions of his friend, logician Hui Shih, into dramatic dialogues and poetic and fables that can be read for their humor and metaphorical brilliance. (A Chinese animator has made Chuang Tzu into a comic book.)[74] They incorporate the logical and linguistic dialectic of the Chinese sophists and analysts of language, while going beyond them to evoke in the reader the changes of viewpoint and consciousness that Chuang Tzu reveals. Chuang Tzu has so mastered the logical and grammatical dialectic of his time that he is not lost or buried in it, like the sophists, but is able to play with it, as he plays with the imagery of the traditional Chinese myths and stories. The use that Chuang makes of the Chinese sophists—borrowing their dialectic while using it to transcend their position—is like the use Plato makes of the Greek sophists, who serve as foils for Socrates's arguments.

Philosophical Taoism was continued two centuries later, though on not so high a level, in the *Lieh Tzu* and the *Huai-nan-tzu*.[75] These works carried on the naturalistic emphases as well as the individualism and hedonism of Chuang Tzu. Chang Chan, who published the *Lieh Tzu* from manuscripts supposedly handed down in his family, also wrote about alchemy and life-extension techniques. This shows that philosophical Taoism and Taoist religion were not necessary disjointed.[76]

Lieh Tzu is of disputed date, but contains passages the same as or similar to those in *Chuang Tzu* (some of which may have been included in earlier, lost versions of the *Chuang Tzu*), along with added material, including a lengthy paean to hedonism. The *Huai-nan-tzu*, a large compilation of the numerous scholars who were patronized by the duke of Huai-nan, incorporates Taoist ideas into a veritable encyclopedia of cosmology, politics, and much else. It also shows that those interested in currents close to philosophical Taoism were also interested in military strategy.

Although the philosophical Taoists were not at all explicitly concerned with religious ceremony, one finds reference to techniques of breathing, meditation, and physical exercise that were to become part of later, popular Taoism. Lao Tzu makes several references to breathing like a baby in the womb. Chuang Tzu refers to "breathing from the heels" and to exercise techniques in which one moves like a bear or stretches like a crane.[77] Both in content and in rhyming and style many of the passages in Lao Tzu resemble primitive, shamanistic hymns, especially songs or prayers for initiates. Many chapters conclude with the enigmatic, "What is it? It is this," which suggests a religious function. The earlier, longer version of the *Chuang Tzu* contained numerous chapters of magic and divination, which were edited out by later commentators and compilers. Some of these passages survive in other works, such as the *Lieh Tzu* and *Huai-nan-tzu* and later commentaries.[78]

The book *Chuang Tzu* contains writings by others besides Chuang Tzu. The first seven chapters, the so-called inner chapters, are agreed to be by Chuang Tzu. Other chapters, although somewhat in the style of Chuang Tzu, are thought to be by writers who shared his antipolitical individualism. One group of chapters (8 through 11) have been called primitivist because of their emphasis on the superiority of the early natural state of humankind. Another set of later chapters (28 through 31) have been called the Yangist because they echo the hedonism originally espoused by Yang Chu.[79] The primitivist chapters are sometimes called anarchist for obvious reasons and include Robber Chih's disquisition on robbers' ethics which mockingly mimic Confucian terminology. The Yangist chapters also include one in which the bandit Chih refutes Confucius, showing his morality to be hypocrisy. There was a tradition (allegedly originated by Yang Chu) of "fostering life," which was criticized by the Confucian Mencius, among others, and was developed by the *Lieh Tzu*, also containing some lost passages from the *Chuang Tzu*. These primitivist and anarchist political tendencies thus do appear within the so-called philosophical Taoism of the *Chuang Tzu*, and link philosophical Taoism to the outlook of the political rebellions that moved Taoism into its place as a major organized religion.

Taoist Religion: Its Birth in Early Peasant Rebellions

Taoist religion burst into history in the rebellions around 200 C.E. that brought about the end of the Han dynasty and several centuries of political fragmentation. In the ensuing rise of local warlords, as well as in the development of the Taoist and Buddhist religions, there are parallels to the fall of the Roman Empire and the early Middle Ages in Europe. In China the period of chaos and disruption ended in the brilliant T'ang dynasty, while the West suffered four more centuries of Dark Ages and never reconstituted its empire the way China did.

We become aware of Taoist religion for the first time in a series of revolts by Taoist political organizations starting in 184 C.E. The Yellow Turbans captured eight provinces of China, two-thirds of the country; however, they were immediately crushed because their leader had such great faith in the automatic coming of the Great Peace (*T'ai P'ing*) that he did not make timely military preparations for self-defense. Another rebellion, of the Five Pecks of Rice movement, was able to hold a province for decades, and its leader was able to surrender on good terms and gain a court position as a religious leader. Another major rebellion among many others was that of Sun En in the fourth century.[80] Sun En's rebellion shows that one must carefully interpret the claim of Sivin that *orthodox* Taoist religion was conservative and accommodated the status quo since the end of the Han, around 200 C.E.[81] Rebellious Taoist sects then become *by definition* unorthodox. It can be granted that institutionalized religious Taoism (as opposed to movements like that of Sun En, which mixed Taoist religion with strong doses of local popular religion) did become supportive of the ruling regimes after its rebellious birth. In debates with Buddhist monks that occurred during the succeeding centuries, the Taoists played down their politically revolutionary origins, and the Buddhists used reference to the early revolts to attempt to sway the listening emperor against Taoism.[82] Even among the early Taoist theocracies, one was conservative and explicitly authoritarian.[83]

The association of early religious Taoism with peasant revolts and secret societies is the component that leads Needham to associate Taoism with radical political revolution. Its antiestablishment individualism, which one finds in Chuang Tzu and in the Seven of the Bamboo Grove, is apparently apolitical but antihierarchical, following the primitivist (6–11) and the Yangist (28–31) chapters in the *Chuang Tzu*.

The ideology of the radical peasant movements that attempted to replace bureaucrats and literati with a more democratic or at least a different theocratic rule was largely expunged from the historical record by the resurgent bureaucrats after the defeat of the revolts. Thus it is difficult to extract the genuine beliefs from the documents available. The text we have of one of the major testaments of the rebellion is believed to have been extensively rewritten by the literati. Nevertheless early Taoist religion showed greater equality of the sexes. All offices except the highest one (the Celestial Master) were open to women, and women served at very high levels in the religious hierarchy.

Huang-Lao, Taoism, and Political Theory

Lao Tzu contrasts with Chuang Tzu in having an interest in the techniques of political rule. The advice that Lao Tzu gives to the ruler seems paradoxical, since rule through nonaction seems to envision a rural peasant utopia in which the people are kept simple by emptying their minds and filling their bellies; however, rule through nonaction was earlier advocated by Confucius in his opposition to published laws and punishments. He portrayed an earlier emperor who simply "faced south," and all went on correctly in his state. Chinese legalist Han Fei, an advocate of published laws and strict punishments, wrote an appreciative commentary on Lao Tzu and was part of the movement of Huang-Lao. That the legalists, who believed in a strong state machinery of laws and punishments, should have allied with the mystical nonaction of Lao Tzu may not be paradoxical. Creel has suggested that the term *nonaction* itself arises in the writings of Shen Pu-hai,[84] wherein it means that the emperor should not be busied with the details of policy, but simply delegate them to able ministers and bureaucrats—not bad advice for an executive. The copy of Lao Tzu found in the 1970s in the Mawangdui tomb puts the division on *te* (virtue or potency) first and the metaphysical portion on *tao* second, the opposite of the later editions. Its order may have been dictated by the legalist interpretation of Taoism. The Mawangdui silk manuscripts are generally taken to present Huang-Lao doctrine in which the Lao is indeed Lao Tzu. It has recently been suggested that the major work, the *Kuan-tzu* (or *Guanzi*) is part of the Huang-Lao tradition, and so its "importance soars."[85]

During the period after the First Emperor of the Qin dynasty's brutal rule, the emperors of the early Han dynasty (second century B.C.E.) had an interest in Huang-Lao doctrine. Also, a number of emperors and royalty in the later Han (first and second centuries C.E.) are claimed to have "greatly loved Huang Lao."[86] Henri Maspero, a great French scholar of Taoist religion of the first half of the twentieth century believed that the god of Taoist religion, Huang-Lao, was a unification of the Yellow Emperor (of later medical fame) and Lao Tzu, as did such later scholars as Wing-Tsit Chan,[87] but later Japanese scholars suggested that the god Huang-Lao had nothing to do with Lao Tzu. Archaeology may settle the issue to some extent.[88]

The first book in English on the Huang-Lao Mawangdui tomb silk manuscripts characterizes the Huang-Lao position as one in which the ruler bases himself on nature and not simply on benevolence.[89] According to Huang-Lao the ruler devises laws (which are grounded according to this account in a natural-law position) based on a normative aspect of nature. Huang-Lao doctrine limits the range of coverage by laws relative to the legalists and limits the power of the sovereign in contrast to the absolutism of the First Emperor of the Qin dynasty. Huang-Lao is an eclectic position that rejects the dependence on the benevolence of the emperor in Confucianism and the nonaction of individualist Taoism, but restricts the scope and rejects the fiat nature of the laws in legalism. Huang-Lao combines Taoist naturalism with the importance of law in legalism. Although Huang-Lao dominated the early decades of the Han dynasty, it was rejected by the emperor Wu, who may have demanded a doctrine that justified more personal power for him.[90] It may be that the *fang shi* (magicians), who gained access to the court offering immortality, were theoretically influenced by Huang-Lao and that Huang-Lao doctrines were conveyed by them into early Taoist religion.[91]

Neo-Taoism: Modern Category for Confucian Literati or Genuine Revival of Taoism?

The term *Neo-Taoism* was originally used in the earlier part in this century to refer to Taoist religion; however, it came to be used by influential historians of philosophy (such as Fung Yu-lan and Wing-Tsit Chan) for the revived philosophical Taoism of the early centuries C.E. Strickmann dismisses the term *Neo-Taoism*, quoting Zürcher's history of Chinese Buddhism.[92] One cannot deny that there was a revival of philosophical interest in Taoism after the collapse of the Han empire and the subsequent discrediting of Confucianism. With the shift away from Taoism centered on Lao Tzu, evaluated solely in terms of its function in advising the ruler, the individualism and aestheticism of Chuang Tzu could be truly appreciated.[93] Even though Chuang Tzu's philosophically sophisticated work did not become popular until the T'ang dynasty of the seventh century, the free-spirited intellectuals of the third to fifth centuries frequently refer to passages from Chuang Tzu.[94]

Components of naturalism and individualism were incorporated by a number of scholarly commentators on Lao Tzu and Chuang Tzu as well as by a number of literary Romantic nonconformists. One tendency was much more metaphysical and systematic, while the other was literary, poetic, and practical. Both developed themes of Taoist philosophy that went against the Confucian emphasis on social organization.

The brilliant Wang Pi (sometimes rendered Wang Bi), dead at twenty-four, wrote metaphysical commentaries on the *I Ching* as well as on Lao Tzu,[95] which is considered by many to be the greatest. His doctrines were called the "dark learning" in that they emphasized the mysterious nature of nonbeing. A major commentary on Chuang Tzu by Kuo Hsiang appeared a half century later and emphasized being and the multiplicity of things.[96]

Some have claimed that *neo-Taoist* is an inappropriate term for these commentators, because in social theory they were very much traditionalists or Confucians;[97] however, their interpretation, emphasizing the unformed, justified the innovations of the new Wei dynasty and replaced the Confucian New Text School of the Han. According to one account, Wang Pi was executed at age twenty-four as part of the reaction against the rejection of institutional and ritual forms,[98] while according to the more usual account he died of a plague. It has been claimed that his "dark learning" played a role in the justification of the novelty of the rule by Ts'ao Ts'ao, the son of one of whose concubines, Ho Yen, was the protector of Wang Pi.[99] Ts'ao Ts'ao was a brilliant military leader, Machiavellian usurper of the throne, a learned commentator, and gifted poet.[100] Whatever the position of Wang Pi—whether Confucian insofar as he supported the literati or Taoist insofar as he rebelled against the previous dynasty and its ceremonies—the commentaries on the Taoist philosophical classics by these legitimizers of the new dynasty's innovations made available intellectual and metaphysical interpretations that were used by intellectuals sympathetic to Taoism over the succeeding centuries. With the incorporation of Taoist nature philosophy into Neo-Confucianism these so-called Neo-Taoist commentators assisted the subsequent massive impact of Taoism on the official Chinese philosophy of nature. Kristofer Schipper, who thinks that the Taoist component of Neo-Confucianism is insufficiently emphasized, agrees with Strickmann in claiming that Wang Pi "was not a Taoist."[101] Wing-Tsit Chan claimed that Wang Pi originated the concepts and language (*li* as principle, with *t'i* as substance and *yung* as function appearing together as a pair of concepts) by which Neo-Confucianism interpreted and absorbed Taoism.[102]

In contrast to the rationalistic school of metaphysical commentators was the sentimentalist School of Light Conversation, which emphasized poetry and elegant, witty conversation on art and sex.[103] When power was usurped by a new chancellor, supporters of the earlier advisors of the Wei dynasty were powerless but openly opposed the puritanism of the new faction. Thus even in the apparently antipolitical poems and behavior of this group, there was a political stance of partially disguised or intellectualized opposition to the values of the ruling faction, maintaining an aspect of Taoist rebelliousness. This was not the revolutionary and millenarian popular rebellion of the beginnings of religious Taoism, but rather an upper-class moral rebellion and evasion of political service in opposition to the regime.

The famous Seven Worthies of the Bamboo Grove were a group of aristocratic nonconformists who in drink, drugs, and nudity attempted to find self-expression and pleasure. Perhaps the most famous of them, Juan Chi, was known for his pose of continual drunkenness by which he avoided the political infighting of the time. Numerous anecdotes describe how he flouted the conventions of dress, mourning, ceremony, and manners.[104] Another of the Seven, Liu Ling, went around naked in his house, and when he was chastised by an unexpected visitor said that Heaven and Earth were his house and his house was his clothing and that the visitor should get out of his pants![105] He also wrote a "Hymn to the Virtue of Wine." The supporters of this way of life could cite Chuang Tzu's passage about the drunken man who does not become injured when falling from his carriage because he is relaxed and not aware that he is falling. They used it as a justification for their use of drink and drugs where Chuang Tzu meant it as a parable.[106] Yet another of the Seven, Hsiang-Hsu wrote a major Chuang Tzu commentary, probably incorporating (or plagiarizing) Kuo Hsiang's commentary, which remains influential to the present.

Romantic Neo-Taoists were for the most part not religious priests, nor were they serious philosophers, as were Lao Tzu and Chuang Tzu and their rationalist Neo-Taoist commentators, such as Wang Pi. Nevertheless some were connected with both Taoist alchemy and Taoist religion. Hsi K'ang (one of the Seven), was a serious philosopher who wrote essays on morals and aesthetics.[107] He was also a pursuer of immortality, and was designated an immortal in religious Taoism after his unjust execution.[108] While working at a forge (showing both alchemical interests and a willingness to work like a lower-class artisan), he snubbed a visiting dignitary by continuing to work and ignoring him, thus gaining his enmity. This led to Hsi K'ang's execution, when the dignitary was in charge of trumped-up court proceedings.[109] Hsi K'ang neatly fits Needham's image of the scientific and egalitarian Taoist.

Alchemy and Physical Immortality

Pursuit of bodily immortality is one of the central themes of Taoist religion, and for some a necessary condition for a doctrine's being considered Taoist.[110] This pursuit led to the use of breathing techniques, exercises, sexual regimens, and alchemy to produce elixirs to lengthen life or even produce immortality. Early (pre-Buddhist) China contrasts strongly with Europe in that the Chinese did not think of immortality in terms of a nonworldly heaven and hell, but in terms of contin-ued physical life in this world, whether on Earth or among the stars. Though nei-ther Lao Tzu nor Chuang Tzu was an advocate of physical immortality, later Taoist religion emphasized it. This quest for immortality surprisingly links Taoism both with politically progressive tendencies in the status of women and with Taoist sci-ence in the form of alchemy.

Taoism recognized both an outer and an inner alchemy (*nei tan*). The outer alchemy involved the chemical transformation of substances. This could lead to the production of elixirs that were to be taken to prolong life and achieve immor-tality. The sexual and breathing techniques were part of an inner alchemy to pro-duce such elixirs within oneself. The connection of alchemy with Taoism has been countered with the claim that the pursuit of immortality was a Chinese tradition before the rise of the Taoist religion.[111] This is true, but Taoism absorbed that tradition and made it central. The other major religions of China—Confucianism and Bud-dhism—did not. Confucianism emphasized worldly duties and status within the state bureaucracy. Buddhism emphasized reincarnation and the freeing therefrom.

Women and Taoism
Lao Tzu reversed the valuations of the traditional polarities, emphasizing the feminine over the masculine. This valuation of the feminine, at least in theory, is strikingly similar to the valuation of the feminine and of egalitarian relations be-tween the male scientist and female nature found in Western Renaissance hermeti-cism and alchemy. Western alchemists were certainly not particularly feminist in personal relations. David Noble has incorrectly used the almost entirely male mem-bership of the Western alchemical fraternity to argue that the alchemists were as antiwoman as the rest of Western science.[112] The theoretical valuing of the fe-male principle by Taoism is at least an improvement over the devaluing of females by Confucianism and Chinese society in general.[113]

Early Taoism also supported various practical, political aspects of female equality. The early *T'ai P'ing* (Great Peace) rebellions prohibited female infanti-cide.[114] In the Yellow Turbans, all offices except that of the highest rank of the divine leader had female as well as male forms of names for the official, showing that these offices were open to women as well as men.[115] This may have contin-ued the tradition of Chinese shamanism in which "the distinction of sex does not seem to have been maintained as a matter of great importance."[116] One scholar claims that the majority of Bronze Age shamans were women.[117]

The early Taoist sects, in contrast to the prudery of most of Chinese Bud-dhism and Confucianism, had group sexual activities as part of their religious ritu-als. Among the Yellow Turbans, as well as in the Sun En rebellion movement, there were mating ceremonies that were recorded and denounced by Buddhist critics.[118] These rituals are often claimed to have died out in a century or two under the pres-sure of Buddhist and Confucian denunciation,[119] but apparently existed locally as late as the tenth century.[120]

The Taoists went further in practices of sexual relations that were positive for women, even if only for male Taoists' selfish reasons. Taoist sexual techniques emphasized the production of female orgasm in order for the male sexual alchemist to absorb the yin energy of the female. For that to occur the male was not supposed to ejaculate. The partner who did not have an orgasm absorbed the energy of the partner's orgasm. This led to the development of techniques for retarding or preventing male orgasm. This functionally encouraged the sexual satisfaction of the woman, though not for altruistic reasons.[121] The art of the bedchamber was traditionally taught from mother to daughter only. Father and son were not supposed to be familiar, so the facts of life were not taught by father to son.[122] In mythology, the teachers of the arts of the bedroom to the sages were women. (Interestingly, some of these women were also said to have taught strategy, relating the military arts to the arts of the bedchamber.)[123] It may be that, since women were the transmitters of the sexual arts, techniques to achieve female satisfaction and philosophical rationales justifying women's orgasm to the advantage of men were taught as well. Chinese sex manuals are the only ancient ones that do not present sexuality solely from the male viewpoint.[124] Overall, Taoism had favorable consequences for women sexually as well as in social and political organization.

Taoism and Popular Religion

Taoism absorbed much of Chinese popular religion. It is true that not all Chinese popular religious rituals or behaviors ought to be called Taoist simply because they are not Confucian or Buddhist, though Taoism absorbed local peasant religious devotions to a much greater extent than Confucianism. Confucianism became integrated with the state religion that had, in very early days, absorbed some features of agricultural and seasonal festivals. By the time of the Han empire Confucianism had become the religion of the ruling elite, not the religion of the masses. Taoism took over the role of integrating peasant beliefs and superstitions into an organized liturgy. Strickmann claims that Taoism defined itself not in opposition to Buddhism or (still less) Confucianism, but against the popular religion.[125] Taoism as an organized religion wished to distinguish itself from the popular cults even while absorbing some of them, but the relation of Taoism to popular religion was close and reciprocal, especially in contrast to Confucianism during the Middle Ages (before Neo-Confucianism and more general education after the Sung dynasty led to Confucianism's revival as both a popular and a state morality).[126] Taoism was certainly not identical with popular religion, but it was close enough to it to encourage one Japanese scholar to identify it as the "unconscious national religion" of China in the way that Shinto is to Japan.[127]

Sivin, who rejects the use of a blanket conception of Taoism, does grant that "orthodox Taoism built largely on beliefs and practices adapted from popular religion."[128] He notes that one might think of the relation of Taoism to Confucianism as that of counterculture to establishment in the modern West,[129] and he means this as a put-down of such a simplistic opposition, relegating it to the realm of clichés, but the parallelism here is more precise and more suggestive than he grants. Certainly Taoism was the view taken up by those out of official favor, as Confucianism was the code of those within the state bureaucracy. A writer on alchemy, Ko Hung, attempted to bring popular beliefs and superstitions to the attention of his aristocratic readers. Sivin uses this to downplay the extent of Ko Hung's Taoism, but in fact it shows the mediating role between popular and elite culture that such an activity might play.

Without being identical to popular religion, Taoist religion (and its more philosophical developments) could have served as the conduit for certain popular beliefs into high culture. It became the mediator of the underground in a way that hermeticist occultism was later in Europe. The association of Taoism with aristocratic dropouts and eccentrics, such as Chuang Tzu or the Seven Worthies of the Bamboo Grove, also suggests Taoism's role in mediating between folk and elite culture.

Taoism, Mohism, and Military Technology

There is a good deal of Mohism in the Taoism of the revolutionary movements. The Mohists opposed war by developing defensive military technology and coming to the aid of cities under siege. Some believe that Mohism was a religion of military engineers. The military interests of the legalists and Mohists stand in contrast to the purely civilian rule envisaged in the Confucian model of the state.[130] Military ranks and terminology were used by the Taoist establishment as an alternative nomenclature to that of the Confucian hierarchy. The military side of Taoism played role in the Taoist contribution to technology.

The association of Taoism with military strategy and technology is unmentioned by either the defenders of its association with science (such as Needham) or by the detractors of such association (such as Sivin and Strickmann). Lao Tzu contains advice about winning by retreating and winning by passivity that resembles both East Asian martial arts and the classics of Chinese military strategy. Confucianism, especially in Mencius and his followers among the medieval Neo-Confucians, was pacifistic and antimilitary. The poor situation of China with respect to its semibarbarian neighbors during the Sung period has been partly blamed on the ascendancy of Neo-Confucianism. Confucian bureaucrat literati generally held military leaders to be inferiors. The downplaying of the military role fits nicely with the social role of Confucianism as the worldview of the nonmilitary bureaucrats. It stands to reason that military leaders would not identify with a Confucianism that held a low estimate of them and might be attracted to views more like Taoism. The *Kuan Tzu* and *Huai-nan-tzu* contain chapters devoted to military matters in books with a Taoist aspect. If military personnel and military strategy were associated with Taoism, might not military technology?

There is an association of Mohist doctrines with early Taoist religion (not with the earlier Taoist philosophy). Mohism was in some ways the first organized religion in China and the major competition with Confucianism in the centuries before the Han empire. Mohists were craftspeople (perhaps military engineers, but possibly even ex-slaves) who formed independent guilds or religious chapters. Mohism included many Western-looking beliefs, including belief in a universe of individual objects, utilitarianism, universal love, formal logic, and quantitative physical science. It apparently died out when craftspeople no longer had independence from rival political powers. During the "feudal" period of warring states, the Mohists could shift alliance from one duke or lord to another and offer their engineering services, thereby remaining an independent force. Once the empire was reunified and centralized as an Asiatic bureaucracy, the Mohists and crafts-people in general lost autonomy and were crushed by elite Confucianism, and the interest in mathematical physics and formal logic largely disappeared.

Wolfgang Bauer notes that many features of the Yellow Turban rebellion re-semble Mohism. Their organization was a military one. They espoused ideas of equality and merit like those of the Mohists. As an organized religion (unlike the official state Confucianism, which lacked local congregations), the early Taoists organized in structures much like those of the Mohists.[131] This affinity and these borrowings may account for the fact that the Mohist canons were preserved in the Taoist canon over the millennia after the eclipse of Mohism, before interest in Mohism was revived by modernizers in the nineteenth and twentieth centuries.[132] Even if the Mohist writings were preserved solely for their usefulness, and not considered themselves Taoist, might there have been a practical tradition that utilized Mohist science via the Taoist establishment in later Chinese military engineering?

Sivin has noted that major scientists of the Sung, Ming, and Ching dynas-ties, such as Shen Kua, a writer on magnetic declination, astronomy, and much else, had associations with Neo-Confucianism. This does not go against the asso-ciation of Taoism with science, since *Neo*-Confucianism differs from earlier Con-fucianism in its doctrines of the natural world borrowed from Taoism.

Conclusions: Continuity and Discontinuity in the Tao

The components in Taoism are connected and justify neither the rejection of Taoism as a useless blanket term nor the debunking of a connection of Taoism to Chinese science. Shamanistic activities contributed to Taoism, both techniques of meditation and of soul flight, and the hymns probably present in Lao Tzu. So did longevity techniques and the search for physical immortality. A strain of primi-tivism and a longing for a simpler, more spontaneous life is found in Lao Tzu's utopia and in the anarchist doctrines found in the middle and outer chapters of Chuang Tzu. Breathing techniques and physical exercises are mentioned in both Lao Tzu and Chuang Tzu. A current of legalist thought with an emphasis on nonaction of the emperor modified and later unified with Taoism in the Huang Lao doctrine. The individualist hedonism of Yang Chu, against which the Confu-cians argued, finds some expression in Chuang Tzu, and much more in Lieh Tzu.

Religious Taoism absorbed popular religion as well as Lao Tzu to develop a personal religion as an alternative to a ritualistic Confucianism divorced from or-dinary peoples' lives and a more indigenous, personal salvation as alternative the newly spreading Buddhism from India. Peasant discontent and millenarian dreams of a Great Peace (*T'ai P'ing*) inspired the Taoist organization that probably uni-fied local Taoist organizations during the rebellions of the Yellow Turbans, the Five Pecks of Rice movement. Lao Tzu's valuing of the feminine and the Taoist sexual alchemy gave positive support intellectually to female qualities and physically to female orgasm. The quest for immortality encouraged the Taoist pursuit of alchemy in both its inner and its outer versions. While Taoist religion spread among the populace after the fall of the Han and subsequent discrediting of its Confucian ideology, there was a revival of Taoist philosophy in commentaries on Lao Tzu and Chuang Tzu. There also was a movement of aristocratic eccentrics and non-conformists who contributed to poetry and who associated themselves or were as-sociated by others with Taoism. Throughout Taoism there was an emphasis on nature and following natural processes and transformations. The political views of Taoism varied from an anarchist, individualist rejection of politics, to a sort of primitivist, anarchist utopia (with an emperor in a characteristically Chinese way), to the political revolutionary movements and theocracy of early Taoist religion, to a Taoist church that reconciled itself with the status quo. Despite these political variations (which are no greater, perhaps less, than those of Christianity), a theme of anarchism and rejection of the state bureaucracy unifies many of the strands of Taoism. The disciplines of breath control, sexual and other exercises, and alchemi-cal preparation of elixirs did not originate with Taoism but were strongly integrated with it. Buddhism also contributed to the techniques of meditation, but Taoism itself transformed Indian Buddhist doctrines as they spread to China into Ch'an or Zen, the quintessential Chinese Buddhism.[133] Finally, during the Sung dynasty (960–1229 C.E.) Neo-Confucianism combined Taoist cosmology with Confucian ethical and social doctrines.

Chu Hsi and Neo-Confucianism

In the high Middle Ages in China, during the Sung dynasty, the philosopher Chu Hsi (1130–1200) synthesized ideas from several revivers of Confucianism into a massive construction that dominated traditional Chinese philosophy for the next six hundred years. Two of the Neo-Confucians whose work Chu Hsi integrated were strongly attracted by Taoism. Chu Hsi incorporated Taoism in order to give a general cosmological and metaphysical framework to Confucianism. He had spent a decade in his youth under the strong sway of Taoist and Buddhist ideas and was aware of the attractions of Zen (Ch'an Buddhism), including its emphasis on medi-tation and concrete experience inspired by Lao Tzu and Chuang Tzu. Chu Hsi fre-quently polemicized against Buddhism on moral and political grounds, while silently incorporating ideas from Buddhism.

Chinese governmental practice, family morality, and social ethics were pri-marily based on the doctrines of Confucius, whose main interest lay in human his-tory, social conduct, and ceremonies. The Neo-Confucians incorporated the Confucian moral and political doctrines into a metaphysical system that included a philosophy of nature largely borrowed from Taoism. The development of this system was largely a response after 1000 C.E. to the Buddhism that dominated China during the Tang dynasty in the seventh and eighth centuries, C.E. The Confucians needed to present a doctrine of the ultimate reality to contrast with the Buddhist doctrine of unreality and cosmic cycles.

Chu Hsi's synthesis of Taoism, Confucianism, and some parts of Buddhism under a dominant Confucian viewpoint resembles in many ways a similar synthesis that occurred in the West around the same period. The synthesis of Christianity, the philosophy of Aristotle, and aspects of Plato is most notable in the system of Saint Thomas Aquinas. Perhaps because of this parallel, Chinese Neo-Confucian thought is often spoken of as Neo-Confucian scholasticism. This borrows the term originally used to describe the combination of Aristotle's logic with Christian theology in the West. Just as the philosophy of Aquinas eventually became the official philosophy of the Roman Catholic church, so too Chu Hsi's philosophy was, for long periods, the official state philosophy of the Chinese empire.

One of Chu Hsi's great impacts on later Chinese culture was to shift the emphasis in preparation for examinations from the ancient Five Classics *(I Ching,* and books of history, rites, and songs) to the Confucian Four Books, which Chu claimed contained the essence of Confucianism: Confucius, Mencius, and the two short works *The Great Learning* and the *Doctrine of the Mean* taken from the *Book of Rites.* Chu Hsi's own commentaries on the classics became official for certain emperors and were used in examination preparation. Since these examinations were taken by anyone who wished to enter the bureaucracy, the impact of Chu Hsi was immense.

The Life and Political Context of Chu Hsi

Chu Hsi was the son of a minor provincial administrator who was removed from his position for opposition to the Sung dynasty policy of appeasement with the northern barbarians. This stood as a warning to Chu Hsi throughout his life. He was strongly opposed to the policy of giving tribute to northern barbarians and refusing to attempt to reconquer territory. In the southern Sung period, the empire consisted of only southern China. Chu Hsi interpreted political appeasement by China as moral weakness. He also considered the strong inroads that Buddhism had made a sign of decay of the traditional Chinese Confucian social organization.

Preceding Chu Hsi's time there was a strong attempt at reform. Wang An-shi was a famous minister who attempted thoroughgoing reforms of the Chinese government. He emphasized law and economic policy and attempted to increase production and eliminate corruption;[134] however, among his own followers and opponents factions developed whose infighting undermined his reforms. It appeared to many that legal and purely institutional reforms were insufficient if the administrators remained morally corrupt. Thus, such Neo-Confucians as the Ch'eng brothers and Chu Hsi put much emphasis on morality.

Shen Kua, who is notable for writing on the nature of the compass and magnetic declination as well as on topics in astronomy, mathematics, pharmacy, and other subjects, was strongly tied to Wang and his reforms. Shen, because of his technical and intellectual talents, was pressed into a high position in government; however, due to his nonpolitical orientation he was discredited by rival factions among both the supporters and the opponents of Wang, being blamed for military losses caused by policies that he himself had strongly opposed. In his retirement at Dream Brook Farm he composed his *Dream Brook Essays*, which include his discussion of the compass and the Earth's magnetism.[135]

Chu Hsi was involved in a many-sided controversy about the appropriate syllabus for the national examinations for the bureaucracy. Wang had emphasized the ancient classics and appealed to historical precedents from an idealized ancient China. In opposition, Ssu-ma Kuang emphasized historical analogies for policy formation based on both ancient and more recent dynasties.[136] A third approach, that of Su Shih, emphasized personal feeling and centered studies on literature, which encouraged a clever, cynical style in the exam essays.[137] In opposition to the historical and literary approaches, the Ch'eng brothers emphasized a fourth, strongly moral, approach.[138]

Chu Hsi believed in the importance of moral principle, but he also believed in the importance of "the investigation of things." He opposed the literary approach and emphasized knowledge of society. In innovative debates before students Chu defended his moral and observational approach and opposed purely internal contemplation.[139] Chu Hsi, unlike some other followers of the Ch'eng brothers and of Chang Tsai, supported the study of mathematics and said, "The study of the calender, . . . astronomy, geography . . . should be paid attention."[140] Chu was notable among the literati for his interests in nature and science. He owned an armillary sphere, although private ownership of astronomical devices was prohibited by the emperor.[141] Chu Hsi's speculations in geology are notable, including his citing marine fossils at high elevations as evidence of the uplift of the land and his vision of mountain building as slow wave motion.[142]

Chu Hsi's Cosmology: The Articulated Philosophical Framework of a Theory of the Magnetic Field

An important feature of Chu's Neo-Confucianism is his absorption of Taoist cosmology. Two Neo-Confucian predecessors had a strongly Taoist orientation. Shao Young wrote the discussion of the hexagrams of the *I Ching* that impressed Leibniz in his study of binary arithmetic, but was not so important for Chu Hsi, because of Shao's emphasis on mathematical or numerological accounts of reality. Another Neo-Confucian, Chou Tun-i, also receptive to Taoism, wrote *An Explanation of the Diagram of the Great Ultimate*, a design from a Taoist priest, on which Chu Hsi then wrote a further major commentary.[143]

At the top is the empty circle of the Great Ultimate, which is also called the Ultimateless. Beneath that are the principles of yin and yang, or movement and quiescence, generated by the Great Ultimate. These in turn generate the five phases of earth, water, fire, metal, and wood. These are five ethers (phases), or versions, of *ch'i*. *Ch'i* is ambiguous in its status and has been interpreted differently, as a kind of fluid or continuous material stuff or as a kind of dynamic force. (Leibniz and the German Romantics will similarly treat matter as made of dynamic forces.) Some writers translate *ch'i* as "ether" while others translate it as "material force." It is similar to a physical plasma, material in nature but carrying electromagnetic force.

From the five phases the *ch'ien* principle is produced, which becomes the male element, and the *k'un* principle, which becomes the female element. These in turn generate all the particular things. This is somewhat like Empedocles's use of the four elements plus love and strife to generate the world, and gives an account of the relationship between the Great Ultimate and the polarities of yin and yang and the development of objects in the universe. One version of this explanation is closer to that of Shao Young's claim that the Great Ultimate gives rise to yin and yang and that yin and yang gives rise to the multitude of things, and is similar to Lao Tzu's statement that "one makes two, two makes three, and three creates the ten-thousand things."[144]

Chu Hsi developed his own explanation of the diagram of the Great Ultimate using the concept of material force *(ch'i)* that was traditional in Chinese philosophy and medicine and fundamental for Neo-Confucian Chang Tsai, whose central concepts were material force and its transformation through yin and yang. Chu Hsi included *ch'i* as a major concept but combined it with *li* (principle) as more fundamental. *Li* is mentioned but not explained in the *I Ching*, and was not used by early Confucians. It appeared in the Neo-Taoist commentary on the *Tao Te Ching* by the brilliant Wang Pi. The Neo-Taoist literature of the early centuries C.E. is the second source of Taoist terminology incorporated into Neo-Confucianism.[145] The Ch'eng brothers' emphasis on principle most influenced Chu Hsi in making *li* central. They applied it primarily to ethics, not cosmology as did Chu Hsi, and had not given an explanation of the relationship of principle to material force.

Chu Hsi's synthesis explains the relationships of principle to material force and of the Great Ultimate to particulars. Chang Tsai, the almost materialist Neo-Confucian, put all emphasis on *ch'i*, or material force, which he identified as the Great Ultimate. The Ch'eng brothers did not explain creation from the Great Ultimate. Ch'eng I claimed that yin and yang have no beginning. Chou Tun-i claimed that the Great Ultimate generated yin and yang but that it was out of tranquillity (yin) that creation occurred, and he did not explain activity (yang). Chu Hsi claimed that the principles of activity and passivity are within the Great Ultimate, which itself is neither passive nor active.

For Chu Hsi, *li* and *ch'i*, are copresent in all things and resemble matter and form in Aristotle, where form does not exist independently as it does in Plato's philosophy. Form for Chu and for Aristotle is always the form of some matter. Chu Hsi denies temporal priority between matter and form, which have always existed, but holds that there is a logical and metaphysical priority of form. Matter could possibly not exist, while form must exist. Nevertheless, they are always complementary. There is duality at the level of appearances but unity at the level of reality. Above the matter and the form of particular things there is *Tai-chi*: the Great Ultimate, a unity containing the principles of all multiplicity and hence the multitude of things. Also, every particular thing contains, in a sense, the Great Ultimate. He uses the metaphor of the Moon whose light illumines all the individual things without the Moon itself being divided. This metaphor is Buddhist, and derives from the Hindu metaphor of the net of Indra in which each diamond in the net reflects every other diamond and its reflections ad infinitum. It closely resembles the hermetic microcosm in which the individual, particularly the human being, reflects the macrocosm, and also the doctrine of monads of Leibniz, who thought of Chu Hsi's metaphysics as isomorphic to his own.

The relationship of principle and material force was not explained by the Ch'eng brothers and was absent from Chang Tsai's doctrine. Chu Hsi made a great explanatory advance by claiming that, although principle and material force are different and separate entities, they are unified within any individual thing and they cannot be spatially separated. This doctrine resembles the Western medieval scholastic doctrine that prime matter (identified with pure potentiality) never exists on its own totally unformed.

Chu Hsi uses his dichotomy of principle and material force to account for human nature. The *ch'i*, or fluid matter, which accounts for both material bodies and minds, has different degrees of clarity and turbidity. Although knowledge of principle is possible for anyone, the preciseness of that knowledge is dependent on being filtered through the *ch'i* of our body, which in different individuals is more or less turbid. The accuracy of our grasp of principles can be blurred by our *ch'i*. This resembles Plato's notion of the receptacle (*chora*) as a principle of vagueness that makes the shapes of the forms when manifest in the world inexact, though for Chu Hsi this is done by the knower, not by space.

Chu Hsi's cosmology contains a view of cyclical process within the universe, borrowed by the Neo-Confucians from the *I Ching* and Buddhism. Earth and Heaven are generated, followed by generation of living things and humans, and finally a period of decay leading to chaos and then new generation. The universe never disappears but becomes successively organized and disorganized. Here, Chu's cosmology resembles that of Empedocles, in which periods of love (attraction and construction) and strife (repulsion and disorganization) alternate in a cosmic cycle.

The cosmic process involves condensations and rarefactions of *ch'i* governed by yin and yang. The active ether is governed by the male principle, yang. When *ch'i* congeals, it is governed by the female principle, yin. The centrality of gender roles in Chinese society, shown in social terms in the importance of the distinct duties of husband and wife in Confucian ethics, illustrated how yin and yang are crucial for the understanding of the cosmos. Similar doctrines are found in the Pythagorean sexualizing of odd and even numbers as male and female, although this latter doctrine did not endure like the Chinese doctrine.[146]

The genesis of the structure of the universe is tied up with the motion and separating out of *ch'i*. The initial separation of Heaven and Earth is done by a kind of centrifugal motion in which the heavy Earth settles to the center and the lighter heavenly *ch'i* whirls off from the center.

Yin and yang do not coexist but rather alternate in the cosmic processes. *Li* governs processes, and yin and yang are instruments of *li*. The word *ch'i* designates individual "instruments" or particularizations. The *ch'i* functions as paradigms for objects, but they are more concrete than the *li*. *Li* is described by Chu Hsi as "above shapes," or nonspatiotemporal, while its instruments are "within shapes," as are the activities of yin and yang.

Li is not active, unlike "active principles" in Western hermeticism and Newton. *Li* seems instead to mirror the Chinese ideal of nonaction in Taoist thought. The Great Ultimate, which contains all the *li* of the many particular things, is infinite and eternal. It is beyond space and time and eternally contains all the principles of the activity of particular things and processes. The Great Ultimate as a unity that contains the principles of multiplicity is like the One in Neoplatonic philosophy, which is likewise unitary and nonspatiotemporal but from which emanate all the lower levels of being including the material world.

This doctrine of the clarity and turbidity of *ch'i* has some resemblance to the doctrines of condensation and rarefaction that we find in Western thought in pre-Socratics such as Anaximander and Anaximenes, in the Stoics with their *pneuma*, in cabalistic thought, and in Spinoza's thought, if one interprets the last in terms of the thickening and thinning of space.[147] This is one of the similarities of Chinese correlative cosmology in its developed form and Western occultist, hermetic, or cabalistic thought. C*h'i* has resemblances to the ethers, fields, and plasmas that played roles in later Western electromagnetic theory. Similarly, the traditional cosmology that fostered the theory of magnetism in China resembles, in its philosophical systematization by Chu Hsi, doctrines in Western hermeticism and the cosmologies of Spinoza and Leibniz that, when made temporal and dynamical by the German Romantic nature philosophers, also contributed to the development of the modern theory of the electromagnetic field.

CHAPTER 4

Social Background of Organismic Thought

"CHINA" IN THREE CLASSIC SOCIOLOGIES

*T*he dominant, atomistic self-image of Western science is reinforced more by the social and economic individualism of Western society than by the actual content of Western science. Examining the social structures and factors that facilitate the development of a holistic and organic science will shed light on the role that alternative versions of science play. Chinese science is perhaps the major alternative to Western science. Of the numerous holistic sciences of indigenous peoples, none is so fully elaborated and recorded as Chinese science. Indian science offers many fascinating alternative conceptions, but is plagued by problems of dating and documentation. The three major accounts of Chinese social structure in the classic tradition in sociological theory are those of Karl Marx, Max Weber, and Emile Durkheim. Marx's theory of the "Asiatic mode of production" is later transformed into the "hydraulic society" theory by China scholar Karl August Wittfogel and ecological anthropologist Marvin Harris. Weber's theory that the role of the city and the family distinguishes Chinese from European social structure is utilized by many sociologists and historians of China. Durkheim's account of language and collective representation as elaborated by Marcel Granet and others within French sinology contrasts Chinese ideographic with European phonetic writing.

Theories of Asiatic Society or Hydraulic Society: Marx and Wittfogel

Extracting the important insights of the theory of Asiatic society or the "Asiatic mode of production" is made difficult by the politically complex career of the theory. It was initiated by Marx but downplayed by Lenin. It was developed in less economic, more geographical, and more ecological dimensions by Wittfogel.

During the period it was rejected by official Chinese and Russian Marxism, Wittfogel turned the theory in an anticommunist direction.[1] More recently it has been used by Harris, and by Rudolf Bahro in the German Green party.[2]

The theory of Asiatic society was developed by Marx in order to account for the apparently nondynamic character of early civilizations, the apparently unchanging features of civilizations such as Egypt and China in contrast with the rapid changes that occurred in Europe.

Oriental despotism, a primarily political concept, arose in French eighteenth-century Enlightenment writings of figures such as Montesquieu, who emphasized the influence of geography and climate on forms of government, identifying despotism primarily with the Asian empires. Montesquieu, like Adam Smith and Marx, followed the account of seventeenth-century traveler François Bernier in claiming that a ruler's sole ownership of the land was the source of the worst despotism.[3]

Enlightenment writers, such as French physiocrat François Quesnay and *philosophe* Voltaire, wrote positively of the Asian empires. An age of enlightened despotism idealized the Asian rulers as models for European enlightened despots.[4] German philosopher Christian Wolff praised the non-Christian religion of the Chinese in a speech that led him into political and religious difficulties.[5] The writers of the major classics of British political economy—Adam Smith, James Mill, and John Stuart Mill—were connected to a greater or lesser extent with the British East India Company. These founders of market economics all wrote about the static character of Indian society. Hegel, familiar with both the French Enlightenment writers and with Adam Smith, presented Oriental society as the first stage of the world history of freedom, the stage in which "one is free" (that is, the despot alone). Marx used ideas of all these writers—the French Enlightenment figures, the British economists, and Hegel—in his concept of Asiatic society and his general social theory.

The Asiatic mode of production is one in which the state extracts taxes from self-sufficient villages or village communes. The internal affairs of the villages are relatively independent of the state. This theory included the development of state-controlled irrigation and flood-control works (later emphasized in Wittfogel's characterization of the "hydraulic society"), the relative weakness or absence of private property, and the self-sufficiency of villages or communes with a small amount of handicraft subordinate to agriculture. Marx applied this concept of the "Asiatic mode of production" primarily to India and China, but also to such non-Asiatic societies as Egypt. Later Marxists and anthropologists would apply it to such societies as the Meso-American and South American civilizations of the Aztecs, Mayas, and Incas.

The major Western reviver and developer of Marx's concept of the Asiatic mode of production was Wittfogel (1896–1990), who applied the concept to China.[6] It is interesting that Wittfogel claims that in his very early days "I was described as a mixture of Karl Marx and Lao-tse."[7] Wittfogel emphasized the impact of environment and natural resources on society, thus the influence of the great rivers—the Nile on Egypt, the Huang He (Yellow River) on China, the Tigris and Euphrates on Sumer, the Indus on India—is made central. The need for flood control (central in China) and irrigation (central in the Tigris and Euphrates valley) gave rise to centralized bureaucracies for mobilizing the force of huge numbers of workers. Wittfogel made the state maintenance of water-control projects central to his theory of Asiatic society, and called it the "hydraulic society." (Ironically, Harris, one of the prime defenders of the hydraulic hypothesis, excluded Mayan civilization a few years before the discovery that the Mayas had an elaborate system of raised islands and canals.)[8]

The story of the concept of the Asiatic mode of production, or hydraulic society, is complicated by political developments. In 1931, just as Wittfogel's first major theoretical book appeared, official Soviet (and hence Chinese and other communist party) Marxism rejected the concept of the Asiatic mode of production.[9] With Soviet emphasis on revolution in Asia, as prospects of revolution in Europe faded, the notion of a static, stagnant traditional Asian society resistant to change was seen as a damper on revolution there. The concept of Oriental despotism was seen as degrading to Asian societies, whose populations were now seen as allies in fomenting nationalistic, anticolonial movements. In 1931 (the year of Stalin's consolidation of power), the Leningrad Conference and the Ban on Factions eliminated the Asiatic mode as a stage in social evolution. The linear, mechanical version of social evolution (with no alternative routes, and minus the Asiatic mode of production) became the official version of Marxism and (given the power of the official communist parties) virtually the only available version of Marxism for the next several decades. A simple, linear sequence is found in Marx's influentially brief and succinct summaries of his doctrine. The Preface to the *Critique of Political Economy*, the most widely cited summary of Marx's views on social evolution, did include "Asiatic" society, slave society, feudalism, and capitalism. Stalin eliminated the reference to the Asiatic mode of production when he quoted it in his "Dialectical and Historical Materialism."[10]

Marx's "Rough Draft" of *Capital*[11] contained a rich discussion of alternative routes of development, but this work was not published until 1941, and the relevant portion was translated into English only in 1964.[12] The century-long delay in the publication and translation, encouraged by Soviet hostility to the Asiatic mode of production and Western hostility to Marxism during the cold war, led people to identify Marxism with linear evolutionary schemas of society and to not recognize the Marxist origin of this theory when used by ecological anthropologists.[13] The political situation was further complicated by Wittfogel's break with communism because of the Hitler-Stalin pact and his subsequent move to an extremely militant anticommunism. During the height of the McCarthy era Wittfogel testified and named names of colleagues in China studies and other branches of university life to the House Un-American Activities Committee with Joseph McCarthy himself present.[14]

Although Wittfogel's earlier (1931) study of China was never translated (presumably because Wittfogel, now a militant anticommunist, did not wish it to be), he presented his ideas in English in *Oriental Despotism*, the very title of which shows the political and anti-Asian tone of the book. Wittfogel's concept of the hydraulic society is expanded in scope but imbedded in an anticommunist political polemic with strong emphasis on democratic Europe versus despotic Asia. Of course this further alienated Marxists from the concept of the Asiatic mode of production, because here Wittfogel, by doubtful extensions of his concept, managed to incorporate Russia, despite lack of irrigation projects, into his model.

U.S. China scholars, including some mainstream moderates, were fired or denied visas during this period and were blamed for the United States having "lost" China. Wittfogel's testimony against his graduate students and such scholars as Owen Lattimore, writer on Mongolia and Central Asia, made liberal academics and China scholars oppose him and his theory. Both orthodox Marxists and liberal U.S. China scholars rejected the theory, though for different political reasons.

Wittfogel's ideas, stripped of cold war polemics, have been fruitfully used by anthropologists of the cultural-ecology school, most notably Marvin Harris, who praises Wittfogel for his contribution to cultural materialism.[15] Another appropriator of Wittfogel's work was Rudolf Bahro, who became an important theorist of the German Green party using Wittfogel's ecological theories. Bahro, while an East German dissident, surreptitiously used Wittfogel's ideas to describe Soviet-style societies.[16] Bahro's turn from dissident communist to ecological activist did not change the theory of geographical (or ecological) determinism he had learned from Wittfogel.[17] Wittfogel's conception thus secretly informs communists, anticommunists, ecological activists, and mainstream anthropologists.

The regulation of agricultural labor and the regulation of water resources, including the prevention of flooding, led to a stratum of nonhereditary bureaucrats who studied the stars and the weather, both as administrators of seasonal agriculture and soothsayers for the emperor. Centralized imperial control and the lack of opposing interest groups encouraged a highly unified conception of society and the cosmos. Needham utilized this hypothesis, along with the thesis of the "weak" city lacking in autonomy (mentioned in Marx but much more developed by Max Weber), and the lack of a strong, independent bourgeois formation of tradespersons and craftspersons, to explain the absence of mathematical, lawful science in China.[18] The relative autonomy of the primitive villages probably reinforced the conception of the coordination of independent elements found in Chinese correlative, organic thinking.

A problem with the hydraulic-society model of Chinese culture is the extent to which it can account for the specifically organic model of holistic thought in China as contrasted with the thought of other hydraulic or Asiatic societies. Ancient Egypt and ancient India are both, like China, classified as hydraulic societies, but their thought differs in significant ways. Can the distinctions in kinds of hydraulic society account for the differences in thought among these societies? India was the paradigm case of Asiatic society for J. S. Mill. The Indian laws of Manu are used by Wittfogel in his account of total social control. Egypt was more centered and dependent upon the river than was India or China, and exemplifies endurance for thousands of years. Neither Indian nor Egyptian thought, however, manifests the extremes of organicism found in Chinese thought.

Indian thought, despite its mysticism, antiutilitarian orientation, and large differences from European thought, is more similar to European thought than is

Chinese thought. Indian philosophy develops a mystical monism with a denial of the reality of perceptible experience. This certainly differs from Western materialism and positivism, but it is more similar to Western religious mysticism and monistic thought (such as that of Parmenides) than it is to the pragmatic, holistic naturalism of Chinese thought. This may be why Indian mysticism has had such a strong appeal in the West during the last century. There are many more Western followers of Indian gurus than there are Western converts to Confucianism. Indian thought also contains philosophies of atomism and materialism as well as schools of logic. This contrasts with the lack of atomistic formal logic in ancient Chinese thought (with the exception of the Mohists). Indian Buddhist missionaries brought logical doctrines to China, but the highly antilogical Ch'an, or Zen, Buddhism is the most characteristic Chinese contribution to Buddhism. Features of societies with the Asiatic mode of production parallel Chinese cosmology. The strong centralization of the state bureaucracy and lack of individualism, combined with the relative autonomy of peasant villages resembles the holistic coordination of separately aligning entities. China's thought shows even more thoroughgoing and pervasive organicism than that of other centralized or hydraulic societies, such as India, Babylon, or Egypt. The Asiatic-mode and hydraulic-society theses do help explain the holism of Chinese thought in contrast to that of Europe, but other factors of social structure and language intensify this contrast.

The Family and the City in China and the West: Max Weber

While hydraulic management was emphasized by Marx and Wittfogel, and the ideographic nature of the Chinese language was emphasized by Durkheim and Granet in contrasting China to Europe, the family and the city are another pair of important factors considered by sociologist Max Weber (1864–1920). Weber claimed that the weakness of the city and the strength of the family were central in distinguishing Chinese from European civilization.

Weber is a founder of modern sociology, along with Marx and Durkheim, and is most famous for his analysis of the role of religion in the development of capitalism, which he utilized to criticize Marx's materialist conception of history. Weber also believed that class-based social analysis did not adequately explain modern capitalist bureaucracy and social life and advocated the importance of such concepts as status and status groups rather than class. Weber used the general process of society's "rationalization" (rather than what Marx called "alienation" from society) to explain the impersonality and meaninglessness of modern society. For Weber, the process of rationalization is the defining feature of modern Western society, and involves the replacement of tacit traditions and skills with explicit rules and the pervasive spread of means/end rationality, including the development of bureaucracy, explicit laws, economics, science, and even the piano.

All observers have been struck by the tremendous role that the family plays in Chinese society. The dominance of kinship and the family, which is characteristic of many prestate formations, has continued through the rise of the Chinese state and its civilization. The centrality of ancestor worship is well known. This may account for the development of the extreme holistic thought maintained through thousands of years of Chinese civilization. The nuclear family and individualism in the West is more easily associated with atomistic philosophy.

In Athens Cleisthenes formed a new constitution in which the four old tribes, founded on gentes and phratries, were ignored. New communal districts, called *demes*, were formed, organized in groups of ten into tribes. These tribes had nothing to do with the kinship-based tribes they had replaced.[19] In the West political and geographical organization replaced kinship as the basis of politics.

In China Heaven and Earth are referred to as Mother and Father, respectively. The emperor is also referred to as *both* mother and father. Chinese metaphysics uses the term *giving birth* instead of *fashioning* or *creating*, in discussing the origins of realities or principles.[20]

Kwang-chih Chang suggests that the earliest Chinese writing found on pottery inscriptions seems to originate from markers or emblems of family lineages or clans. Chang contrasts this family-emblem origin of writing with the origin of writing in Sumer, which is hypothesized to stem from the Sumerian accounting system.[21] The significance of this contrast, if true, is immense: that Chinese writing was originally developed for genealogy while Western writing originated for economic tabulation.

There is evidence of ancestor worship even in pre-Shang, Neolithic China. Paul Wheatley suggests that the earliest cities about which we have archaeological knowledge of sequence arose from religious centers, and that the Chinese city originated from the place of worship of the ancestor cult.[22] Ancestor worship led Chinese religion and mythology in a very different direction from the rest of the world, and involved an absence of creation myths (the recorded Chinese creation myths are later developments). Ancestor worship has a relatively small mythic potential,[23] although China has a rich legendary history. Chinese culture has perhaps the premodern world's most highly developed emphasis on historical record and historical sense. In contrast to the presentism of the early Greeks (whose great political historian Thucydides said that nothing much of interest had happened before his own generation), the earliest lengthy Chinese writings were historical records. Confucius continually referred to earlier legendary or semilegendary historical precedents. The writers of the newly founded Chou (Zhou) dynasty (1100–256 B.C.E.) invented a history of the Hsia (Xia) (2200–1760 B.C.E.) dynasty, before their own predecessors, the Shang (1760–1100 B.C.E.), in order to justify Chou usurpation of Shang rule with a history of how the Shang had previously ousted the Hsia.[24]

In the system of ancestor worship the ancestors participate in cycles of rituals that are highly predetermined, while the gods of the ancient Middle East and of the Greeks were unpredictable. Thus the Chinese objects of worship are subsumed under a higher order, while the Western gods control or disrupt that order.

The holism of Chinese thought is characterized by Needham as "organic." Donald Monro suggests that the family, rather than the individual biological organism, might be the appropriate unit to compare with Chinese thought structure.[25] The tremendous strength of the Chinese extended family, and the relative weakness of the city as an independent unit, is used by Max Weber to account for the lack of capitalism in China. Weber examined the influences of the great religions of the world on the economic life of various societies and is most famous for his theory of the role of Protestantism on capitalism and the role of other religions in India and China for the absence of a rise of capitalism in those societies. Weber's project of comparative sociology of religion was part of an even broader one of developing a social account of the rise of rationality in the East and West. Although Weber mentions science only occasionally, this broader theme of rationality clearly applies to the rise of formalized, natural science, as well as to other rationalistic phenomena, such as the capitalism and bureaucracy on which he focused.

The major units of Chinese society below that of the emperor were extended family groups. Just as religion centered on ancestor worship, power and organization within the village and the city was based on family groups. Military organizations apparently grew out of extended family groups through co-optation of recruits as honorary members of the family. Something like this in modern Western society is the Mafia "family." Family is clearly the central organizing principle of both, although many nonbiological "kin" are incorporated into the resulting organizations.

As in many cultures Chinese villages grew out of extended families; however, what is unusual is that the legal status of city dwellers was that of belonging to their family and to the village where the temple of their ancestors was located.[26] Cities consisted of village districts and did not function as unitary corporate bodies. In ancient Greece, the transition from family organization to citizenship in the city was made through public cult meals in the prytaneum. The primordial family activity of the common meal was literally transferred to the public city fraternal cult.[27] In China, nothing like this ever happened. As Weber wrote, the "fetters of the kinship group were never shattered."[28]

The notion of citizenship that originated with the independent city-state was unknown in China. In contrast to both ancient Greece and medieval and Renaissance Europe, independent cities were very rare in China and nonexistent once the empire was founded.[29] Cities were ruled by mandarins who were agents of the emperor. Marx had, in passing, sketchily foreshadowed this theory of Weber's with his parenthetical remark that in Asia "the great cities are to be regarded merely as princely camps."[30]

There was no dichotomy in China between the public state and the private family as in the West. The Chinese city did not have an inner political life.[31] There was no city cult or city god, so important in the ancient Greek city. The Chinese city, by virtue of being controlled by the federal government, lacked the independent life of free cities in Europe. In China all cities had the nonautonomous status that capital cities (for example, Washington) in many modern nations have by virtue of their functions for the federal government.

The tremendous (and for an advanced civilization, unique) strength of the family in China allowed extremely holistic thought forms present in precivilized family-based societies to be maintained in a civilization of the highest complexity and organization. The lack of Chinese autonomous cities and dominance of the emperor offers the strongest contrast to contentious ancient Greek intellectual life.

Collective Representations and Language: Emile Durkheim and Marcel Granet

Besides the Marxian theory of Asiatic society, and the Weberian theory of the family and the city, a third tradition of social explanation of Chinese thought is that based on the work of Emile Durkheim. A major figure of French sociology and a contemporary of Weber in the late nineteenth and early twentieth century, Durkheim gave an account of the social origin of cosmological concepts and the collective ceremonial origin of religion.

Durkheim, having absorbed nineteenth-century theories of the crowd, claimed that the origin of supraindividual religious concepts resided in the phenomenon of group assembly. Primitive individuals participating in a group sensed a dimension of reality beyond themselves through those emotions collectively held by the group. This was the origin of the "sacred" for Durkheim.[32] Society itself is the real object toward which emotions concerning the sacred are directed.[33] Durkheim called these shared, emotionally charged beliefs the *collective conscience*, or *collective representations*.

Another relevant part of Durkheim's work is *Primitive Classification*, written with his cousin, Marcel Mauss.[34] Cosmological classificatory systems are claimed to be collective representations, both as concepts involved with social forms of life and as cosmological concepts representing the structure of society. This was the origin of the account of cosmologies as either modeled on or unconsciously mirroring the structure of the society in which they arise.

Durkheim and Mauss emphasize cosmologies of Australian aborigines, presuming them to be particularly primitive ones, representing primitive social structures. They also include a chapter on the civilized Chinese, alleging that the Chinese have a classificatory system similar to that of the supposedly much more primitive aborigines. For our theme of the Chinese invention of the compass and its connection with holistic thought, it is of interest that Durkheim and Mauss report that the four cardinal points are correlated with colors and animals, and the eight compass points with the eight trigrams of the *I Ching* and the eight powers of nature.

The particulars of Durkheim and Mauss's work have been much criticized; however, one of their followers, Marcel Granet, wrote that their investigations "mark a date in the history in sinological studies." Discreetly refraining from criticizing the inaccurate details of their account, Granet paid tribute to his masters' framework of social analysis.

In several important works, especially *Chinese Thought* (*La Pensée chinoise*), Marcel Granet developed a basically Durkheimian account of Chinese culture with great richness and knowledge of the Chinese texts. Other scholars, such as Jacques Gernet, a historian of China, have, in turn, followed Granet. Granet's accounts of the origins of Chinese peasant religion in peasant festivals in *Festivals and Songs of Ancient China*[35] and the unity of Chinese cosmology with Chinese society and politics in *Chinese Thought* may be successful because China fits particularly well certain (Durkheimian) notions of society which form the bases of his analysis. The close linkage of the social order with the cosmology in China is particularly explicit as is the close linkage of religion and politics. What in other societies would be a matter of reinterpretation and unmasking is in China a matter of explicit statements in the classic texts.

In his book *Festivals and Songs*, which is based on the ancient Chinese *Book of Songs* (or *Book of Poetry*, or *Odes*), Granet claims to trace the primitive festivals in the accounts of the songs. He describes festivals in which bands of young men and women are brought together after having been isolated in the towns of their families. As he summarizes part of his account in *The Religion of the Chinese People:* "The bands opposed in the contests were made up of young people who must not either be from the same village or the same sex, . . . who before that had never met. We might not be able to imagine what their emotions were; they had such potency that on each occasion the young people burst into poetry . . . when they faced one another in . . . a duel of dance and song."[36]

Out of these festivals, Granet derived the religious beliefs as collective representations. For instance, in the spring festival was the crossing of the stream by young people before their first unions. "Shuddering at contact with the running water, the women felt as though they had been penetrated by the souls floating on the sacred springs; . . . it seemed like the coming of the spring had freed their waters from the underground prison where the dead season had enclosed them. From these images and feelings, was born the idea that the souls of the dead, . . . broke loose along the vernal springs from the subterranean retreat in which death had shut them up."[37] One could hardly find a better exemple of the kind of collective emotion to which Durkheim attributes the origin of the sacred.

In *Chinese Thought*, Granet develops the theme of the connection between cosmology and society along with an account of the influence of Chinese language on thought. "The notions to which the Chinese attribute a categorical function depend essentially upon the principles upon which the organization of their society rests; they represent a kind of institutional base for Chinese thought."[38] Granet associates the yin and yang with the emblems of the rival choruses mentioned above as well as with the seasonal movements of men's and women's labor: women weaving in the huts in winter and men going out to the fields in summer. The Tao, Granet claims, is a concept of a less archaic period. He identifies the *Tao* as "road," or "way," with the actual route of the circulation of the lord.

The act of circulation through the various regions of the realm mirrors the movement of the Sun through the seasons. The Sun of Heaven progresses through the kingdom, "bringing the seasons and the months into harmony."[39] This universal circulation can also be achieved through movement within a single building. The Ming T'ang (the Hall of Light, or Hall of Distinction) was divided into nine rooms.[40] Its pattern allegedly reproduced the well-field system of Chinese agriculture, as well as the provinces of the kingdom and the months of the year. By clothing and food as well as location within the building, the king could identify his life with the order of the universe.

Granet's account of the Chinese language presents it as unable to express true abstractions. Confucian *jen* is not "altruism," and Mohist *ch'ien ai* is not "universal love." Granet sees the language heavy in value judgment and rich in practical efficacy. The language consists of what Granet calls "emblems" (a term also used by Durkheim for things such as national flags). Granet asserts that "almost all words [in Chinese] connote singular ideas . . . the dominant desire for specification, particularization, the picturesque. . . . The words . . . seem to correspond to conceptual images . . . united, on the one hand, to sounds that appear to be endowed with the power of evoking the . . . image, and, on the other hand, to signs that represent the gesture which is noted by the motor memory as essential."[41] He sometimes goes so far as to claim that there is no word for *old man*, in general, but many different words for various social aspects of age. There is no socially neutral term for *death*, but words for specific ranks and rituals.

Granet claims that rhythm replaces syntax. The position of a word in a sentence is not essential although the rhythm of the sentence is. (Some qualifications were made in the 1970s and 1980s on the alleged lack of grammatical word order in Chinese, but Granet's claim still finds strong support in a less absolutistic version.) The rhythm evokes in the reader something of the collective emotion of the festivals. Although this overemphasizes, as all of the studies do until recently, the degree of transferability between the verb and noun placement in a sentence, it does correctly emphasize the degree to which rhythms, especially with respect to parallelisms of analogies, are used not only for poetic purposes, but to clarify the syntax and logic of complex presentations in Chinese. Such contemporary writers as A. C. Graham, Derk Bodde, and Chad Hansen, who disagree on the degree of logical argumentation in Chinese philosophy, agree on this point, as we shall see in the next section.[42]

The separation of the written language, which is common to all, and the spoken language, which varies from region to region as strongly as neighboring European languages, helps maintain a constancy of meaning. Pronunciation can change without changing meaning. This is an issue we shall examine in more detail when we consider contemporary accounts of the relationship between the written and the spoken Chinese language.

Granet made major generalizations about Chinese thought, such as the following, cited in a contemporary work on the influence of language upon thought: "The Chinese have no taste for abstract symbols. They see within time and space only a collection of occasions and locations. It is the interdependencies, the solidarities which constitute the order of the Universe . . . Law, the abstract, and the unconditional are excluded—the Universe is one—as much with respect to its societal as with respect to its natural aspect."[43]

Granet offered many early speculative sociological and linguistic explanations of the Chinese worldview and thinking that differ from those of the modern West. Despite the brilliance of many of his points, his attempt, following Durkheim, to give a grand social theory of Chinese thought and language led his work to be neglected by unimaginative and socially unreflective sinologists, who were more careful with the philology of classical texts. Granet's work is sometimes contrasted as reckless speculation with the more cautious scholarship of Henri Maspero.[44] He was often thought of by sinologists and experts of Chinese literature as a sociologist, while he was ignored by most sociologists because his field of interest was ancient, not contemporary, China. As a young graduate student, C. Wright Mills was virtually alone among American sociologists in his interest in Granet's ideas which he attempted to disseminate within a few years of the publication of *La Pensée chinoise.*[45]

Despite Granet's relative lack of direct effect upon Chinese studies outside France, he greatly affected indirectly the theoretical framework of sinologists studying Chinese religion and thought. His work on structures of kinship in China was sympathetically developed by Claude Lévi-Strauss in *The Elementary Structure of Kinship*[46] and structural anthropology eventually had a major impact on Chinese studies. Similarly, Granet's influence extended to the studies of comparative primitive religion by Mircea Eliade, whose works have inspired many scholars of Chinese religion.[47]

The Relation of Chinese Language to Thought

The Chinese language has a propensity for treating the world as a continuum, and for denying separate, internal ideas or external, abstract propositions. The lines between poetry and prose, art and science are blurred by the ideographic nature of Chinese writing. The first and last of these characteristics of Chinese, especially, make a holistic approach to reality easier than in Indo-European languages.

The early-twentieth-century work of Durkheim and Mauss concerning the linguistic shaping of thought, including Chinese schemes of classification, was not systematically followed up until the mid–twentieth century. Benjamin Lee Whorf, whose name is attached to the thesis that language shapes thought, never mentioned Durkheim.[48] The popularity of the Whorf hypothesis in the 1950s was replaced by an equally fashionable dismissal by the 1980s. Nonetheless, Chad Hansen has made a number of illuminating points about philosophical implications of differences between Chinese and Indo-European languages.

The Whorf Hypothesis
Whorf investigated languages of the indigenous American peoples, especially of the Hopi. He claimed that the structure of Hopi language was so different from Indo-European languages that it produced a totally different worldview, leading the Hopi to conceive of the universe in terms of process.[49] Edward Sapir formulated a weaker hypothesis, that grammatical differences make it easier or more difficult to formulate certain concepts, and the general claim is often called the Sapir-Whorf hypothesis.[50]

The Whorf hypothesis has become a less frequent point of reference in philosophy and much of linguistics, in part because of Whorf's often naive ways of formulating his claims. Two philosophical developments have led many philosophers and linguists to sidestep or denigrate the Whorf hypothesis. First and foremost is Noam Chomsky who developed a formal, transformational grammar based on the claim that there is a "universal" grammar underlying all languages. In the formalization grammars of various languages are claimed to be surface variations or transformations of this supposedly universal, deep structure, and it follows that there are no ultimate, root differences. Chomsky claims that recent work (in terms of his Minimal Program) shows that Chinese is very close to English. Those who used Chinese as a counterexample for the supposed universal-grammar thesis took their examples primarily from written, classical Chinese. Against critics of the universal-grammar hypothesis, Henry Rosemont has questioned whether written classical Chinese is a direct transcription of the grammatical structure of ancient spoken language.[51] If classical written Chinese is not a transcription of the spoken language, then these counterexamples fail, because written classical Chinese is not a "language." However, even if at the deepest level of structure all languages are the same, there are strong divergences in surface structure that could bias conceptual schemes by making some formulations easier in one language than in another.

An example of a Chomsky-based attack on the Whorf hypothesis based upon this new consensus is Steve Pinker's.[52] He identifies Sapir's more modest hypothesis with Whorf's. Pinker then concentrates on Whorf's thesis about vocabulary, emphasizing recent studies that debunk the assertion that the Eskimo have many words for snow. This debunking itself focuses not so much on Whorf's study as on the exaggerated versions of it in popular writings and elementary textbooks,[53] which do not deal with the effects of grammatical structure as opposed to vocabulary. Pinker also suggests that Whorf's hypothesis, as commonly applied to Chinese, is racist.[54] Of course, claims that different cultures have language-shaped differences in frequency of certain patterns of thinking need not show the superiority of Western thought to Chinese. Rather, the emphases might be advantageous in some fields and disadvantageous in others. If Chinese organicism is enhanced by the nature of Chinese grammar, the superiority of the early Chinese understanding of magnetism to that of Europeans is a case in point, as are precocious biological advances, such as biological pest control and hormone therapy. If process and relational thinking is aided by Chinese grammar, early Chinese recognition of evolution in the heavens (sunspots, supernovas) would be examples of its advantages over Western reification.

Another philosophical idea that seemed to undermine the Whorf hypothesis is W.V.O. Quine's concept of the indeterminacy of translation.[55] Quine argues that the precise specification of sameness of meaning or of the containment of one meaning in another is prevented by the indeterminate nature of meaning. Radical translation is impossible because of our inability to distinguish between the different means of classification involved in pointing to an object or entity. Quine suggests that an informant pointing to a rabbit might be identifying the substance rabbit, or the parts of a rabbit, or an event, or a rabbit-appearance. If we cannot distinguish between these metaphysical schemes implicit in object identification, then neither can we assume that the radically foreign language works the same as or differently from ours. In these cases, Whorf cannot sensibly make claims about different metaphysical schemes of various languages.

While denying the extreme cases of incommensurable conceptual schemes that are not mutually translatable, we can still grant Sapir's weaker form of the Sapir-Whorf hypothesis.[56] It is quite possible to formulate in English the sorts of insights that the Hopi language makes easier to present. Critics of Whorf are fond of pointing out that the process view of the world that Whorf attributes to the Hopi language is found in some Western philosophers such as Henri Bergson and Alfred North Whitehead. Process philosophy is a minority view, however, which has arisen only recently and is notable for its obscurity of expression, precisely because of the substance bias of European languages. Whorf himself attempts to convey what he presents as the Hopi worldview in articles written in English, showing that the Hopi concepts are, at least partially, expressible in English. It does not show, however, that they are naturally or easily so formulated. Donald Davidson criticized Whorf noting that Whorf presented Hopi sentences in English translation,[57] but Graham notes that "there is no paradox here; Whorf would hardly have denied that bilingual speakers would be clearer about the divergence with an equally sophisticated Hopi account to compare with his."[58]

Ideographs, Kircher, and Leibniz

One of the major areas of dispute concerning the relationship of the Chinese language to thought concerns the "ideographic" nature of the written Chinese character. Westerners who originally came into contact with Chinese thought that Chinese characters represented ideas rather than sounds. Athanasius Kircher, an early Jesuit polymath, claimed that Egyptian hieroglyphics and Chinese ideographs directly represented concepts, while Western languages represent sounds. Kircher produced large and sumptuously illustrated baroque texts full of symbolic diagrams that supposedly explained things by virtue of the symbolic imagery itself.[59] He also published a large text concerning Chinese ideograms and their relation to Egyptian and other picture writing.[60] The philosopher and scientist Leibniz was interested in Chinese as something he thought approximated his notion of an ideal language of thought, where each symbol precisely represented a fundamental concept.

Both Kircher and Leibniz, who emphasized the ideographic nature of the Chinese language and who considered it superior to a phonetic language, are important also in the development of concepts of magnetism and field theory, respectively. Kircher concerned himself with the nature of the magnet, with the magnetism of the Earth, and with magnetism in astronomy. Leibniz formulated an organic view of the universe that influenced later Romanticism and was an important proponent of the continuum view of nature and of the relational nature of space and time.

As more was learned about Chinese, and as twentieth-century linguists attempted to integrate their accounts of Chinese with those of other languages, a strong reaction against the view of Chinese as ideographic set in. In fact, Michael D. Coe makes Kircher into the villain in his story of the decoding of Mayan glyphs.[61] For Coe, Kircher is the creator of "the myth of the ideograph" that misled generations of linguists concerning the nature of Egyptian, Chinese, and Mayan languages.[62] John DeFrancis, the author of major textbooks of Chinese, wrote polemics in several works against the "myth of the ideograph," claiming that all languages are speech based. In the case of Mayan language, Coe states that decoding was delayed for up to a century by the refusal to relate Mayan inscriptions to the spoken language of the indigenous peoples of the area. DeFrancis is motivated by his admirable concern for the alphabetization of Chinese in order to increase literacy among the Chinese; however, these concerns can be pursued without denying ideographic features of Chinese.

Coe ridicules "the dead hand of Kircherian thought," basing himself on an elementary, secondary account.[63] Kircher has been unjustly made the whipping boy of Egyptologists, many of whom have not read him, rather than their honorable founder.[64] Kircher correctly hypothesized that ancient Egyptian was related to modern Coptic (a conjecture that later scholars, to their detriment, rejected) and was the first to determine the phonetic value of a hieroglyph resembling waves of water related to the Coptic word for water, *mu*, as the sound "m." Kircher's compilations of inscriptions were valuable down through the eighteenth century as reference material for others.[65] The source for Kircher was the work of Horapollus in late antiquity, who claimed to reveal the meaning of the hieroglyphs, but is now seen as having simply made up his meanings. His work was mostly erroneous but was based on some correct information. The later Egyptian priests, after the hieroglyphics died out in the culture beyond the temple, manipulated the images for incantations. Any correct information Horapollus handed down concerning the meanings of hieroglyphs was from scribes during this late period.[66]

Criticism and Defense of the Ideographic Thesis

Those who completely reject the notion that Chinese is ideographic emphasize that one component of many Chinese graphs is phonetic. There are 895 phonetic components that appear in Chinese graphs (mostly on the right). Opponents of the ideograph claim that these allow the reader of Chinese to correctly pronounce two-thirds of Chinese graphs without further knowledge. DeFrancis and Coe are correct to reject the simplistic notion that Chinese is *wholly* ideographic, but they illegitimately move from the fact that Chinese has a major nonideographic component to the claim that Chinese is not ideographic at all. A third of the Chinese graphs still cannot be read in terms of sound even with full knowledge of the phonetic components.

Coe and DeFrancis are right to emphasize that Chinese ideographs are not pictographs and often serve a phonetic function; however, Western phonetic languages are not entirely phonetic (that is, they do not, for the most part, directly represent the sound structure of the words). Jacques Derrida correctly points out that there is a "myth of the phonetic" (which is parallel to Coe and DeFrancis's "myth of the ideogram") in that the arbitrariness of language eliminates any direct tie between the spelling and the sound in a phonetic language, just as it eliminates a direct, intuitive, Kircherian tie between the ideograph and the object. Derrida claims that genuine phoneticism is not achieved even in phonetic spelling because of spaces and punctuation marks. Kircher's baroque claim of the superiority of the ideograph to the phonetic was replaced by the claims of Rousseau, Hegel, and later linguists of the superiority of the phonetic to the ideographic. Rousseau correlated the sequence from pictographic to ideographic to phonetic with the sequence of progress from savagery to barbarism to civilization.[67] The supposedly transparent revelation of thought by speech makes the phonetic writing system superior.[68] Coe's and DeFrancis's attacks upon the ideographic interpretation contain a tacit residue of belief in the cultural superiority of the phonetic—what Derrida calls "phonocentrism." To counter the privileging of the phonetic Derrida has appealed to hieroglyphic writing, called "a new Egyptology."[69] He says of nonphonetic writing, and presumably of the philosophical approach he builds upon it, "It is to speech what China is to Europe."[70]

Another fact that goes against the close identification of written Chinese with spoken Chinese is that the dialects of spoken Chinese are as far from one another as are European languages from one another and are no more mutually comprehensible. A speaker of Mandarin can no more understand Shanghai or Cantonese dialect than an English speaker can understand German. The written language, however, is common to all eight or more spoken languages, or dialects, that make up Chinese. Chinese movies shown to Chinese audiences have Chinese subtitles, comprehensible to all dialects.[71]

Though some two-thirds of Chinese characters can be read correctly in terms of sound if one knows only phonetic elements in the graphs, these sounds can be radically different in the different dialects. Written Chinese has an interlanguage, or translation, function. Written Chinese already performs the function with respect to the mutually incomprehensible spoken dialects of Chinese that ideas, meanings, or ideal propositions perform in Western philosophy. The theory of ideas was not needed. The language that refers to mental entities rather than things, "mentalese" (to use Wilfrid Sellars's term), was unnecessary for Chinese philosophers.[72]

Those who oppose the notion that Chinese is ideographic often do so for the politically admirable reason of opposing racist theories of the inferiority of Chinese to European languages. Coe ridicules the nineteenth-century evolutionary theory of language that claimed that language evolved from pictograms to ideographs to syllabic (one symbol per syllable, as in Japanese) to alphabetic symbols. He calls it one of many pseudo-Darwinian theories that were used for the purpose of justifying European racial superiority and colonial rule.[73] Coe and DeFrancis, however, tacitly grant the very evolutionary theory that they claim to oppose. They think that to show that Chinese is not primitive or simpleminded they must show that it is not ideographic. In fact, DeFrancis asserts that all languages arose as pictographic languages[74] and that Chinese is "abysmally bad" as a phonemic writing system.[75]

Some features of classical Chinese suggest that the written language was more detached from the spoken language than are syllabic or alphabetic languages.[76] For one thing, classical Chinese is extraordinarily terse, and modern Chinese seems relatively terse compared to modern European languages. Hansen suggests that if the ancient Chinese spoke with the terseness of written classical Chinese they would be far more laconic than Gary Cooper.[77]

The interlanguage function of Chinese can be seen to the extent that its symbols were borrowed and used for other purposes in East Asian languages, such as

Japanese, Korean, and Vietnamese. Japanese *kambun* texts present classical Chinese works in Chinese characters but with added markings to identify Japanese sentence order and grammar.[78] This shows that written Chinese functions on an even broader scale than as translator of Chinese dialects. DeFrancis wrote a satire called "The Singlish Affair"[79] in which he presented a supposed report of the transcripts of meetings of a group of Japanese, Korean, and Vietnamese scholars who were involved in a project to force Americans to write English in Chinese characters. Many readers, including professional linguists, believed the story. DeFrancis maintained that this showed the ignorance and ideographic bias of experts in Chinese. What it shows is that written Chinese *could* function in this manner.[80]

Another feature of spoken Chinese that shows the separation of written from spoken Chinese is what Graham calls the "phonetic poverty and semantic wealth" of Chinese. If one disregards tones, Chinese contains some three hundred distinct syllables. If one adds the four tones of Mandarin or the eight tones of Cantonese, this leads to a thousand to two thousand syllables distinguishable by sound alone. Given that a great many Chinese words are monosyllabic, with the proportion of monosyllabic words to the total number of words increasing greatly as we move further into the past, there are four thousand words in a minimal vocabulary and eight thousand to fifteen thousand words in a more technical vocabulary. This means that the average spoken monosyllabic word will average some four different meanings distinguished by context. Written Chinese distinguishes these words by adding to the phonetic component a radical that gives the class of things to which the ideograph refers. If the English word *bat* had a wood radical attached, we would know it referred to a baseball bat, not a flying mammal (whose symbol would have an animal radical).

Derrida, Grammatology, and the Defense of the Ideographic Nature of Chinese

Derrida criticizes the presumption of priority of spoken to written language. Part of the support for this presumption is an excessive focus on Western alphabetic and syllabic languages to the exclusion of ideographs and hieroglyphics.[81] In the criticism of this thesis held by linguists following Leonard Bloomfield, Derrida's point is neglected, even by A. C. Graham, David Hall, and R. T. Ames, who discuss Derrida in considerable detail on other topics. Graham refers to Derrida with respect to fundamental polarities underlying our syntactical or grammatical structures. Polarities of light and dark, male and female, right and left, and so on, which we find in Chinese cosmology as well as in other premodern thinking, are claimed by Graham, following Derrida, still to underlie modern philosophical reasoning.[82] Graham does refer to the possibility of applying Derrida's deemphasis on the spoken in favor of the written, but claims that classical Chinese dispute was primarily oral, and so this does not apply to the early period. Hall and Ames make use of Derrida's ideas on deconstruction, but do not cite Derrida's equally important critique of the assumption of priority of spoken language over written language.[83]

Derrida traces this assumption back to Socrates's distrust of writing as portrayed by Plato at the end of *The Phaedrus*.[84] This relates to the transition from an oral to a written culture in which written words become detached from the intentions and answerability of the author. Derrida also notes Rousseau's claim that writing is a corruption of the authentic, primitive transparency of spoken language as part of the general corruption of civilization. Derrida traces these attitudes into twentieth-century theorists of language, such as Ferdinand de Saussure and Claude Lévi-Strauss. Bloomfield and his polemical follower DeFrancis exhibit a particularly naive version of this assumption of the priority of spoken to written language.[85]

Derrida's defense of writing as in a sense prior to spoken language depends on a broad, highly metaphorical conception of writing. This conception includes all those aspects of language that are undecidable and open to further indefinite sequences of interpretation. On a more literal and historical level, a case has been made by science-reporter-turned-anthropologist Alexander Marshack that very simple forms of written symbolic expression appear simultaneously with the earliest forms of spoken symbolic expression. Marshack cites patterns of dots and sequences of drawings of what appear to be phases of the Moon on early bone tools. Marshack claims that these are early written lunar calendars.[86]

Some Consequences of Ideographic Writing for the Relationships of Language and Painting, Logic and Poetry

Hansen notes that both the Chinese and the Westerner first exposed to Chinese are struck by the pictorial aspect of the Chinese language even if, in fact, it contains phonetic elements. This ideographic aspect of the Chinese language probably accounts for the relative closeness of writing and painting in Chinese art. In China, calligraphy is a major art form and one's level of culture is estimated at least as much by one's handwriting (calligraphy) as by one's grammar. In the West, conversely, handwriting counts for little.

The fact that Chinese paintings often inscribed with a poem as part of the visual beauty of the painting, as well as functioning as a verbal description or supplement, shows the closeness between poetry and painting in Chinese art. In China, later owners and connoisseurs of a painting would often write tributes or encomiums on the painting itself. This may seem strange to Western connoisseurs of art, who would consider such an action on a classic Western painting as defilement and vandalism.

Poetry and prose are less separate as well. Chinese spoken syllables are distinguished not only by their vowels and consonants but also by tones. Given the relatively few syllables, and hence monosyllabic words, distinguishable solely by vowels and consonants, spoken Chinese would be much more ambiguous and context dependent than it is without differences in tone. The standard example in Elementary Chinese is the word *ma* which can mean "mother," "horse," "hemp," or "scold." In spoken Chinese these words are distinct because each is given a different one of the four tones of Mandarin Chinese (high, rising, falling, and falling then rising), clearly distinguishing the spoken words. Cantonese has eight such tones that allows for even greater disambiguation. Despite the tones, Chinese spoken words still have multiple meanings; however, the tones augment three hundred or so basic syllables to twelve hundred to twenty-four hundred distinct syllables with tones. This leads to an interesting contrast between the use of tone and its relationship to emotion. In English, a rising tone is used for a question. (California "Valley Girl" speech applies this rising tone universally, giving a hesitancy to all statements.) In English we use different tones of voice for doubt, commands, and flirting. Since Chinese has already preempted certain uses of tones for the actual meanings of the words, explicit words must be added to designate functions such as doubting or questioning. In English the statement of fact is the paradigm, while the emotive or persuasive uses of language are considered outside the formal grammar. In contrast, in Chinese philosophy, as in Western literary analysis, it is natural for emotive tones and attitudes to be just as much an object of formal analysis as the factual content of the statement.[87]

Another feature of Chinese that brings poetry and logic closer together than in English is the extreme simplicity of the basic grammar. Languages can make grammatical distinctions by modifying the core words (in inflected languages, such as Latin), adding suffixes, infixes, or prefixes (as in agglutinative languages, such as Turkish), or by word order (as in analytic or isolating languages, such as Chinese). English is the most isolating of Indo-European languages, but makes some use of other devices. Chinese is the most isolating language in the world. Unlike French or Latin, and even more so than English, any changes in word order in Chinese will change the grammar. Furthermore, Chinese puts adjectives, adverbs, and subordinate clauses before the word or phrase they modify. Modern Chinese allows a construction for possessives and subordinate clauses that permits long, complex sentences; however, if there are three or more levels of subordinate clauses and modifiers, it becomes very difficult to follow such a sentence, and the sentence itself sounds highly unnatural. These complex constructions are themselves "a response to Western pressure."[88]

Classical Chinese writers often used parallel constructions in order to give hints to the reader of the grammatical structure, which might otherwise be ambiguous. The less ambiguous sentence would supply the grammatical order, and the parallel sentence would be interpreted as having the same grammatical order as the initial sentence.[89] This led to the use of poetic devices, such as paradigms, as Roman Jakobson would call them, involving analogy and metaphor as a way of giving information about the grammatical structure of the sentence. Poetical structure in classical Chinese philosophy is itself part of the grammatical structure. Along with tones this makes emotive uses of languages and poetic metaphors in classical Chinese very much part of the syntax and logic of argument. The use of analogy is central to much Western thought concerning magnetism from the Renaissance through Maxwell.

Mass Terms, the Continuous Universe, and Holistic Metaphysics in Chinese

One of the most important suggestions concerning the influence of Chinese language on thought is Hansen's claim that Chinese nouns in general behave like "mass nouns."[90] Most ordinary English nouns are "count nouns." These nouns can be preceded by numbers, such as "one pencil, two pencils," "one woman, two women." Mass terms, such as water, sand, or grass, are not properly preceded by numbers. One cannot say "two waters" or "two grasses" unless one means *kinds* of waters or species of grasses. One can say "one glass of water," "two glasses of water," or "one bushel of grass," "two bushels of grass." Similarly, although count nouns admit singular and plural, mass nouns do not admit singular or plural. Again, "two sands" does not make sense (unless one is speaking of *kinds* of sand), but "two buckets of sand" makes sense. In addition, the indefinite article does not apply to mass terms. One does not say "a water," but "a glass of water," or "a bucket of water."

The hypothesis that Chinese nouns are mass nouns has several interesting consequences with respect to Chinese metaphysics or the implicit worldview of the Chinese language. Entities treated as mass terms would not be considered built up from discrete individuals, as in the count-noun paradigm. In Chinese the word for "a human" and for "humans" in general (either as a class or as an abstraction) is the same. One has to say something like "two head of humans" the way one says "two head of cattle" in order (unnaturally) to get the count-noun effect.

Hansen's claim accounts for the atomistic bias of Western thought (in which count nouns are typical and paradigmatic for nouns) and for the continuist or holistic bias of Chinese thought (in which mass nouns would be the standard).[91] The world presupposed by Chinese is not an aggregate of things, but a continuous stuff (say, *ch'i*) that is divided up in various ways. The principle of division is relative to our purposes, giving the sort of perspectivalism that one finds in Chuang Tzu. To reject an absolute perspective and to recognize different ways of classifying things for different purposes need not be a nihilistic, self-refuting relativism, but may rather be pragmatism.

Hansen's thesis was criticized by one of the foremost grammarians of Chinese, Christoph Harbsmeier,[92] who argued that more recent studies of Chinese nouns (such as his own) have shown that there is a distinction between count nouns and mass nouns in Chinese. Hansen somewhat modified his position in response to the criticism. He still argues that Chinese nouns in general *behave* like mass nouns in English rather than like count nouns in English, even if Chinese has a subtle implicit distinction built into its usages. Chinese lacks singular versus plural and lacks the indefinite article. Abstractions in Chinese are generally denoted by the same word that denotes an individual.

Hansen's revised, weakened thesis still effectively captures the tendency of Chinese to treat the world as continuous stuff that is then divided in various ways for various purposes, as opposed to an aggregate of atoms or individuals.[93] Graham writes, "The most interesting conclusion for Chinese philosophy drawn by Hansen from his mass noun hypothesis was that China tends to divide the world down into variously divisible and countable parts, the West tends to assemble it from individuals. This still holds if we reclassify philosophical terms as generic rather than mass nouns (few would be count nouns)."[94] Graham thinks that Davidson assumes a count-noun-based model of the world as "obviously" an aggregate of individuals in his critique of the Whorf hypothesis. Davidson writes,

"We cannot attach a clear meaning to the notion of organizing a single object (the world, nature, etc.) unless that object is understood to contain or consist in other objects"[95] Davidson's thought presupposes a language that sharply contrasts singular and plural, and he uses examples that show he assumes wholes are made up out of prior, independent parts, rather than being continua (a closet full of objects but not the ocean).[96]

In summary, despite recent reaction against traditional claims in this regard, the structure of the Chinese language, particularly the written language, would seem to predispose speakers and writers to certain emphases different from those of speakers of Indo-European languages. These include a closer unity of written language and picturing (such as painting); the role of written Chinese as an inter-language means of communication; eliminating the need for abstract propositions or mental "idea"; disassociation of the written form from the phonetics of spoken language; and a tendency to treat the world as continuous rather than discrete, as viewed from different relative perspectives, and divided into different classifications for pragmatic purposes. All these features of the Chinese language support or predispose to continuism, holism, and field theory.

CHAPTER 5

Interlude: Peter the Wayfarer and the Western Reception of the Compass

———❧———

The first certain Western reference to the compass is found in Alexander Neckam's book *On the Nature of Things*, written around 1190 C.E. The direction-finding quality of the lodestone was unknown to Adelard of Bath in his encyclopedia in 1117 C.E. There are references to the compass in poems by William the Clerk and by Guwot de Provins; however, the dating on the latter is very uncertain.[1]

Neckam writes, "The mariners, . . . when in cloudy weather they can no longer make use of the sun, touch the needle with the magnet which is then swirled round in circles, and when the motion ceases, the point of the needle looks towards the North."[2] Although the transmission of the compass to the West has often been conjecturally attributed to the Arabs, Arabic references to the compass are all later than those in the West. The first mention is in a Persian anecdote collected by Muhammad al-'Awfi in 1232 C.E.[3] The first extensive reference is made by Bailak al-Qabajaqi in 1242 C.E. Bailak writes: "When the night is so dark as to conceal from dark the stars . . . , they take a basin full of water . . . ; they then drive a needle into a piece of wood or corn stock so as to form the shape of a cross, and throw it into the basin of water prepared for the purpose, on the surface of which it floats. They afterwards take a lodestone of sufficient size . . . ; bring it near the surface of the water, give to their hands a rotary motion; then they suddenly and quickly withdraw their hands when the two points of the needle face north and south."[4]

The lateness of Arabic references suggests that the compass may have reached the West by an inland route. Stories of magnetic islands predate the references to the compass. A late version of these stories appears in *Arabian Nights* about an island that pulls the nails out of ships.

In 1269 C.E., *The Letter on the Magnet* by Petrus Peregrinus (Peter the

Wayfarer) appears. This extraordinary work is at a level of experimental analysis unequaled until the seventeenth-century. William Gilbert, in 1600, in his book *On the Magnet*, borrowed his methods largely from Peregrinus.[5] We know exactly when *The Letter* was written as it is dated August 8, 1269. The author was in the army of Charles of Anjou, at the siege of Lucera, a town in southern Italy.[6]

In contrast to our precise knowledge of the date of the letter, we know virtually nothing about its author. He is sometimes known as Pierre de Marincourt, with reference to his hometown in France. His Latin name, Peregrinus, is variously translated as "Wayfarer," "Traveler," "Pilgrim," and even "Stranger." We would like to know if his travels and pilgrimages extended to the East, but this knowledge eludes us.

The experimental clarity and accuracy of the treatise makes it one of two highly advanced experimental works of the Middle Ages, along with Theodoric of Freiburg's treatise on the rainbow. *The Letter on the Magnet* has no known experimental precursors on the subject leading up to it. This might suggest knowledge on Peregrinus's part of researches in Asia regarding the magnet. *The Letter* has no methodological or philosophical reflections. It is primarily experimental and descriptive, with a few speculations. Petrus Peregrinus says of the researcher: "He must himself be very diligent in handicraft also, in order that through the operation of this stone he may show wonderful effects. For by his carefulness he will be able in a short time to correct an error which in an age he could not possibly do by means of his knowledge of nature and mathematics, if he lacked carefulness in use of hands."[7]

In a series of short, paragraph-length chapters, Peregrinus shows how to recognize a lodestone, how to determine its north and south poles, how to induce magnetism in a piece of iron, how to reverse the poles, and how to produce little magnets out of a big magnet.

Peregrinus also devises several instruments, among them two designs for an astrolabe. The magnet is used to determine the north direction, and the astrolabe is used to determine the angle of a heavenly body from north on the horizon. He also proposes a perpetual-motion machine based on magnetic attraction. He points out that the compass does not point directly to the North Star; the compass needle points to the poles of the heavens rather than to the poles of the Earth or the Pole Star.

Thus, Peregrinus, suspiciously like the Chinese, explains the orientation of the compass by its relationship to the universe as a whole, rather than to its pointing to a particular object. He specifically criticizes the view that the compass points to a deposit of iron at the North Pole by noting that there are lodestone deposits all over the world and one might as well point out that there are large deposits of iron at the South Pole as well. In addition, he criticizes the view that the magnet points to the Pole Star by noting that the star does not always coincide with the exact direction of the compass.

Roger Bacon, a well-known proponent of science and technology and contemporary of Peregrinus, writes a great tribute to Peregrinus as his teacher and guide, whose knowledge outreached that of anyone of his age and time. He states, "for he does not trouble about discourses or quarrels over words, but follows the works of wisdom and keeps quietly to them. And so, though others strive blinkingly to see, as a bat in the twilight, the light of the sun, he himself contemplates it in its full splendour, on account of which he is a master of experiments and thus by experience he knows natural, medical, and alchemical things, as well as all things in the heavens . . . he has inquired into all operations of metal-founding, and the working of gold and silver and other metals, and of all minerals; and he knows all things pertaining to the army and to arms. . . . He has already labored three years on one burning mirror [set for?] a fixed distance and, by the grace of God will soon complete it."[8]

PART II

Renaissance Occultism

CHAPTER 6

Occultism and the Rise of Early Modern Science

*R*enaissance occultism, although pursued primarily by educated, elite figures, had connections with peasant thought. The clearest example of direct contact of this sort is Paracelsus.

Paracelsus and the Paracelsians

Renaissance occultism includes hermeticism, natural magic, Neoplatonism, cabalism, and alchemy, a major current of which is Paracelsianism. Theophrastus Philippus Aureolus Bombastus von Hohenheim—Paracelsus for short—was an extraordinary figure in Renaissance chemistry and medicine. Paracelsus (1493–1541) has often been called the "Luther of Medicine" both because of his time and place—the Germany of the Reformation—and because he challenged medieval medicine based on the works of Aristotle and the Roman physician Galen around the time that Luther challenged the medieval Catholic Church.

Paracelsus means "above Celsus." Celsus was an ancient medical author whose elegant Latin prose impressed the humanists as well as the physicians. Thus, Paracelsus's own name was a boast, and he himself was not shy about boasting. In speaking rhetorically to the traditional medical practitioners, he wrote, "Let me tell you this: every little hair on my neck knows more than you and all your scribes, and my shoebuckles are more learned than your Galen and Avicenna, and my beard has more experience than all your high colleges."[1]

Paracelsus was truly a figure of the underground. Among the Renaissance theorists of nature he was the closest both in his way of living and in his theories to peasant life and beliefs. The son of a country doctor who practiced alchemy, he studied under a famous alchemist and worked in the mines. Though he was not of a very low social class, he did live with lowlifes and was a social outcast in several respects. Paracelsus rejected the traditional Aristotelian and Galenic medicine and sought to replace it with his own theory and practice. He also rejected the Latin language of scholarship exemplified by his namesake Celsus. The sources of much of Paracelsus's writing are in "German folklore, and local traditions long since lost."[2]

Paracelsus lectured and wrote in his native German, causing consternation among the traditional medical scholars. One of his defenders later wrote, "It is true that Paracelsus spoke often in German rather than Latin, but did not Hippocrates speak Greek? And why should they not both speak their native tongues?"[3] Because of his lectures in the vernacular and his coining of new terminology for medicine, students in Basel, goaded on by traditional members of the faculty, ridiculed him as "Cacophrastus."[4] Paracelsus's followers also wrote in the vernacular.

Paracelsus frequented taverns and enjoyed drinking bouts with the peasants. From his mining experience, he wrote the first treatise ever on occupational medicine, dealing specifically with diseases of miners, with whom he sympathized. He charged high fees from the rich and low fees from the poor. He wrote, "There are not many rich people endowed by nature with blessed generosity." He described them as "choking with vulgarity, cruelty, and avarice and utterly lacking in understanding . . . stupid, arrogant, proud men who imagine that . . . they are masters of heaven and earth."[5]

Paracelsus sided politically with the poor and was sympathetic to the Anabaptist religious movement, which was opposed by both Catholics and Lutherans in Germany. He supported religious reformation and redistribution of wealth.[6] Heer has written: "The spiritualist socialism and communism of Paracelsus combined the main elements of Franciscan spirituality and Anabaptism";[7] "Paracelsus rejected charity, in the sense of organized benevolence, as a false solution and a self-deception. The tasks facing contemporary society were too great for charity and could only be solved by rational planning and social legislation. There must be obligations of service and enforced labour for idlers. . . . It might be necessary to murder tyrants for the sake of the common good."[8]

Paracelsus wandered about Europe throughout his life. He worked in the medical position (lowly as it was at that time) of surgeon and invited barber-surgeons to attend his lectures. His followers in medicine were primarily lower-class apothecaries, not upper-class physicians. The Galenic physicians attempted to control prescriptions of drugs, but the apothecaries ignored the rules during the years before the English Civil War. Paracelsian doctrines opposing those of the physicians and ridiculing establishment medicine found fertile ground among the rebellious apothecaries.[9]

Paracelsianism was in vogue among supporters of the radical currents of the Puritan Revolution. The antirationalist doctrines of Paracelsus fit the turbulent times as well as the religious enthusiasms of the Puritans. John Webster combined the Paracelsian worldview with radical political ideas and plans for educational reform. Defenders of the mechanical philosophy attacked the natural magic of the Paracelsians, their radical politics, and their conception of science or natural magic as solely in the service of society.[10] The Royal Society specifically eschewed in the proposed charter of 1663 "meddling with Divinity, Metaphysics, Moralls,

Politiks"[11] in order to disassociate itself from the identification of radical Puritan political activity with science during the Puritan Revolution.

Paracelsus emphasized nonelite sources of knowledge. He wrote: "Not all things the physician must know are taught in the academies. Now and then, he must turn to old women, to tartars, who are called gypsies, to itinerant musicians, to elderly country folk and many others who are frequently held in contempt. From them he will gather his knowledge, since these people have more understanding of such things than the high colleges."[12] Here we see Paracelsus open to the traditions of women (soon to be called witches) as well as to Eastern knowledge.

The view of the universe as a living organism, holistically relating microcosm and macrocosm, is one we have seen in Chinese thought. Paracelsus had no contact with the Far East but refers to Central Asian "tartars" as a source of his knowledge. He actually did travel to Russia and was a prisoner of Tatars after their siege of Moscow. He studied Tatar customs, including the medicine and magic of shamans. Later he accompanied a Tatar prince on a diplomatic mission and learned about an opiate from a magus he met on this journey.[13]

Similarly, the concept of a holistic living universe was part of the traditional peasant medicine of women healers, presumably stretching back to pagan, pre-Christian times. The view of Earth as female and mother was replaced by the view of it as a lifeless object of male exploitation with the rise of the mechanistic view of science.[14] This went along with the replacement of female healers by male physicians aided by the persecution and slaughter of thousands of so-called witches.

The alchemists' image of the relation of scientist to nature was one of "coition," or sexual intercourse of male scientist and female nature.[15] In contrast, the Baconians and spokesmen of the Royal Society presented the relation as one either of forceful seduction—date rape in modern terms—or else of chaste communication.[16] The witch trials may have provided the metaphors of probing and torture that were used to describe the new science. The egalitarian sexual congress of scientist and nature among the alchemists was replaced by an unequal relationship of exploitation of female nature by the scientist. The sexual imagery of alchemy was replaced by "the chaste and lawful marriage of Mind and Nature" for Bacon.[17]

It is notable that one of Paracelsus's earliest writings, which combines theology with ideas about human biology, elevated Mary to the status of a quasi goddess and also argued against the Aristotelians that both sexes have an equal role in the creation of the form of the child. In Salzburg, where the residents chafed under the combined religious and political rule of an oppressive prince-archbishop that eventually led to the Peasant Wars, Paracelsus initially supported the establishment Catholics. In defending the cult of Mary against quasi-Protestant attempts to downgrade her status, however, Paracelsus went so far as to give her a higher goddess counterpart (probably influenced by peasant Earth goddess beliefs). The Church attacked Paracelsus's doctrines as heretical, which drove him to ally with the radical Anabaptists. In the course of further polemics, Paracelsus argued against the exclusive role of sperm in producing the form of the child and claimed that both sexes equally supplied a "liquor of life" that formed the child. Thus Paracelsus's early association with political and religious radicals was also tied to his elevating the role of the female principle, both in theology and in medicine.[18]

Paracelsus's theory was tied to the alchemical tradition of viewing the universe as a unity of microcosm and macrocosm. The human body as microcosm reflected the cosmic astral arrangement. Paracelsus applied the theory of sympathy and antipathy to describe the relationship of microcosm and macrocosm. Allen G. Debus notes that the Paracelsians were among the first defenders of William Gilbert's *On the Magnet*: "In contrast to Aristotelians, who insisted on action through contact, the Paracelsians found no difficulty in accepting action at a distance."[19] Like the Chinese, they were not puzzled by the phenomenon of magnetism. The alchemist Robert Fludd, who like the Chinese was a devotee of geomancy, also defended "magnetic" remedies, such as the weapon-salve cure, in which the weapon rather than the wound is treated. (The weapon-salve doctrine was itself based on a spurious, or pseudo-Paracelsus, text[20] but was defended by Paracelsians.) In analogy with the magnet, Fludd called such cures by sympathy "magnetic." A visiting physician, William Maxwell, wrote: "Among other things, he was able to tell me of the wonders of a magnet . . . ; it had such power of attraction that when he applied it to his heart, it drew him with such force that he could not have held out for long. For the Fluddian magnet is nothing other than desiccated human flesh, which certainly possessed the greatest attracted power; it should be taken from a body still warm, and from a man who has died a violent death."[21] This magnetic analogy was central to the later, more qualified defense of the weapon salve by Jan van Helmont, who also later used it in his treatise on the plague.[22]

In his own work, Paracelsus was analytical in investigating the body. He concentrated on local areas of the body (such as the liver), which he correlated with cosmic influences, but he also proposed specific cures. Ironically, although he emphasized the relation of the individual bodily parts to the cosmos, he did not emphasize the holism of the individual body itself to the same extent as some later medical theorists.

Paracelsus did not merely speculate, he also proposed specific chemical cures, including the use of mercury as a remedy for dropsy, which "is due to a morbid extraction of salt from the flesh, a chemical process of solution and coagulation. . . . This process does not depend at all on the quality and complexion, but is a 'celestial virtue' endowed with its own 'monarchy.' . . . Mercury will drive out the dissolved salt, which has a harmful corrosive action on the organs, and preserve the solid—coagulated—state of the salt in the flesh, where it is needed to prevent putrefaction. Mercury will affect the cure specifically in everybody. . . . Neither vomiting or sweating—the universal cures of the ancients—are therefore the curative factors."[23]

The movement he spawned within medicine is often referred to as *iatrochemistry*. Followers of Paracelsus opposed treatment by bloodletting that was advocated by the traditional Galenic physicians. Paracelsus replaced herbal medicines with new chemical medicines. In addition, he replaced the Galenic doctrine that "opposites cure" with the folk doctrine that "like cures like."[24] Paracelsus also advocated smaller doses of medicine than the traditional physicians. For instance, in the case of syphilis, mercury was again used but in smaller amounts than was traditionally administered.[25]

Paracelsus advocated a new system of elements to replace the Aristotelian system. In the place of Aristotle's earth, air, fire, and water, Paracelsus introduced mercury, salt, and sulfur. The sulfur and mercury came from Islamic alchemy. Paracelsus added salt to the soup, and sometimes used the four Aristotelian elements, sometimes his own three, but the very introduction of the three-element system threatened the whole Aristotelian worldview. The Aristotelian elements had come to be correlated in the Galenic and medieval systems with the four humors—blood, phlegm, yellow bile and black bile—as well as with the four temperaments—sanguine, phlegmatic, choleric, and bilious. The Paracelsian elements, like those of the Chinese in this respect, were processes rather than substances.

Paracelsus shows most clearly the links among the Renaissance occultist philosophy and science of nature, the beliefs of peasants, the radical political movements of peasants, and the positive pagan valuation of the female principle in nature. Paracelsianism also used the notion of the magnet as a general model for causal influence, or action at a distance, which would encourage later investigations of magnetism. Though Paracelsus (emphatic as he was about the influence of the astral macrocosm) was ignorant of technical astronomy, his doctrines had a major impact on the new cosmology by helping Tycho Brahe reject the crystalline spheres of the medieval world picture.

Tycho Brahe and Paracelsus: The Smashing of the Spheres

Paracelsus typifies much of the new Renaissance thought, such as the doctrine of the microcosm and macrocosm, the revolt against traditional, Scholastic learning (including Aristotle in science and philosophy and Galen in medicine), and the influence of peasant and women healers' beliefs on literate thought. Despite his great contributions to medicine and chemistry, however, Paracelsus had little to say about rigorous astronomy and physics. Paracelsus's thought was militantly qualitative. Nevertheless, one major astronomer, Tycho Brahe (1546–1601), was influenced by Paracelsus to make a revolutionary revision in the traditional view of the solar system involving the smashing of the crystalline spheres previously believed to carry the planets and the fixed stars in their orbits around the Earth. His theory was intermediate between earlier Earth-centered astronomy and newer Sun-centered astronomy. Further, Tycho supplied the data for Kepler's magnetic theory of the solar system.

Paracelsus's own astronomical ideas contributed nothing directly to scientific astronomy. He was primarily concerned with the possible influences of the stars on humans and, conversely, of humans on the stars. Like many theorists of nature of his time, Paracelsus denounced the astrologers while presenting ideas very close to those of astrology. For example, he claimed that astral powers influenced the capacities of the ideal physician as well as the condition and diseases of various parts of the organism. Paracelsus claimed, however, that the highest power was not that of the visible stars but rather of what later would be called astral bodies.[26] He denied that physicians were totally controlled by stellar influences, and claimed that they could overcome negative influences of the stars by free will. Nevertheless, the condition of wounds in the body as well as the effectiveness of medicines was tied up with the *astra*, or astral bodies.[27]

Tycho Brahe developed more precise observational astronomy than any of his predecessors and of a sort that would not interest Paracelsus, though throughout his life Tycho accepted ideas of Paracelsus. One of his major teachers at Rostock was a professor, Levinus Battus, who taught medicine in relationship to astronomy and alchemy. Battus wrote on alchemy as a follower of Paracelsus.[28] A friend of Tycho by the name of Severinus attempted to systematize Paracelsus's wild and disorganized thought using ideas of the Neoplatonists.[29] Tycho Brahe's later observations on the comet of 1577 were conducted in close association with Severinus.[30]

Tycho became known to posterity for his precise, naked-eye astronomical observations and his elaborate astronomical observatory and equipment, as well as for his less-advanced astronomical theory. He also practiced medicine and performed alchemical experiments throughout his life, particularly for the two years immediately preceding some of his first major astronomical writings. At that time, Tycho lived with his uncle, who was the only family member who appreciated Tycho's scientific interests.[31] Even later, when Tycho was fully occupied with a massive program of astronomical observations at his island castle in Denmark, he continued alchemical experiments. He also dispensed free medical treatment to local peasants, raising the objections from local physicians.[32] Tycho sent one of his eminent assistants to be trained in chemistry as well as astronomy.[33] He saw his empirical and mathematical astronomy as part of one great science that included the relationship of the stars to the Earth and human beings through astrology and alchemy. His scientific astronomy was tied to astrology as a necessary foundation for a future, more precise, astrology.[34]

After Tycho had become famous and respected because of his tract on the new star of 1572 (what we would call a supernova), he gave a series of lectures to members of the university at Copenhagen in support of astrology. These lectures defended astrology against those Christian theologians who objected that it left no place for free will. The lectures were filled with, and served as a defense of, Paracelsian doctrine.[35]

Tycho's teachers and friends shared Paracelsus's rejection of Galen's traditional medical doctrines and Aristotle's traditional philosophy and science. His major change of the physics of the solar system was justified by the Paracelsian claim that the heavens were filled with fire, a physical element, rather than ether, a substance unique to the heavens according to Aristotelians. In accordance with this belief, Tycho rejected the existence of the crystalline spheres, which according to Aristotelian and Scholastic astronomy, carried the planets in their orbits.[36] Tycho was fascinated enough with the new star of 1572 to issue a pamphlet in which he described his observations and prophesied upheavals and new religious and social arrangements, including the ruin of the pope and Antichrist, as well as the defeat of the leader of Moscow.[37] Tycho's later prophecies concerning the comet of 1577 were taken up by the English revolutionaries.[38]

Tycho considered the physics of the heavens to follow that of Earth. This, too, came from the Paracelsian rejection of a strict dichotomy between the nature of the heavens and the Earth. A genuine, physical astronomy would not arise until the work of Tycho's assistant and collaborator, Johannes Kepler, who was not at all a follower of Paracelsus (or of the Cabala, hermeticism, or Rosicrucianism).[39] Kepler was, however, sympathetic to a weakened version of astrology, possibly overcome by free will, similar to that of Paracelsus.[40] Tycho's Paracelsian approach to these physical considerations allowed him to be a philosophical as well as a scientific collaborator of Kepler, despite their differences over whether the Earth or the Sun was the center of the solar system, and conflicts between a great observational astronomer with limited mathematical ability (Tycho) and a theoretical astronomer of genius who needed the observational data (Kepler).

Certainly, the Paracelsian influence was not the only one on Tycho's momentous rejection of the heavenly spheres. Tycho followed another Renaissance philosopher, Geronimo Cardano, in believing that comets were celestial objects rather than weather phenomena. He analyzed the great comet of 1577 as an astronomical object rather than an earthly, atmospheric object in the traditional Aristotelians manner.[41] Tycho's animistic, Paracelsian side was evident in his speculation that the comet was a sort of monstrous birth, part of the general births and growths of celestial bodies.[42]

Renaissance Occultism, Hermeticism, and the Egyptian Hermetic Corpus

The standard story of the rise of modern science is one of the rise of atomism and the mechanical worldview, accompanied by empiricism—basing knowledge upon direct sense perception. In this history, magic was banished and occult properties (those not accessible to observation) were ridiculed as remnants of the Dark Ages.

The Yates Thesis, Hermeticism, and Renaissance Occultism

Although some historians of alchemy and of Renaissance Italian thought have emphasized the role of the occult and of hermeticism in early modern science, the standard account of its rise was not challenged on a wide front until the 1960s and the Yates thesis. Frances Yates suggested that the "hermetic tradition" was a major stream in the transition to modern science.[43] In particular she reinterpreted Giordano Bruno as a magician and advocate of a return to ancient Egyptian religion. Previously, Bruno had been presented as a martyr of science who died for preaching the Sun-centered solar system and infinite universe. In the nineteenth century, liberal nationalists and anticlericals erected statues of Bruno in many Italian towns opposite Catholic churches as a symbol of their opposition to Catholic domination and in support of scientific enlightenment.[44] Russian Marxists embraced Bruno, Spinoza, and the French vital materialists (such as Diderot) as part of the tradition culminating in dialectical materialism. Bruno was indeed a martyr to a view of the world opposed to Christianity, but not to modern science. Yates showed that Bruno had no knowledge of the technical details of the Copernican theory and used it for occultist, geometrical diagrams. The Inquisitors were uninterested in Bruno's philosophical and scientific theories.[45] When Bruno himself raised the issue of Copernicanism, the Inquisitors ignored it and moved on to question him whether he had praised heretical princes.[46]

Central European writers have long emphasized that the astronomer Kepler's cosmology contained elements of mathematical mysticism and astrology. Historians of chemistry Allen Debus and Walter Pagel stressed the importance of such alchemists as Paracelsus and van Helmont. With the extension of the Yates thesis, the giants of the physical and astronomical revolutions—Copernicus, Kepler, Leibniz, and Newton—were also recruited into the hermetic tradition or its combination with alchemy that Yates identified with Rosicrucianism.[47] This leads to a radically different vision of the heroes of early modern science as, at least in part, natural magicians, alchemists, and occultists. Bruno and Copernicus become Sun worshipers, while Newton becomes "the last of the magicians."[48]

Needless to say, there was a counterattack against this shocking reconceptualization. Historians sympathetic to the traditional view of science have argued vociferously against drafting Copernicus and Newton into the hermetic tradition. Self-proclaimed defenders of philosophical reason contested the Yates thesis in favor of more orthodox philosophies of science. Bruno was closer to later scientific thought than Copernicus and Kepler (who both placed the Sun not just near the center of the solar system, but near the center of the universe) in *not* placing the Sun at the center of the universe. Bruno drew the full consequences of Copernicanism, in denying that the Earth was at the center, since nothing could be at the center of an infinite universe. He was the first to view the Sun as a star and stars as suns, and accepted the many worlds of the ancient Greek materialists. He was the first to take the cosmology of Lucretius's epic poem of materialism seriously.[49]

The hermetic tradition has often been understood as the broad movement of a variety of magical and mystical views that contributed to ideas and techniques now considered part of science. In the widest sense, the hermetic tradition simply equals the occultist or magical tradition. In a much more narrow and proper sense, the hermetic tradition refers to a specific body of works, the hermetic writings (*Corpus Hermeticum*). Yates talks about the hermetic-cabalist tradition, thereby including the Cabala, a body of Jewish, mystical texts that were Christianized by some Renaissance writers.[50]

In the broad sense of the hermetic tradition, historians of science have included not only hermeticism proper and the cabalistic tradition, but also alchemy, natural magic, and sometimes Rosicrucianism. Few now deny that Bruno made much use of hermetic writings, or that van Helmont and Gilbert show elements of the animistic, alchemical, and magical worldview. Many recent scholars admit that Kepler adheres to mathematical mysticism and an idiosyncratic astrology. The extent to which Copernicus was motivated by hermeticism has been strongly questioned, while the extent to which alchemy motivated Newton's concept of force in physics is controversial; nevertheless, no one today denies that Newton wrote more on alchemy and biblical chronology than he did on physics. Thus, Yates's work encouraged an interpretation of early modern science in which occultism plays an important positive role.

Yates's identification of the *entire* early modern occult tradition with the hermetic tradition was criticized even by those who, following her, now grant the important influence of occult or alchemical ideas on early modern science. Besides the strictly hermetic tradition there is the Cabala, as Yates emphasized in her later writings, and medieval magic, not dependent on hermeticism, was absorbed in the Renaissance. The tradition of natural magic and animism in Renaissance natural philosophy also does not depend on the Hermetic Corpus. Furthermore, even for those who support strictly hermetic doctrines, a distinction must be made between historical appeals to the *Corpus Hermeticum*, in order to gain prestige for doctrines through claims to antiquity, and actual use of hermetic ideas as sources or motivation for scientific or pseudoscientific activity. The identification of one's doctrines with "ancient theology" (*prisca theologia*) to bolster claims, as Newton does, need not be the same as building ones' theories on those ancient doctrines.[51] Although influenced by alchemist ideas, Newton appealed to the wisdom of Moses and Pythagoras in setting precedents for his atomism and gravitation, respectively, without actually using ideas of either in those theories.

Hermeticism in the Strict Sense and the Afrocentrism Controversy

The hermetic writings are now thought to have been largely composed and assembled in the second century C.E. They are attributed to Thrice-Great Hermes (Hermes Trismegistus), who is also identified with the Egyptian god Thoth. The composition is a product of the mystical and desperate atmosphere of the Roman Empire, and manifests the reverence for ancient Egypt that had come to permeate the continuers of Greek philosophy. Most Platonists and Neo-Pythagoreans were magical and superstitious and had a strong belief in astrology by this period.

The hermetic writings were attributed to an extremely ancient period, many thousands years B.C.E., in order to raise their prestige. The Renaissance rediscoverers of the hermetic writings accepted the claim to vast antiquity at face value, and believed the writings were older than the Bible and much older than the great Greek philosophers of the fifth century B.C.E. After all, Hermes was the same as Thoth, who gave the Egyptians law and writing. When the hermetic writings in Greek were brought to Italy from the Byzantine Empire after the fall of Constantinople, the wealthy patron Cosimo de' Medici was so excited by their aura of antiquity and wisdom that he ordered them to be translated before the works of Plato![52] Since Plato was, then as now, considered by many the very greatest of philosophers, this suggests the prestige in which the hermetic writings were held.

The hermetic writings contain religious wisdom as well as magic. The resemblance of certain of the hermetic accounts, such as the creation story, to the Old Testament, as well as the similarity of certain statements to the doctrines of Pythagoras and Plato, led many Renaissance writers to believe that the hermetic works were the source of the philosophies of Pythagoras and Plato, and were contemporary with Moses and the writing of the Book of Genesis. The similarity to the Bible allowed Renaissance hermeticists to call Hermes a pious writer and to defend the magic of Hermes.[53]

The reputation of the hermetic corpus as early Egyptian wisdom was undermined by Isaac Casaubon in 1614. Using the textual analysis of his day, Casaubon showed similarities between the hermetic corpus and the writings of Saint Paul, the Psalms, and Plato's *Timaeus*. Far from being Egyptian precursors to the Bible and Plato, he claimed these passages were lifted from these writings. Yates and others saw this as the death blow to hermeticism.[54] (Isaac's son Meric opposed the Puritan Revolution by attacking claims to divine inspiration, such as hermeticist and mathematician John Dee's supposed conversations with angels.)[55]

Yates notes that some writers, such as Robert Fludd and Athanasius Kircher, simply ignored Casaubon's textual criticism. Further, Ralph Cudworth, a Cambridge Platonist, argued that although the hermetic texts themselves were compiled in the early Christian era rather than a millennium before Christ, they contained genuine ancient Egyptian knowledge mixed with more recent material. Although many scholars through the mid–twentieth century followed Casaubon, other writers agreed with Cudworth. Yates herself notes at the end of *Bruno* that hermetic writings in Coptic have been found,[56] strengthening the case for considering the hermetic works to be of genuinely Egyptian (as opposed to Hellenistic Greek) origin. The discovery of a Coptic version of the hermetic corpus further suggests that these writings were not simply later Greek forgeries, but derived in part from genuinely Egyptian doctrines, some of which go back to Egypt prior to the Greek and Roman conquests.[57]

The discovery of the Nag Hammadi library Coptic version of some hermetic writings associated with Gnostic works has forced reevaluation of the relationship of the Hermetica to Egypt. The greatest scholar of the Hermetica, A.-J. Festugière, dismissed the Coptic texts and their possible impact on his massive study as mere contents of a vase (punning with *la vase*, "slime" in French) from Egypt.[58] For him the Egyptian elements in the Hermetica were "local color . . . touches of the exotic" which "have little more importance than ibises or palm trees in a fresco at Pompeii."[59] The Coptic works, as well as Armenian texts, support the claim that Egyptian wisdom literature was embedded in commentaries by Greek-speaking Egyptians.[60] The "popular" or "technical" (magical) and "philosophical" Hermetica were parts of a single educational or indoctrination process, and cannot be easily separated. The Egyptian influence is more frequent in the magical materials than in the philosophical works, but the latter supplied a theoretical framework for the former and for the more spiritual stages of the initiates' ascent.[61]

In the course of arguing the case, in *Black Athena*, for African-Egyptian priority to and impact on ancient Greek thought, Martin Bernal collected the claims of a number of twentieth-century writers for the ancient Egyptian authenticity of many passages in the hermetic writings. Bernal's main goal was to undermine the Eurocentric arrogance of the traditional history of civilization and to strengthen the claims of non-Europeans, in particular black Africans, as major contributors to the sources of science and philosophy in ancient Greece. These roots are usually attributed to Aryan inhabitants of Greece who immigrated from Central Asia. Classicists, especially the most respected German classicists of the nineteenth and early twentieth centuries, were very anxious to expunge and discredit any claims that non-Europeans were the sources of the classical Greek culture, which was allegedly the font of all rationality, science, and democracy. Anti-Semitic German scholars worked particularly hard to expunge claims about Middle Eastern Phoenician influence on Greek civilizations, as described in Bernal's chapter entitled "The Final Solution to the Phoenician Problem." Most important is the extent to which Western classical scholarship during the age of imperialism and racism worked hard to discredit a millennia-old tradition stemming from the Greeks themselves—namely, that the Egyptians were the source of the science and philosophy of ancient Greece. With the decline and conquest of Egypt by the European powers, as the status of Egyptian civilization dropped rapidly in the eyes of its European colonial masters, the view that Egypt was the source of Western wisdom was rapidly reversed. While such figures as Shakespeare and his contemporaries looked upon Greeks as generally untrustworthy, writers of the nineteenth century elevated ancient Greece to the highest point of civilization, art, and culture.

Additionally, Egyptian civilization, which cannot be denied to have been older than that of Greece or of most others, had to be hermetically sealed off from black Africa. The colonization, conquest, and enslavement of black Africa, coinciding with the rise of modern racism, necessitated that claims to civilization had to be absolutely denied to this area.

A leading Egyptologist, James Henry Breasted, did protest too much in his concern about the race of the ancient Egyptians. Breasted claimed that outside of "the Great White Race . . . the Mongoloids on the east and the Negroes on the south . . . played no part in the rise of civilization." Furthermore, "On the *south* . . . lay the teeming black world of Africa, as it does today. . . . Sometimes the blacks of inner Africa did wander along this road into Egypt, but they came only in small groups. The Negro peoples of Africa were therefore without any influence on the development of early civilization."[62] This is somewhat ironic, given that Bernal counts Breasted as one of the more enlightened figures in Egyptology because of Breasted's attribution to the Memphite theology abstract thought usually claimed to have originated with the Greeks.[63]

Christian writers from the late seventeenth century to the present, in opposition to the anti-Christian Egyptianism of the hermetics and the Masonic movement, labored to undermine the alleged antiquity of the hermetic works. In defense of greater age of the Hermetica, modern scholars have pointed to elements that suggest non-Greek and very early sources of portions of these texts. Names for Thrice-Great Thoth in demotic Egyptian texts have been found in the second and third centuries B.C.E., several centuries earlier than Casaubon's date. Also, the astrology in some hermetic texts dates back to the period of Persian rule over Egypt.[64]

Flinders Petrie, around the turn of the twentieth century, noted that the pessimistic *Lament* in the hermetic corpus must have referred to persecution of Egyptian religion in the Persian period and that it referred to Indians and Scythians as typical foreigners. In addition, the hermetic writings mention an Egyptian king who could be no later than 340 B.C.E.[65] Scholars of the early twentieth century argued against Petrie's claims on the grounds that this would discredit the originality of Greek thought and on grounds of the superior prestige of classical studies over Egyptology.

Non-Egyptian sources also mention Hermes, or Thoth. In the famous passage in the *Phaedrus* in which Socrates criticizes learning from written documents (much celebrated by Derrida),[66] Plato discusses Thoth as the inventor of writing. Additionally, Thoth is claimed by some biblical scholars to appear in the Book of Job. There is also the ancient Phoenician cosmology of Sanchuniaton, which was long dismissed, like the hermetic writings, as a late forgery but which has more recently been noted to parallel Middle Eastern texts from the thirteenth century B.C.E.

If indeed the Yates thesis is correct—that the hermetic writings were a great stimulus for Renaissance science—and the thesis that the hermetic writings contain a core of genuinely early Egyptian material is also correct, and the thesis that Egyptian culture can be considered at least in part to be a black African culture is correct, then the two major scientific breakthroughs of Western history (that of the ancient Greek philosopher-scientists and that of the Renaissance philosopher-scientists) can be attributed to a black inspiration. Bernal's work, controversial as it is, primarily limits itself to the controversial claims concerning the black Egyptian impact on early Greece—hence the title—however, the black Egyptian thesis concerning the hermetic writings plus the Yates thesis would allow Afrocentrists less controversially to lay claim to an African role in the origins of early modern science.

One area on which the Hermetica impinge that is part of the background of genuine science is alchemy. Egyptian temples at Memphis were famous for alchemy. Zosimus, the first alchemist whose writings are recorded and a hermeticist, visited a temple in Memphis to see a special furnace. His associate in alchemy was a priestess, Theosebia, who had a circle of alchemical followers.[67] (Interestingly for the female associations of the alchemical worldview, Zosimus refers to his predecessor, Maria the Jewess, the earliest named alchemist.)[68]

There is the further issue of to what extent Egypt can be considered black. Reactions to the claim of black Egypt range from dogmatic assertion to equally dogmatic ridicule. Opponents of the claim that Egypt was black can rebut the claim that, for instance, Cleopatra was black, but the blanket denial that Egypt was black is misleading. As Egyptologist Frank J. Yurco, no supporter of the Afrocentric thesis, wrote, Egypt was a multicultural society with inhabitants who ranged from very dark skinned to very light skinned and that had hardly any racial prejudice.[69] The Egyptological advisor of *Archaeology*, in an editorial comment, notes that we simply do not know the skin color of many of the major historical figures in Egypt. There are pharaohs with negroid features, judging from their statues. Nubia, a black kingdom south of Egypt, was a powerful and influential civilization in the seventeenth century B.C.E., and it ruled Egypt in the eighth century B.C.E.[70] (The last great center of genuine Egyptian demotic and hieroglyphs was in the far south and was linked to Nubians.)[71] Murals show black Egyptians, sometimes in servile roles, sometimes as rulers receiving homage or collecting tribute from white-skinned people.[72]

Similarly, one can attack straw (or actual) persons who make outrageous or erroneous claims for Egyptian influence on Greek or early modern science. For instance, classicist Mary Lefkowitz uses the example of an Afrocentric speaker at Wellesley who claimed that Aristotle plundered and robbed the library of Alexandria for his ideas.[73] She can point out that the library at Alexandria did not exist until after Aristotle died,[74] but this hardly settles the issue of the extent of either genuine Egyptian influences on early Greek beliefs (Bernal's claim) or the influence of native Egyptian religion and culture on late Hellenistic thought in Egypt (where very clearly there were major borrowings).[75] Interestingly, in discussing the hermetic documents, Lefkowitz claims that Casaubon settled the issue that the hermetic documents were forgeries.[76] She cites, but does not seem to have attended to, studies of Egypt in late antiquity published in the three decades since Yates's work that suggest that parts of the hermetic corpus are genuinely ancient Egyptian, while other parts are Greek forgeries. Critics of Afrocentrism who argue against straw women—such as a black Cleopatra—tend to emphasize that the Greek and the Egyptian inhabitants of Alexandrian Egypt lived in hermetically sealed communities, but more recent studies of Egypt in late antiquity suggest that Alexandrians spoke both Greek and Egyptian.[77] Lefkowitz portrays the whole Masonic myth of the "Egyptian mysteries" as the solely harmful source of historically false Afrocentric accounts of Egyptian wisdom; however, George Sarton, the founder of American history of science as a profession, was influenced by Masonry and Egyptianism to the extent that he named his journals *Isis* and *Osiris* after Egyptian gods and in his scholarly surveys of ancient science emphasized the Egyptian impact on Greek science.

The issue of whether Egyptian science influenced Greek science, let alone the issue of whether ancient Egyptian science had any role in the *science* of the early modern founders of physical science, is separate from that of whether the genuinely Egyptian elements of the Corpus Hermeticum *philosophically* or *religiously* influenced the founders of early modern science. Robert Palter has written a long criticism of Afrocentric claims concerning the origins of science.[78] He carefully criticizes claims by Bernal and others concerning the advanced state of Egyptian mathematics and medicine and the supposed advanced mathematical and astronomical knowledge encoded in the structure of the pyramids (pyramidology). When Palter comes to criticize claims concerning African influences on Newton's *science*, however, he shifts his target from Bernal to another, much less well known, writer who has made claims concerning the African origins of Newton's physics.[79] Palter makes ironic use of the notion of counting "Newton in the ranks of the Afrocentrists" and heaps scorn on the idea. Even though claims that Newton was directly influenced in his mechanics by ancient Egyptian wisdom are totally without merit, the claim that Newton appealed to an ancient wisdom that he identified with the Egyptians (as well as with Pythagoras and Moses) can hardly be denied. Palter himself notes the hypothesis that Newton's use of the "ancient wisdom" was part of his search for the cause of gravitation.[80] Here the distinction between the use of hermeticism as historical precedent or justification for one's position and theories, the use of hermeticism as an inspiration or motivation for one's activity, and the use of hermetic doctrines in one's actual science is surely relevant. Africa has been denied *any* role in the rise of "Western" science and rationality. There-fore ancient Egyptian texts, stemming from a society in Africa that at crucial times was ruled by black Africans and throughout its history was a mixture of races—black, white, and everything in between—had important motivational and justificational roles in the formation of the worldviews of major early modern sci-entists. This is significant in itself, regardless of the truth or falsehood of any stron-ger claims.

Occultism, Science, and Rationality

To many defenders of traditional history of science as enlightenment, the claim that hermeticism or occultism contributed positively to the development of science is anathema. Edward Rosen concludes his polemic against the claim that Copernicus was an hermeticist with, "Out of Renaissance magic and astrology came, not modern science, but modern magic and astrology."[81] One can doubt the significance of Rosen's quantitative content analysis, which concludes, "The hermetic association amounts to about 0.00002 of the *Revolutions* [Copernicus's major work],"[82] in terms of counting sentences concerning hermeticism without regard to qualitative or conceptual significance. One critic has noted that Rosen's attempts to purify Copernicus's science of all such influences "comes close to hagiography: the temptations of Saint Copernicus."[83]

Devotees of magic, the occult, and witchcraft are almost universally con-sidered to be pursuers of the irrational. Students of hermeticism and the occult in the development of science are considered themselves to be advocates of the irra-tional, by a kind of guilt by association. Historian Herbert Butterfield (mainly re-sponsible for both the terms *scientific revolution* and *Whig theory of history*) wrote of students of alchemy that, "like those who write on the Bacon-Shakespeare con-troversy or on Spanish politics, they seem to become tinctured by the lunacy they set out to describe."[84]

There are several confusions here. First, to depict a movement is not to ad-vocate its aims or to approve of it. Many state universities have opposed the exist-ence of departments of religion because the public does not distinguish between the study of religion as a historical phenomenon and the advocacy of a religion. During the 1990s school textbooks were criticized for leaving out the role of reli-gion in history (such as the role of religious beliefs in the Pilgrims' journey to America and in the thought of Martin Luther King, Jr.) on the mistaken belief that to study such a causal role of religion in history would undermine the separation of Church and State. Historians and philosophers should not be guilty of such a simple confusion; however, the vehemence with which certain historians and phi-losophers of science have criticized the study of the hermetic tradition suggests that, subliminally at least, this error is being made.

A surprising case is that of Paulo Rossi. In an anthology, *Reason, Experi-*

ment, and Mysticism, Rossi rails against what he considers the overemphasis on the role of hermeticism in the scientific revolution. Rossi usefully developed ideas of Zilsel concerning the role of the artisans and attitudes toward artisan activity in the scientific revolution, as well as the study of the role of hermeticism and of magic in Francis Bacon.[85] Rossi cannot be opposed to the study of these issues as such. His objections appear to be primarily against such contemporary figures as Paul Feyerabend, who criticize rationalism. Rossi's concern is with opposing contemporary irrationalist movements and with Theodore Roszak's *Making of a Counterculture*, which claims to be based on the work of Thomas Kuhn. Rossi wishes to emphasize what is new in modern science and to downplay the continuity of modern science with its precursor.

Besides the confusion of depiction of a historical movement with contemporary advocacy of it, there is the more subtle confusion of the characterization of rationality itself. A. Rupert Hall, who has presented the more traditional view of the scientific revolution and opposed Marxist and social historians of science, simply asserts that we know what rationality is. He writes, "The word *rational* has a definite and useful meaning; by this I mean that criteria can be established by which propositions about nature can be divided into three categories: the rational, the irrational, and the doubtful. And these criteria have been accessible to all men [*sic*] and all times."[86] Hall ignores that there have been various notions of rationality, some of which are narrower than others, and which thus would exclude propositions that other approaches would accept as part of rational discourse. One thinks of Cartesian rationality, which ties rationality very closely to the intuitively self-evident "clear and distinct ideas"; of technological, instrumental rationality, which would identify rationality with manipulation and control, and of the setting of means to ends without evaluation of the ends, but which would be considered too narrow by critics of technocracy; of economic rationality or "rational, economic man," in which rationality has to do with maximizing utility or profit; of formal rationality, which identifies rationality with logic and calculation; and the contrast of these with the transcendental Reason of Romantics (such as Schelling) or of Hegel, in which intuitive or contradictory modes of reasoning are considered higher than mere calculative rationality.

When one moves from Hall's dogmatic assertion to an attempt to characterize these criteria, difficulties arise. Recent philosophy and history and science have undermined simplified models of the scientific method. The criticism of the logical positivist verifiability criterion of meaning and of Popper's falsifiability criterion of scientificity makes the simple demarcations of single propositions or theoretical systems, of the sort Hall assumes, as acceptable or unacceptable quite problematical. Larry Laudan writes that "rationality is a many splendored thing,"[87] but this seems to be putting a good face on the admission that we have no general characterizations of rationality.

Mary Hesse has written several essays grappling with the issue of rationality in scientific explanation in relation to hermeticism.[88] She concludes that there are no agreed-upon criteria for rational, scientific method. Hesse notes that the simple induction model of science has been widely criticized and rejected. This model holds that the growth of science occurs automatically from the accumulation of pure observations unbiased by theories or expectations. The hypothetico-deductive model of science, in which hypothesis are tested by consequences deduced from them, does not exclude the role of irrational factors in the initial creation or invention of the hypothesis. Thus, Hesse further concludes after a lengthy examination of the alternatives that there is no general context-free set of criteria to distinguish rational from irrational method in science. Nevertheless, she thinks that historical investigations of particular incidents or periods in science will allow such distinctions to be made on an ad hoc basis. Hesse ends up appealing to our intuition to distinguish the irrational from the rational in science. She seems to reintroduce induction in her approach to the history of science; that is, she generalizes from a few case studies (which are supposedly self-evident with respect to their manifestation of rationality) to a general characterization of science.

Hesse frames her investigation in terms of an attempt to distinguish internal from external factors in the history of science. Internal factors, most narrowly, are those involving the inner logic of science itself. External factors are those involving the effect of nonscientific ideas as well as nonconceptual psychological and social factors. The internal approach has been broadened in the last half century to include the impact of philosophical and other intellectual ideas on science, but still excludes the social factors. Further, then, Hesse maintains that due to the open questions about the logic of scientific method, no general criteria are available to distinguish internal from external factors in the histories of science.[89] Nonetheless, Hesse wishes to claim that one can distinguish the mechanistic from the hermetic tradition. This can be done in terms of debates between their respective partisans. Hesse's distinctions depend on an intuition that is not rationally justified.

The distinction between science and magic looks straightforward today. Science is a large, institutionalized and formalized system, while magic is associated with parlor tricks or with localized fringe activities. The main distinction that is made retrospectively is simply that science is identified with that which works and magic with that which does not work; however, at the time of the birth of modern science, such institutional and pragmatic means of distinguishing magic from science were not available.

Traditional technology and magic were often closely intertwined. Ceremonial or superstitious activities were often just as much a part of the rules of technological procedure as were the physical operations that we accept as actually yielding the practical results. For instance, for a traditional Japanese sword maker, even today, various prayers and invocations are as much a part of the sword-making procedure as are the various metallurgical operations. Even today, such workers as coal miners and sailors include good luck charms and the warding off of jinxes with their routine activities. The introduction of a systematic method of trial and error might teach us which factors are essential to the technological process and which are extraneous; however, in highly dangerous, practical affairs such experimentation might be seen as irresponsible. This sometimes happens, as in the case of a post–World War II, indigenous South Pacific canoe navigator who purposely made a long voyage without performing the usual ceremonies in order to test the efficacy of the rites, and found them unnecessary for success. The distinction between magic and technology in terms of which components work and are replicable presupposes the very scientific method that was then not yet developed.

An interesting example of some of the difficulties concerning the notion of rationality and the appeal to technological success is to be found in an essay by Newton-Smith on Isaac Newton and rationality.[90] His essay has the merit of dealing with the issue of rationality while taking seriously the importance of alchemical activity for Newton's self-conceived scientific program. His position is rare for a philosopher of science, who usually consider almost solely Newton's work in mechanics and optics, revered by contemporary scientific standards, ignoring Newton's alchemy and leaving that to the historians—rather the way the positivists left the context of discovery to psychologists and sociologists. Newton-Smith appeals to technological success as his criterion of scientific success.

He rightly denies that there is any specifically scientific kind of rationality, and says there is just rationality. In this vein he praises the marchioness of Francesco Algarotti's dialogue in *Newton for the Ladies* (1739) for attempting to apply Newtonian rationality to a theory of love. Apparently Newton's alchemy was just as rational as his physics, it just did not succeed practically. One could object that some aspects of alchemy had technological-chemical payoff (gunpowder in China, and particular results of Western alchemy in chemistry) and that Newton's alchemy contributed to his idea of gravity. Newton-Smith denies the applicability of specific scientific methods, such as those of Popper and Mill. He also rejects complete externalism, which explains the holding of scientific theories in purely sociological terms as absurd (on the basis of the argument that the success of science would be a miracle if scientific theories were accepted simply on the basis of social or ideological demands). He says that it would be odd if following dictates of church or party rather than nature leads to technological success, but he neglects the extent to which technological success is socially defined. Richard Rorty, in an ambiguous article on science and success, notes that just as philosophers of science often attribute the success of technology to scientific theory, Khomeini could attribute the success of his revolution to Islamic religious belief, and the rulers of the Soviet Union could attribute the success of the communist revolution to Marxist theory.[91] (Rorty simplistically attributes the success of Marxist revolutions outside of Russia to the KGB, but does not seem to think we can distinguish success based on theory from luck or brute force.)

The term *magic* itself includes much we would later call science. So-called natural magic in the Renaissance often included what we would call scientific and technological operations. Giambattista della Porta's book *Natural Magick* contains a mixture of strange superstitious claims and scientific observations and operations. Even later, in the seventeenth century, one work of mathematics was entitled *Mathematical Magick*.[92] Today, such a title is taken totally metaphorically; however, this was not always so in the early modern era.

In large areas, such as alchemy and early chemistry, the proportion of the mixture of magical and religious elements to the chemical operations was very large. The term *occult properties* grades from its medieval definition as properties that cannot be directly observed through its literal association with the realm of the occult, as we should call it (as in Agrippa's *Occult Philosophy*) to its partial return to referring simply to properties inaccessible to observation in Enlightenment thought (while retaining its antiscientific connotations).[93]

There are two areas where the contrast between the rational and the irrational may be more clearly made. One is the role of inspiration in the hermetic tradition as opposed to the methods followed by mechanistic physics; the other is the publicity of the scientific results in contrast to the secrecy of results in alchemy and magic.

Hermetic philosophers, sometimes quite explicitly, rejected the role of reason in favor of inspiration and intuition, as Fludd's polemic against Kepler. Certainly one can contrast intuition with discursive reason, or reason proceeding sequentially through a number of stages. Intuition, however, may have an essential role to play in the process of invention and discovery itself.[94] Even if intuition is not logically analyzable, insofar as it is a necessary stage in the process of discovery, it has its place in a rational methodology. Methodologies for invention or handbooks of heuristics now often include a role for intuition and unconscious incubation as a useful procedure. A late-twentieth-century astronomer's claim that imagination is necessary for scientific method (which, according to the author is not accounted for by either Baconian induction or Aristotelian deduction) is Allan Sandage's praise of the American writer Edgar Allan Poe's work "Eureka": "Poe dismissed the methods of both Bacon and Aristotle as the paths to certain knowledge. He argued for a third method to knowledge which he called imagination; we now call it intuition. . . . Imagination, or genius, or intuition, lets the classification start."[95]

What the occult tradition and hermeticism contributed to later science is to be found primarily in certain models or conceptions or root metaphors. Notions, such as that of force, powers of active matter, and of action at a distance, are among these contributions. They are at the level of models and metaphors, not of methodology. They may be contributed from intuitive sources, but later science found it useful, indeed necessary, to incorporate them into its structure.

In drawing the contrast between hermetic and Paracelsian appeals to intuition and inspiration against logic, we should be careful not to misestimate the role of logic in the new science of the Enlightenment. Formal deductive logic was primarily identified with the moribund medieval Scholasticism by advocates of the new mechanistic or empirical sciences. Formal, deductive logic was not associated with progressive science as it is, for instance, in the modern categories "logic and philosophy of science" or "logic of science." Both Bacon and Descartes, the founders of empirical inductive science, and of mathematical, rational science and philosophy, respectively, rejected formal, deductive logic. Bacon rejected Scholastic deductive logic for the logic of observation and induction. Descartes (despite being the alleged source of the much touted French logical mind) rejected formal deductive logic in his *Discourse on Method*. On the one hand, Descartes identifies it with the moribund Scholastic, syllogistic logic, which is useful for ordering ideas already possessed, but useless for discovering new ideas. On the other hand, Descartes identifies formal logic with the "Lullian art" (the method of combining concepts following the cabalistic permutation of letters in meditation developed by Raymond Lull), which Descartes dismisses as largely fantasy. (Descartes had earlier shown a considerable interest in the Lullian art, listening to an aged and garrulous mnemonist who could discourse for an hour on any subject and on twenty different subjects in succession.)[96] The method that Descartes proposes is based in part on the constructive and creative intuition of geometry, which he contrasts with sterile Scholastic logic. The main interest of the early moderns was in the method of discovery, not in methods of justification. Indeed, the two were identified both in Baconian inductive method and Cartesian mathematical method.[97]

Thus the polemics concerning the irrationalism of hermeticism and even of history of science that emphasizes the roles of hermeticism and occultism in the birth of modern science involve at least two major confusions. First of all, to study the role of religion, occultism, or whatever in history is not the same as to advocate it or support its claims to truth.

Second, the dismissing of all hermetically based research and theorizing in early modern science as irrational often derives from a narrow, inadequate conception of rationality. Just as philosophers of science today do not reject theories simply because they are not based on simple, enumerative induction, and realize that theories can come from a variety of sources, so philosophers of science ought not reject the hermetic sources of conjectures or models in early modern science, or assume that if a theory has a component inspired by or modeled on hermetic or occultist conceptions that the theory itself must be irrational.

The other contrast of the occult as esoteric with the publicity of science is a more significant and useful distinction. Alchemical and much of other occultist knowledge was purposefully made obscure and difficult for the reader, so that only the pure adept would receive the true message, and the vulgar would only be confused. Recent social historians of science have rightly emphasized the transition from the occultist esotericism to the publicity of modern science. Such writers as Galileo and Paracelsus purposively communicated in the vernacular language of their culture rather than in Latin to appeal to the learned layperson rather than the narrow-minded late Scholastic.[98] Galileo's *Dialogue Concerning the Two Chief World Systems* is read in literature courses in Italy. John Dee, an Elizabethan occultist and student of all the sciences, wrote a preface to Billingsley's English translation of Euclid, emphasizing along with the Neoplatonic role of mathematics in describing the spiritual realm the usefulness of mathematics for navigators, surveyors, and others seeking mastery of nature. The printing press also contributed to the availability of cheaper editions of books and to the uniform reproduction of scientific illustrations. Anatomical and mechanical illustrations often lost their original features in manuscript copying, as the copyists lacked the scientific knowledge to realize that certain minor details were significant. Similarly, the formation of scientific societies and journals in the seventeenth century led to the public communication and replication of results. Newton exemplifies here, as in so many other areas, a figure on the cusp between the esoteric occultist knowledge (present in his unpublished alchemical writings) and the public knowledge embodied in his published *Principia*. Newton, as he moved into the British establishment, forsook his private alchemical laboratory activities for the public role of government bureaucrat (at the mint) and public national scientific figure.[99]

The ideal of public knowledge, and open publication of results for testing and criticism by others, is central to the normative conception of science of philosopher Karl Popper and sociologist Robert King Merton.[100] Works by Popper and Merton on this topic were products of the Depression and World War II. Merton claims that the values held by the scientific community include "organized skepticism" and even "communism," in the sense that results are open to criticism and knowledge is a public good which, once published, can be used or built on by anyone. Popper claims that the very nature of science consists of openness to criticism (which he elsewhere links with the "open society" of democracy). Merton, who during the 1930s wrote classical studies of the social basis of early modern science and published in the Marxist journal *Science and Society*,[101] retained provocatively the name "communism" for the public property in information, while Popper, anti-Marxist from his young manhood, identified science with the "open society" to which he contrasted the Soviet Union.

This normative conception of ideal science certainly contrasts most strongly with the secretive and esoteric nature of occultist and alchemical knowledge. Whether in Popper's epistemological and logical form or in Merton's sociological form, it also contrasts with much that actually goes on in science as social practice. Science in the descriptive sense, that is, science in the form of activities of individuals identified as scientists by credentials and positions in scientific institutions, often does not manifest the publicity and openness to criticism that Popper and Merton attribute to science in the normative sense. Obviously a person who has all the degrees and positions that make him or her a scientist in the descriptive, institutional sense may lack the intellectual honesty and openness that are part of the ideals of science. Recent cases of scientific fraud, especially in biomedical research and the psychology of IQ, make that obvious.[102] Outright fraud is probably not significant for the structure of science in general, and negligible in physics. More significantly, "big science," a large-scale enterprise dependent upon military and industrial funding and grown into a professional system with its own economic and political interests, may often not even manifest the theoretical ideals of science. Military and corporate secrecy, whether in patented or closely guarded secret industrial processes, military technology secrets, or codes, clearly involves the denial of the norms of publicity and "communism" that Merton described.[103] Furthermore, the growth of all sorts of new claims to intellectual property through copyrights and patents during the 1980s promises to undermine the free development of scientific techniques of experiment in molecular genetics in such areas as gene sequences and gene cloning. The strong reaction of scientific professionals and scientific professions and their publications against criticisms of hereditarian and psychometric research; criticism of dangers of nuclear testing, nuclear power, or genetic engineering; research on the safety on environmental pollutants; and military funding and direction of research all show that openness to criticism, when it comes from outside of science or even from outside of a professional subspecialty of science (such as criticism by geneticists of psychological IQ heritability studies),[104] is no longer welcomed. The growth of the format of the scientific research paper, and the conventions (or lack of literary style) of scientific prose, shows that the scientific paper is written for a very small circle of specialists, and not for the general, even general scientific, public. The openness to testing and replication of scientific results (granting for the moment that such results can be understood outside the narrow specialty) is in many areas prohibited by the high cost of the technology needed. This is most true in the area of fundamental particle physics, where the requisite equipment may cost billions of dollars and, in the case of the superconducting supercollider, can become so expensive that even the national government refuses to build it.

In the mid–twentieth century some of the most fundamental areas of physics were shrouded in a veil of secrecy due to the involvement of many of the world's leading physicists in military nuclear technology. Papers on nuclear physics, on cryptography, and much else have been classified because they involve, or are conveniently claimed to involve, military secrets. Furthermore, the complexity and difficulty of contemporary mathematical scientific theories makes competent criticism by outsiders very infrequent. This difficulty, combined with the counterintuitive and mysterious concepts of contemporary physics, the technological wonders of biomedical and computer technology, and the search for public media attention by scientists, leads to coverage of science in the media in terms of technological magic and mysterious, less-than-half-understood theories of the universe. The difficult nature of the scientific theories themselves combined with the military and corporate secrecy that surrounds much research has led to an account of "Scientists as Magicians—Since 1945."[105] Experts on nuclear deterrence policy have been called the "Wizards of Armageddon."[106]

Are these mere metaphors, not to be taken seriously? Certainly the esotericism of modern professionalized big science, where it exists, is different from the esotericism of the occultist natural philosophers and alchemists. Within the relevant subcommunity or specialized subdiscipline, replication goes on. The publicity of science does not extend to the general public as much as it did in the founding days of Galileo's or Darwin's science, but within the discipline, and at times between disciplines, criticism continues. There are elements of craft knowledge in science, which are taught in laboratory practice and not available in published descriptions, as Thomas Kuhn and Michael Polanyi have rightly emphasized;[107] however, the ideals of publication and peer review are pursued. Peer review has its limitations in cases where networks of colleagues or adherents of the same school of thought approve one another's work and dismiss that of opposing schools, but the very criticism of publication and of the failure of peer review attests to remnants of serious support of the traditional norms of science.[108]

Twentieth-century scientists have a much more tightly knit, structured community than did Renaissance natural magicians. That community itself, however, although having its own internal norms of rigorous criticism, is less open to the criticisms and contributions of outsiders than was science before its professionalization, bureaucratization, and militarization. Although contemporary science is certainly less intentionally esoteric than the occultism and alchemy of the past, it is more esoteric than early modern experimental science, primarily by virtue of its complexity and abstractness, but also in part because of its professionalization and role in military and other social policy. Proliferation of jargon and formalism also helps keep outsiders out.

Esotericism is today widely considered a highly negative attribute of the alchemical and occultist tradition. Remarks by science policy experts and historians of technology (if not by some more traditional science popularizers) that characterize scientists and technologists as "wizards" or "magicians" are meant as criticisms, not compliments. There is, however, one area in which the relationship of science to alchemical and hermetic adepts is described in a positive manner, contrasting the alchemists favorably with contemporary scientists and even holding them up as a model in this respect for the improvement of contemporary science. This is the case in the areas of the moral motives of the scientist, commitment to society, and self-awareness of scientific activity as a way toward self-understanding.

Alchemists were not simply investigating nature, but also engaging in activities to improve themselves psychologically and spiritually. Many seventeenth-century alchemists, of the period of the English Civil War, were concerned with science as a way to social betterment of humanity, through education, technology, and natural religion. These aspects of science as a self-conscious activity devoted to improvement of oneself and of society in general are sometimes held up today by those critical of value-free science and scientists who deny social responsibility for the consequences of their actions.

Theoretical biologist Brian Goodwin has compared alchemy favorably to contemporary science, at least with respect to the alchemists' using scientific research activity as a means to find out more about themselves as well as a means to discoveries about nature.[109] Goodwin claims that the alchemists "attempted to fuse knowledge and meaning by combining *scientia*, the study of natural process, with morality, man's attempt to realize his own perfectibility and self-fulfillment, itself a continuous process."[110] (Goodwin is a student of embryologist C. H. Waddington, who studied Arabic alchemy in secondary school with his chemistry teacher Eric John Holmyard, who was one of the few English-speaking experts on the topic at that time.)[111] Jerome Ravetz also has used alchemy and hermeticism as models for the political and moral improvement of contemporary science. He reminds the reader that "the dominant traditions of academic science have developed out of conflict with other styles of scientific work." He notes that the fact that scientific success in our terms has sometimes been achieved by these alternative traditions has "embarrassed" traditional history of science, and he recognizes "a similarity of doctrine and style, and sometimes a linking tradition, as far back as the Taoists of ancient China, through Saint Francis of Assisi, to Paracelsus, William Blake and Herbert Marcuse." Ravetz concludes with Bacon, whose science is dedicated to social improvement. He had earlier referred to Barry Commoner, who attempted to direct science toward ecological improvement and at the same time involve the public in contributing to scientific surveys and research and is seeking to link the social-reform aspects of the hermetic tradition to ideas of a socially responsible "science for the people" today.[112] Implicitly this would involve broadening the rationality involved in science to include critical self-consciousness and social rationality.

Thus in their commitment to self-conscious self-improvement as well as to the betterment of society in their scientific research, hermeticists and alchemists held a notion of rationality, if only in embryo, resembling that of the Frankfurt School critical theorists, without either the latters' excessive contempt for or excessive faith in the self-enclosed power of science. Frankfurt School science is characterized as part of instrumental reason as a method of prediction and manipulation, very much along the lines of logical positivist Morris Schlick's characterization of science. This is then contrasted with Max Horkheimer's "critical reason," which is related to the traditional metaphysical philosophies of Plato and Aristotle, and which for Jürgen Habermas (at least in his earlier version) had a different, emancipatory interest from the purely manipulative interests of science. Thus science itself is characterized as intrinsically noncritical and manipulative.

In an article with the apparently paradoxical title "Is Science Rational?" Marx Wartofsky attempts to relate some of the critical theory of rationality to the question of the rationality of science.[113] He characterizes the irrationality of science used for ends destructive of the species as scientific reason becoming dysfunctional. The study of the functional uses of science for species survival and well-being he characterizes as itself a social science, a metascientific science, or what others have called a science of science. In this Wartofsky would differ from the Frankfurt critical theorists who claim that the reason to be applied to science is not itself scientific and that science itself can be very narrowly characterized in terms of a positivist instrumentalism. Insofar as the hermeticists can be thought of as doing science both in their chemical experiments and in their self-improvement and benefiting of society, this position on a kind of science of science to make science foundational to human ends could be seen as a contemporary characterization of the sort of reason in science with the aim of human betterment sought by the hermeticists.

CHAPTER 7

Occultism and the Social Background of Science

PEASANTS, ARTISANS, AND WOMEN

—————

$\mathcal{T}$he relationship of radical Puritanism to alchemy and occultist science in the seventeenth-century English Civil War is a prime example of the relationship of holistic, correlative doctrines to peasant thought and the underground of history. Many philosophers of science, such as Mary Hesse, have wished to distinguish social issues from the hermetic way of life and the positive doctrines (such as the powers of matter) that hermeticism contributed to physics. In contrast, I wish to investigate the role of the social context in the development of the doctrines we are tracing, doctrines that were accepted and then later rejected for political and religious as well as scientific reasons.

Head and Hand: Artisans and Literati in the Rise of Science

A more general issue than the relation of radical Puritanism to English hermeticism is the issue of the relationship of nonliterate thought to Renaissance science. Edgar Zilsel has argued that the origin of modern science in the early seventeenth century was a product of the collaboration of the literate scholar and the illiterate artisan.[1] In non-Western cultures and prior to the late Renaissance in the West the two groups were separate.

In China, the Confucian literati were usually separate from the lower-status artisans, although some literati who oversaw military projects did have contact with craftspeople and builders. The Mohists, followers of Mo Tzu, were the only school in ancient Chinese thought that developed systematic interest in optics and physics, as well as holding a view of reality as built from individual substances and a geometry based on pointlike minima. The Mohists were craftspeople, including military engineers who offered their services to cities under siege. Mo Tzu's inelegant and crude writing style suggests a lower-class background. The Mohists, artisans capable of producing a philosophical and scientific school, disappeared with the consolidation of the empire. Taoist religion first entered history in a series of peasant revolts. The philosophical Taoists who went to live in the countryside to return to nature had further contact with peasant life and beliefs and absorbed earlier folk religious doctrines and exercises.

Among the earliest scientist-philosophers of ancient Greece in the sixth and fifth centuries B.C.E. were Thales in Asia Minor and Empedocles in Italy who are claimed to have performed engineering feats. Thales is said to have diverted the course of a river for King Croesus of Lydia as well as to have developed methods of navigation and of measuring the distance of ships at sea. Empedocles was claimed to have made the marshes near a city less insalubrious by diverting a river into them.[2] With the rise of slavery as a major source of manual and crafts labor in ancient Greece by the fourth century B.C.E., the prejudice against manual labor discouraged gentlemen of the time of Plato and Aristotle from dirtying their hands with work fit for slaves. Thus mining, metallurgy, pottery, and other crafts were not absorbed into theoretical science in ancient Greece. One exception to this almost total rejection of hands-on activity by the elite was medicine, where the handling of and operation on human bodies was necessary. Thus, the Hippocratics and Aristotle had some input of manipulative technique into their conception of the world. Aristotle defends the study of even "lowly" living creatures in the *Parts of Animals;*[3] however, even he thought that the highest form of intellectual activity and pleasure was the contemplation of the stars and of philosophy.

A partial exception to the premodern separation of head and hand was in the medieval monasteries, where some literate monks performed manual labor to ensure the economic self-sufficiency of the community. Mechanical inventions, such as the clock, were developed in the monasteries although without a practical, physical theory of such devices. Although medieval monasteries were relative self-sufficient and may have been the source of some mechanical devices, many of the inventions absorbed in the late Middle Ages came from China or central Asia.

Zilsel notes that the fusion of craftsperson and intellectual was not complete even during the Renaissance. He argues that full fusion occurred only in the early seventeenth century under the pressure of the demands of commerce and industry stimulated by incipient capitalism. As Robert Lopez has suggested, the characteristic figure of the Renaissance, the artist-scientist, was a product of the breakdown of barriers between social classes that resulted from the Black Death and the economic depression during the Italian Renaissance.[4] Previously, literate scholars were for the most part insulated from the manual crafts. Similarly, craftspeople were mainly illiterate and ignorant of learned theorizing about the world.

During the economic and social turmoil of the Renaissance, déclassé intellectuals wandered the roads and were thrown together with illiterate artisans. Out of the interaction of these two groups emerged the scientifically knowledgeable artist. Figures such as Leonardo da Vinci, who combined scientific and technological investigations with studies of perspective, anatomy, and pigments for his paintings, and Leon Battista Alberti, who designed buildings in the classical

Roman style and wrote about perspective in painting as well as architecture, the family, and government, are considered to typify the Renaissance. Zilsel notes that the idea of progress was more common among the technological artisans than it was among the purely literary humanistic scholars. The humanists looked upon ancient Greece and Rome as a golden age whose perfection could never be re-achieved, while the artisans projected the progress of practical techniques into an incipient philosophy of general historical progress.[5]

The very phenomenon of the intellectual Renaissance is a product of the absorption of underground currents of practical thought into the higher literate culture. Zilsel emphasizes the cooperative social relations of the artisans' workshop in contrast to the isolated cell of the monk or scholar-humanist. He also believes that the experience with forging mechanical devices and with practical construction on the part of the artisans encouraged the mechanical and causal view of the world, in contrast to the magical view. This point could be supported by the Chinese case of the Mohist artisans' physical and individualistic utilitarian view in contrast with the correlative cosmology coming out of peasant thought and made official by the primarily Confucian scholar-officials at the court. Zilsel slightly overgeneralizes this account, however, for instance in his treatment of Gilbert. True, Gilbert was strongly influenced by sailors and instrument makers as well as by metallurgists, and over a third of his book deals with navigation and metallurgy, but Zilsel's claim that this association dispels occultist thinking from the core of Gilbert's work and that Gilbert's "animistic metaphysics is nothing but the emotional background of his thinking and does not affect the empirical content of his science,"[6] neglects the extent to which metaphysical background or metaphysical presuppositions can have a role in selecting problems and in directing and elaborating research through strategies for formation of auxiliary hypotheses or models.

The contact with the artisans and their practices was a crucial aspect in the development of early modern science, especially for mechanics. The intelligentsia's contact with peasants and peasant woman healers was another source of input for the rise of science in the Renaissance. Both Paracelsus and Agrippa assert that one must absorb the skills and knowledge of soothsayers and old healing women. In the case of magnetism and the doctrines of forces and fields, this contact with peasants' knowledge was at least as important in supplying a metaphysical framework and model in which the phenomena can be considered as was contact with the artisans in developing experimental instruments.

Puritanism, Paracelsianism, and the English Civil War

We have noted the connection of Paracelsian doctrine to surgeons and apothecaries, who were the lower strata of the medical profession. There is also the larger complex of issues of the relationship of Puritanism to science. Robert K. Merton, a sociologist, extensively developed the thesis of the positive contribution of Puritanism to science as a movement and way of life, which had been suggested earlier by Max Weber. Merton's concern was with Puritanism in general, and with mainstream early modern science as traditionally conceived.[7] Merton was primarily interested in the science of the early Royal Society, one of the first and per-haps most important of the scientific societies founded with the birth of modern physical science. The science of the Royal Society was expressed in the great main-stream atomistic and mechanical chemistry and physics of Robert Boyle and Isaac Newton. Merton also examined the contribution of practical technological con-cerns, such as mining and warfare, on the rise of modern science. It is the topic of Puritanism and science, however, that attracted almost all interest to Merton's early work.[8]

The prevailing view of the history of thought as one of struggle between progressive science and backward religion dominated the decades in which Merton's work appeared. The contrary thesis that religion could have a positive effect on science was considered shocking and controversial to followers of posi-tivism and the Enlightenment, such as America's first professional historian of sci-ence, George Sarton. Sarton was from Belgium and from the strongly anti-Catholic, Masonic, and Comtean positivist tradition. He named his two history of science journals *Isis* and *Osiris* (names of Egyptian gods). Sarton, who considered him-self a strong defender of the rationalist scientific tradition, took from his Masonic, anti-Catholic background the Egyptianism that traces back to Rosicrucianism and the hermetic tradition itself. Like the early hermeticists, he was, at least in his choice of journal titles, pitting Egyptian religion against the religion of the Catholic church. The notion that religion has contributed positively in various ways to the thought of such major scientists as Newton and Kepler is now widely accepted as part of intellectual history.

Merton made impressive quantitative tabulations to show the purported Pu-ritan affiliations of members of the Royal Society and of early modern British sci-entists. Since this work, critics have argued that Merton defined either Puritanism or science or both too broadly and elastically, and specifically that he identified Puritanism with Protestantism in general. I. B. Cohen, summarizing a half cen-tury of debate, concludes, however, "Merton has conclusively demonstrated that the growth of seventeenth-century England was not due to the chance concentra-tion of a handful of geniuses but was conditioned by social and economic forces and a religious ethos."[9] Merton also has argued that the value structure of Puri-tanism contributed to the emphasis on practical activity and social utility. The view of science in the service of social betterment, which was propagated by Francis Bacon and the members of the Royal Society, was closely connected to the Puri-tan ethic.

Merton concentrated primarily on the sort of science propagated by Bacon, Boyle, and Newton, underemphasizing the hermetic conception of science. The social correlate of hermetic science during the English Revolution was with the more radical elements within the Puritan movement. Boyle and Newton, for in-stance, were allied with moderate or latitudinarian tendencies within Protestant-ism and were opposed to the radicalism of some of the hermetic reformers.

Radical Puritan reformers, such as John Webster, were allied with hermeti-cism and Paracelsus. Webster was a surgeon and sometime chaplain in the New Model Army, which contained many radical religious sectarians, including Levelers,

Diggers, and Ranters.[10] Webster attacked Scholasticism as did the future members of the Royal Society; however, he wished to replace Scholasticism with hermeticism rather than mechanistic philosophy. He also attacked the universities in the name of a more practical education. The science that Webster and others advocated was to be tied to universal education and medical care. His science resembles some of the radical science movements in recent decades, but it lost out to a science that explicitly allied itself with the rulers and the propertied classes. Both Seth Ward and John Wilkins, future founders of the Royal Society, attacked Webster on both scientific and social grounds. They rejected both his magical worldview and his radical reformism and defended the mechanical worldview in the service of greater private profit and national power.

Regarding the number of Puritan members of the Royal Society, various definitions can lead to divergent results. One (extreme) example is the work of Lewis Feuer, who includes all the Restoration nobles who were basically honorary members of the later Royal Society and concludes that a majority were hedonists and not Puritans (the latter determined by very narrow criteria of lifestyle). Despite such gerrymandering, it seems evident that a number of important early members were Puritans. The vast majority of the scientists who made what are generally accepted as the advances in early modern British science were recruited to science during the Civil War. Work has also been done (compilational and not theoretical) on the disproportionate contribution of Quakers to British science, far higher, in proportion to the total number of members of the faith, than is the Puritan one.[11] (This is of interest in considering Spinoza's association with British Quakers in the Netherlands after his expulsion from the synagogue.)

Beyond these disputed issues of head counting and definitions, there are the further issues of the role of the politically radical sects in the origins of British science. Gerrard Winstanley, the leader of the Diggers, or True Levelers, perhaps the most notorious group in the radical wing of the revolutionaries, proposed general education that contained considerable science. (This science emphasized medicine and practical knowledge relevant to agriculture and navigation as well as astrology, but definitely included much material that would today be universally considered scientific, such as astronomy.)[12]

Although P. M. Rattansi has argued that the hermeticism of the late sixteenth century was more conducive to the unity of religious aspiration and scientific research than was the Puritanism of the next century,[13] Charles Webster has emphasized the importance of the overlap between the Paracelsian alchemical movement and the hopes for universal renewal during the English Civil War. Max Weber, and Robert K. Merton elaborating on application of Weber's famous Protestant ethic thesis to science, emphasized the role of work as a means to personal reassurance of salvation predestined in Calvinism (soteriology).[14] In contrast, Charles Webster in his studies of the period has emphasized the role of hopes for the Last Judgment and collective salvation (eschatology).[15]

The expectation of the millennium and the Last Judgment became associated with expectations of social utopia on earth. The "saints" would have dominion over the natural world and live in health and harmony. The Fifth Monarchist

Mary Rand (or Rant) prophesied that alchemical making of the philosophers' stone or gold would be common by 1661.[16] Seven years later the reformer Samuel Hartlib interpreted this prophecy in a more moderate fashion as concerning use of alchemy for public benefit.[17]

The more radical of the Puritans were dismissed by A. Rupert Hall in their relation to science by the claim that they were not "able, tough-minded, and professional" but "soft-headed, amateurish or incompetent," using contemporary notions of scientific professionalism to characterize the religion- and politics-saturated era of the Puritan Revolution.[18] Even if this characterization is true, the Puritan millenarians formulated major institutional structures for science and technology, including programs for improved public health services (through medical training of the clergy as a national health service, modernization of medical doctrine replacing Galen with Paracelsus, and education of the city poor in herbal medicine).[19] Puritan millenarians also formed scientific fraternities, culminating (in more restrictive and deradicalized form) in the Royal Society. They proposed improved practical education, modernization of the universities, the economic organization of technology, scientific agriculture, collection of economic statistics, improvement of the lot of the poor, and incentives for inventions and discoveries. The Puritan revolutionaries proposed social and economic planning for technological development far beyond anything instituted in the Restoration—or since. These notions were hardly envisaged even as goals again until the twentieth century.

But beyond these contributions to the support and propagation of science, as well as the recruitment and motivation of scientists, the Puritans contributed substantially to the hardest mathematical sciences.[20] People such as John Graunt and William Petty contributed to quantification and to the general quantitative approach to all questions (even if their most famous achievements are in social science areas such as population and economics—and even here Graunt and Petty have been praised as pioneers of mathematical population ecology).[21] John Wilkins was a major figure in seventeenth-century scientific and mathematical communication, as well as being in the nucleus of the Royal Society.

The founders and leaders of the Royal Society conducted intellectual warfare on several fronts. For one thing, they wished to reject the old Aristotelian Scholasticism in the name of the new empirical science. Additionally, they fought the radical reformers and popular hermeticists with the mechanical philosophy. They also attempted to contain the dangerous materialistic and atheistic implications of the pure mechanical philosophy, however, by adding to it certain notions of spirits and active powers, which they borrowed in part from the hermeticists.

Even the publication of Newton's *Principia* has been tied to political maneuverings. The usual story of the publication of the *Principia* was that Edmund Halley, the astronomer of Halley's comet fame, practically dragged it out of Newton in response to finding that Newton had calculated the consequences of the inverse-square law. This suspiciously echoes the story of how William Harvey's biological work had been brought to publication. Newton may well have not been aware of the surrounding political motives in the publication of his great work; nonetheless, Halley's toadying letter to King James II and his ode to Newton couched in the philosophical language of court circles of the time suggest that the publication of Newton's work was a linchpin in the attempt of supporters of the Church of England to gain influence in the court of King James II.[22] Newton, who had closet alchemical and hermetic interests behind his mechanical and empirical science, played a central role in the new mainstream worldview propounded by the Anglican church. The appeal to spiritlike, active powers as well as to passive matter allowed physics to dissociate itself from atheistic materialism.

The absorption of concepts of forces and powers surreptitiously brought from alchemy and hermeticism into the orthodox and mainstream Newtonian system of the world, which became the ruling cosmic paradigm for the next two centuries, is a prime example of the absorbtion and co-optation of the intellectual ferment of the underground by the establishment.

Women, Witchcraft, and the Witch-Craze

A tantalizing and difficult question is the connection of learned Renaissance hermeticism to popular peasant magic and witchcraft. Magic had led a continuous, underground existence during the Middle Ages. The literate, hermetic Renaissance magicians strongly disassociated themselves from black magic. They claimed to pursue a good and angelic magic as opposed to a black and diabolical one. The pursuit of magic was generally frowned upon by the Church, and figures like Marsilio Ficino were careful to distance themselves from black magic.

Nevertheless, the period of Renaissance hermeticism overlaps with the witch craze that accompanied the rise of modern rationalism. The beginning of modern physical science was accompanied by frenzied persecution of witches on the basis of largely fabricated testimony and forced confessions. Important figures in the Royal Society, bastion of modern science, were strong believers in witchcraft. Joseph Glanvill, a Fellow of the Royal Society, published a defense of the reality of witchcraft in 1666, the *Annus Mirabilis* (because of the number 666 in the Book of Revelations; it was also the year the young Newton, sent home from college because of the plague, worked out his major discoveries).[23] Part of the support of persecution of witches among followers of modern science may have been the violent attempt of a new and insecure reason totally to exclude its opposite. Foucault has made a claim of this sort with respect to the confinement of the insane in Paris at the birth of the Age of Reason.

In England and Germany the great majority of putative witches burned were women. Even though in some areas, such as Estonia and Iceland, and some regions of France, so-called warlocks were the majority of those executed, it can hardly be denied that the more explicitly sexual tortures publicly imposed upon females in England and Germany tie the witch mania to an attack on women.[24]

Feminist writers, such as Carolyn Merchant, Evelyn Fox Keller, Barbara Ehrenreich, and Brian Easlea, note that the persecution of witches and the rise of modern rationalism and mechanical science involved a downgrading of the status of women in society and of the female principle in cosmology. Acceptable occupational roles for women were restricted, and terms such as *applewife, alewife,* and *oysterwife* dropped out of the language. The only occupational word of this sort that survived in its literal occupational form was *housewife*.[25] *Fishwife*, it might be noted, also survives today, but now is primarily associated with the aggressive use of abusive and vulgar language by women, rather than with the selling of fish.

King James I ordered the clergy to preach against "insolencies of our women."[26] Literature and songs warned women against masculine behavior. Women folk healers were displaced by the new male medical scientists. The struggle for power in medicine probably played a role in the persecution of witches.[27] English medical men petitioned Parliament in 1421 to ban any practice of medicine by women. In 1511, Parliament did ban medical practice of women who were thought to "partly use sorcery and witchcraft."[28]

Advocates of modern science, such as Henry Oldenburg, argued for a "masculine" philosophy.[29] Bacon is famous for this theme, as well as for his language of the scientific knowledge of nature as forceful seduction of feminine nature by masculine scientists: "For you have but to follow and as it were hound nature in her wanderings and you will be able when you like to lead and drive her afterwards to the same place again. Neither ought a man to make scruple of entering and penetrating those holes and corners when the inquitistion of truth is his whole object." This quote is given by feminists as illustrative of Bacon's sexism.[30] Thomas Sprat, in his 1667 *History of the Royal Society*, writes, "The wit that is founded on the arts of men's hands is masculine."[31]

Henry Oldenburg, correspondent with Robert Boyle and scores of European scientists, proposed "to raise a many philosophy" and make sure that "what is feminine . . . be excluded from the [Royal] Society's true philosophy." Joseph Glanvill, chronicler of the Royal Society, not only defended the reality of witches, but advocated a masculine philosophy for the society, fostering the "manly sense" and avoiding the deceit of "the women in us."[32]

We know little about the actual beliefs of the peasant women accused of being witches. What we know of the doctrines of these alleged witches comes solely from their persecutors, as the women themselves did not leave written works. There are three possible sources for the alleged doctrines of witches. One is the systematic construction of demonology by late medieval writers and by the witch-hunters themselves. Another is peasant folk religion and pagan pre-Christian survivals in peasant customs. A third is the psychopathology of the "witches" themselves, perhaps accentuated by torture. It is more difficult in the case of witchcraft to separate the beliefs of the "witches" themselves from the doctrines of their persecutors and opponents than it is even in the case of religions competing with Christianity or of Christian heresies during the Roman Empire. At least in the case of religions, such as those of the Gnostics or Christian heretics, we have documents written by the proponents of the competing religion. Even in the case of some heretics and competing faiths where no original documents are available, we still have quotations from the originals embedded in Christian tracts directed against them. In the case of so-called witches, we have no such authentic documents. The few documents that exist are presumably forgeries or else extorted under torture. The systematic exposition of demonology was most likely produced by the medieval

Scholastics and not by witches. Nevertheless, it is a mistake to dismiss the notion that elements of peasant pagan religious beliefs occurred in the practices of some persons persecuted as witches. One of Britain's leading historians, H. R. Trevor-Roper, in his index cites "Murray, Margaret, her 'vapid balderdash,'"[33] but does not bother to cite or reference her book. Murray's mistake was to attempt to reconstruct the pagan religion directly from the witch-hunters' reports containing standardized theological descriptions of witch behavior to which the accused witches were forced to confess. Murray attempted to justify every detailed accusation of the witch-hunters in the form of actual peasant beliefs. This mistake (plus sexism) led to Trevor-Roper's contemptuous dismissal.

More recent writers, such as Carlo Ginzburg, emphasize an illiterate peasant culture that had reciprocal interactions with high culture. Ginzburg amusedly recalls that he had been called a "Murrayite" simply because he argued that Murray was correct insofar as she claimed that there were real pagan rites (not insofar as she described them solely on the basis of the witch-trial reports).[34] Ginzburg makes use of the Inquisition records concerning a semiliterate miller, Menocchio, and his cosmological and political views. Ginzburg writes: "In Menocchio's case, it is impossible not to think of . . . oral transmission—from generation to generation. This hypothesis appears less improbable if we think of the diffusion of these very years . . . of occult and shamanistic currents. . . . Menocchio's cosmology fits into this little explored area of cultural relations and transmissions."[35]

Ginzburg makes a strong case for the relationship of peasant culture to beliefs of those persecuted as heretics and witches. He notes: "The Catholic Church at this time was engaged in a two-front war: against high culture . . . that wouldn't conform to Counter-Reformation patterns, and against popular culture. As we saw, subterranean convergences could develop between these two very different opponents."[36] Ginzburg cites neither Heer nor Zilsel despite the potential usefulness of their work to his.

Besides the doctrines imposed by the Scholastic persecutors and the peasant culture that was a source of pagan beliefs and heresies, there is the possible psychiatric origin of the witches' reported beliefs. Some writers (such as Paul Feyerabend and H. R. Trevor-Roper) made such psychopathological interpretation of the witches' beliefs. Feyerabend and Hans Peter Dürr present the literal interpretation and the psychiatric interpretation as equally valid, incommensurable accounts.[37] Trevor-Roper, among others, uses the psychiatric interpretation as the "true" interpretation. There is a remarkable consistency in the reports by the "witches" that does not seem to be fully accounted for by the confessions demanded of them by their torturers. Some witches went beyond what was demanded and give remarkably consistent accounts of phenomena such as flying to the witches' coven, having sexual intercourse with the Devil, and feeling his ice-cold penis. Many psychiatrists have taken these reports as a form that mental illness took during this period. One of the difficulties in dealing with the phenomenon of the witch trials is distinguishing fantasies and hallucinations of a psychopathological nature from actual traditions and ceremonies in these accounts. The phenomenon of flying can be given an obvious Freudian interpretation in terms of sexual intercourse, but it can also be related to flying in shamanism. Siberian and Central Asian shamans were noted for their ability to go into a trance, then claiming to fly to different places. Some of the stories of Pythagoras concerning his ability to travel rapidly or even to be in two places at once suggest the connections of the religious aspects of the Pythagorean religion with shamanism.[38]

Despite the fact that the major hermeticist authors strongly disassociated themselves from black magic and witchcraft, there is evidence of connection in that some learned critics of witchcraft attacked witches by attacking occultist philosophers. For instance, Jesuit Martin Del Rio attacked the Renaissance philosopher Cornelius Agrippa as part of his general attack on magic. Significantly, Agrippa had also written a defense of the superiority of women.[39] Jean Bodin, an important French philosopher of the period, believed in witchcraft and wrote a work called *Démonomanie* which attacked the evils both of witches and of Renaissance philosophers such as Giovanni Pico della Mirandola and Agrippa.[40] Bodin wrote a ferocious reply to Johann Weyer, who doubted the reality of witchcraft. In the eyes of some opponents, witchcraft and popular magic were associated with the intellectual works of the Renaissance hermeticists and cabalists.

Weyer was a student of Agrippa, who himself was often accused of black magic. Weyer wrote a humane account of witchcraft that in many ways was a forerunner of the psychiatric account. He claimed that the phenomena in the witches' confessions were hallucinations caused by illness or by actual demons; however, the witches themselves were victims rather than evil. Weyer recommended Christian charity and medical treatment for witches, rather than torture and burning at the stake.[41]

Bodin was perhaps the most advanced and sophisticated political philosopher of his day. He was called the Aristotle and would be called the Montesquieu of the sixteenth century and wrote on comparative history and the quantitative theory of money as well as political philosophy.[42] It is shocking to find Bodin's demanding burning at the stake both for witches and for those who, like Weyer, denied that witches were genuinely evil. Bodin claimed this was necessary for social order, which in turn depended on religion for its maintenance. Bodin scoffed at the Copernican theory that the Earth traveled around the Sun, claiming that this went against the senses and all of physics, yet accepted all of the most ridiculous claims made by the witch-hunters.[43]

The shift in the attitude toward magic in the intellectual classes from the Middle Ages to the Renaissance is tied to the growing impact of the lower social strata on literate thought. In the Middle Ages, black magic was practiced but was generally frowned upon and suppressed by the Church.[44] During the Renaissance, magic became intellectually respectable and was supported by learned philosophers, scientists, and political patrons of the arts. This shift shows a weakening of the power of the traditional ideas of the Church and a strengthening of paganism. The rebellious resurgence of paganism included not only the revival of the Greek and Roman classics, but also the revival of the Egyptian religion of Hermes and of popular magical traditions as well. The threat of this resurgence of the worldviews of the lower classes—tied to religiously and politically radical, communal movements— may explain, in part, the ferocity of the persecution of presumed witches and the denunciations of hermeticism by the supposedly serene rationalists of the new mechanical science.

Just as in the case of witchcraft, the issue of the existence and content of peasant culture is problematical. Given that the peasant culture was illiterate, we have very few documents attesting to it. Some artists of the late Renaissance and early modern period (such as the writer Rabelais and the painter Brueghel) incorporated peasant culture into their works. Thus, Mikhail Bakhtin, in his path-breaking work, made use of the writings of Rabelais in connecting numerous indirect references from the literary tradition to reconstruct features of peasant culture and their reciprocal interactions with the high culture.[45] Similarly, Lucien Febvre, of the French *Annales* school, made use of the works of Rabelais to delineate the "mentality" of the period.[46] Clearly, depending on the literate works of highly educated people does not give us direct entry into the world of peasant culture but rather into one filtered through the beliefs and stylistic conventions of the high culture. Some historians have denied that we can have any access at all to the peasant culture of past ages, analogous to the way that some critics of Freud treat the unconscious. Foucault ran into this sort of problem in his early work on the history of madness and retreated to the position that it was impossible, in principle, to investigate the totally irrational. As Ginzburg writes, "At this point, Foucault's ambitious project of an *archéologie du silence* becomes transformed into silence pure and simple."[47]

Some historians deny any interaction between aristocratic culture and folk culture or that popular culture is simply a degenerated, trickle-down product of aristocratic culture. The thesis of this work, like that of Heer, Zilsel, Ginzburg, and Bakhtin, holds that illiterate popular culture has indeed affected the high culture at least as much as it has been affected by it. As Ginzburg writes, the case of the miller with his cheese cosmology shows "the popular roots of a considerable part of high European culture both medieval and postmedieval . . . hidden but fruitful exchanges moving in both directions between high and popular cultures. The subsequent period was marked, instead, by an increasing rigid distinction between the culture of the dominant classes and artisan and peasant cultures. . . . The decisive crisis had occurred . . . with the Peasants' War and the reign of the Anabaptists in Muenster. . . . The necessity of reconquering, ideologically as well as physically, the masses . . . was dramatically brought home to the dominant classes."[48]

Various forms of impact of the lower-class culture on the high culture continued even after increased class barriers. Social movements of the lower classes (such as those of the French Revolution in the enghteenth century, and subcultures such as the Bohemian in the nineteenth and twentieth centuries, as well as Marxist revolutionary political movements) have mediated between popular culture and high culture.

Some of Ginzburg's critics searched for parallels between the cosmology of his barely literate miller and that of the upper classes. Menocchio stated, "All was chaos . . . and out of that bulk a mass formed—just as cheese is made out of milk— and worms appeared in it," and angels "were produced by nature *just as the worms were produced by a cheese.*"[49] Some scholars have attempted to find precedents for the cheese and worms cosmology in the works of literate intellectuals in the same period; however, although they have found putrefaction of cheese mentioned in Pietro Pomponazzi without cosmogological implications, and angels emerging from chaos in the hermetic *Pimander*, they have not found anyone connecting these elements.[50] An exact statement is to be found in a Paracelsian tract, *Philosophia ad Athenienses*. Neither Ginzburg nor his reviewers or critics have noticed it. "All things are said to have come from the traditional prime matter . . . called by Paracelsus the *Mysterium Magnum*—the great generating substance. . . . It stands to more particular substances in the same way that cheese may be considered the mystery for worms or milk the mystery for cheese or butter."[51] The passage does not prove that Menocchio's cosmology was borrowed from the Paracelsians or that the Paracelsians' analogy was borrowed from peasant thought, but lends support to the notion that Paracelsian cosmology absorbed many peasant beliefs.

The problem of the relationship between learned hermetic philosophy and peasant magic and witchcraft is similar to the problem of the relationship between Chinese Taoist philosophy and popular Taoist religion. In both cases, there is a certain parallelism between the intellectual and the popular doctrines. At the same time, there are clear differences between articulate, philosophical systems and inchoate peasant superstitions.

In the case of Taoism, this has led some scholars, such as Herrlee G. Creel, to deny that there is one common Taoism that embraces both the popular religion and the philosophical writings. Creel's reaction is understandable, since the doctrines of the popular religion and of the profound philosophers often differ significantly. On the other hand, there are close connections between the rebellious and skeptical doctrines in philosophical Taoism and the social rebellions launched periodically under the banners of secret societies embracing religious Taoism. Certainly Taoism, with its anarchistic and naturalistic strains, contrasts with the devotion to authority and indifference to physical nature found in Confucianism. Hermeticism, in late antiquity, was also a "two-tier" religion, involving both learned and popular versions, "with superstition for the masses and true knowledge or *gnosis* for the elite."[52] Frances Yates strongly contrasts learned magicians, such as Agrippa, Pico, and Ficino, with popular magic; however, the issue arises regarding the connection between the reaction against witchcraft and the attack on hermeticism and cabalism.

The usual explanation for the end of belief in and persecution of witches accredits the rise of mechanism and the spread of science which made impossible the supposed role of demons. This view is found in standard histories of the rise of rationalism in Europe, beginning with Richard Bentley's Boyle Lectures and continuing through William Lecky's history of the relation of religions and science. Keith Thomas in 1971 espoused this view as well claiming that "the absurdity of witchcraft could henceforth be justified by reference to the achievements of the Royal Society and the new philosophy," and citing Richard Bentley as evidence.[53]

It is certainly true that the persecution of witches began to decline in the late seventeenth century as the mechanical worldview began to develop. In addition, acceptance of the mechanical worldview excludes a role for spirits as such (in its materialistic version) or for spirits other than human and divine (in its Cartesian version). The relationship of the proponents of the experimental method and the mechanical worldview to the belief in witches is much more complex. Nicholas de Malebranche, a close follower of Descartes who went further than his teacher in denying a causal role of the mind on the mechanical body, nevertheless asserted that witches exist, but gave a mechanical explanation of how they operate.[54] The British Royal Society was not active in opposing the witch craze, and several of its leading members supported the belief in witches, as Charles Webster has shown.[55] The contemporary writers who summarized the achievements of the Royal Society and sang its praises did not list opposition to the belief in witchcraft or to witch-hunting as an accomplishment of the society. These writers surely would have mentioned it if they had thought it would enhance the prestige of the society.

Robert Boyle, a major figure in the history of chemistry and a proponent of the atomic theory, instigated the publication of the book *The Devil of Mascon*, which discussed diabolical goings-on. Boyle accepted demonology and did not deny the existence of witches. He wrote to Joseph Glanvill that the book on the devil would help in combating atheists.[56]

Glanvill, a leading propagandist for the Royal Society, was also a very strong believer in witches, as were several other figures associated with the society, such as John Aubrey. Glanvill's major work was *Saducismus Triumpantus*, which was probably the most influential English work on the topic of witchcraft. Glanvill strongly defended the existence of demons and devils. His prestigious association with the Royal Society is illustrated by his name on the title page of his book on witchcraft being replaced with "A Member of the Royal Society." Other members engaged in scientific investigations of such phenomena as fairy rings and "non-Adamical Men." These phenomena were denied by Robert Plot, although Plot himself believed that evil spirits did exist. No major figure in the Royal Society wrote to combat Glanvill and his belief in witches. One minor figure, John Wagstaffe, opposed the existence of witchcraft and was ridiculed by Royal Society members not only for his arguments and sources but for his "dwarfish appearance."[57]

The only major reply to Glanvill was by John Webster, whom we have already met as a radical reformer and Paracelsian natural magician. It was not the proponents of the mechanical philosophy who did the most to combat belief in witchcraft but rather it was followers of the occultist and alchemical traditions, such as Weyer (a student of Agrippa) and John Webster. Eminent members of the Royal Society and successors to Newton's chair supported the reality of witchcraft, while Weyer and John Webster, understandably eager to deflect accusations of witchcraft against themselves and other hermeticists, pioneered the "psycho-analytic" explanation.[58]

The belief in witchcraft had severely declined by 1700, although some leading scientific figures continued to defend it. In 1737, William Whiston wrote on the power of casting out demons. Whiston held Newton's Chair of Natural Philosophy and authored *A New Theory of the Earth*, which claimed that the Earth was formed from a nebulous comet and that Noah's flood was caused by the near approach of a comet.[59] Whiston claimed that demons were as well confirmed as Boyle's experiments on air pressure or Newton's experiments on gravitation.[60]

A theme of Keith Thomas that has been presented by physicist Steven Weinberg, citing a talk by Trevor-Roper, is that witch-burning declined because of belief in the laws of nature;[61] however, Bodin (who was a major figure in the development of juridical law, which according to Zilsel was carried over into laws of nature) as well as members of the Royal Society (who associated with lawgivers such as Boyle, Hooke, and Newton) believed in and defended the belief in witches. Apparently belief in laws of nature was quite compatible with belief in witches. It was not so much the content of lawful science itself that led to the decline of witchcraft, but rather it was the success of science and scientific medicine as institutions that no longer necessitated the discrediting of female peasant healers and their more respectable male hermetic partisans, at least in educated and elite circles. In the wider social context, it would seem that changes in theology and judicial matters had at least as much to do with the defeat of the witch-persecution craze as did the rise of specifically scientific beliefs. Robert Mandrou has documented how changes in theology and changes in judicial law were the major factors in the decline of witch-persecution trials.[62]

Much of the persecution of witches in the sixteenth century was fostered by magistrates in an atmosphere of threat and crisis involving the rise of the new Protestant churches. The Catholic church was threatened by these new heresies and attempted to suppress any sort of behavior that was apparently opposed to the Church. The Protestant churches saw the Roman church as the Devil and attacked what they considered irreligion, whether Catholic or pagan. Meanwhile, the crisis of authority among the peasants may have led to dissatisfaction with the official religion and the institution of ceremonies (identifiable with witchcraft) that based themselves on tradition but inverted the authority structure (as occurred seasonally in carnivals but did not occur more frequently in locally institutionalized form). Catholics tended to associate witchcraft with the Protestant heresies. Protestants identified witchcraft with the Devil of Rome and threats to their existence. Manichaean dualism of good and evil (itself a medieval heresy) permeated the worldview of both sides.

After the period of religious wars and the rise of even more radical Protestant sects, the magistrates and theologians changed their strategies of control. After the Council of Trent, there was a movement within Catholicism to eradicate pagan vestiges in the Church. The former persecution of witches as genuine agents of the Devil was now considered by the Church equally pagan as the peasant belief in witches. Both needed to be eliminated. The purification of Christianity played at least as much of a role in the growing opposition among theologians as did skepticism. Also, the strategy of accusing opposing churches of being in league with witchcraft or Devil-worship was rejected by magistrates as counterproductive. Because supernatural arguments had been made use of by politically revolutionary or reforming Protestant groups, there was a turn to naturalistic arguments.[63]

In summary, while the contact between illiterate artisans with technical knowledge and literate scholars with philosophical and mathematical theory gave rise to experimental science, the heritage of peasant beliefs in a holistic universe of occult influences was incorporated into more sophisticated hermetic and occultist natural philosophy and science. Paracelsianism and hermeticism were important in the radical doctrines of Puritans in the English revolution. Witchcraft was at least in part tied to genuine pagan survivals among the peasants. The enemies of witchcraft were defending the new masculine philosophy of experimental and mechanistic science as well as opposing the threat of religious and political peasant rebellions. Some of the believers in the reality of witchcraft attacked not only witches but occultist and hermeticist intellectuals. Support of the mechanistic philosophy did not in itself lead to the denial of witchcraft, but it did lead to persecution of witches and opposition both to the peasant beliefs and to such intellectual beliefs as hermeticism.

Rosicrucianism and the Thirty Years' War

During the Thirty Years' War (1618–1648) that ravaged central Europe, Descartes had his vision of the reform of philosophy and the sciences while resting in a warming hut as a soldier of fortune. The war was a central event in the political and religious reflections of great systematic theorists of philosophy, religion, and science of the middle and late seventeenth century—such as Leibniz, who wished to harmonize the competing religions and governments of Europe to prevent the bloodshed that had destroyed much of Germany and Bohemia. After all, the war had wiped out, primarily through disease and starvation, as much as one-third of the population of vast regions in Germany and central Europe.

The beginning of the Thirty Years' War was one of those overdetermined events which, like the opening of World War I, appear to be minor incidents but happen in a context in which any threat or disequilibrium triggers massive hostile forces that fear each others' moves. The event was the election of Frederick of the Palatinate as ruler of Bohemia. The Holy Roman emperor was ordinarily chosen by seven electors, ecclesiastical and temporal rulers of German lands and central Europe.[64] A member of the Austrian Hapsburgs had been named emperor. Central Europe was split between Catholic and Protestant populations, but three of the electors were Catholic and four Protestant. The swing vote was the elector from Bohemia. In this election, the Protestants had no suitable candidate, so all the electors were willing to accept Ferdinand, an Austrian Catholic who had been made ruler of Bohemia and whose representative voted for him to be Holy Roman emperor.[65] Just as the electors were unanimously electing Ferdinand as their emperor, too soon for the news to reach the electors in Germany, the Czechs revolted. In the famous defenestration of Prague, two of the Catholic representatives along with their secretary were thrown out of a window by the mob.[66] The angry knights and lords of Bohemia elected Frederick of the Palatinate, who had recently married Elizabeth, the daughter of James I of England. It was thought that he would bring Protestant allies to the defense of Bohemia against the Catholic Hapsburgs' attempt to retake Prague. The young king and his queen arrived in Prague to general applause and celebrations, but personal indiscretions[67] combined with misplaced compensatory attempts at prudery rapidly led to loss of their popularity.[68]

Meanwhile, James I, who later was considered incompetent as ruler, and some of whose arbitrary actions led to the tensions behind the English Civil War, pursued a policy of compromise with the Catholic forces in Europe. James's compromising policy contrasted with the reputation of Elizabeth I for opposition to the Catholic states, although she herself had actually been somewhat more cautious than supporters of a war with the Catholics remembered. Likewise, the German Protestant rulers, while promising some money that never arrived, were unwilling to commit troops to the defense of the new king and queen of Bohemia. After ruling less than a year the "Winter King" of Bohemia was defeated at the Battle of White Mountain by Catholic troops, and Ferdinand retook and sacked Prague and the surrounding regions. This defeat and the ensuing attack on the United Provinces (the forerunner of the Netherlands) by Spinola, an Italian general, rallied the Protestant countries to belated action, and the Thirty Years' War began.

In *The Rosicrucian Enlightenment*, Frances Yates presents circumstantial evidence that behind the so-called Rosicrucian Manifestos were the marriage celebrations and the initial arrival in triumph of the Winter King and his queen along with the exaggerated hopes that many Protestants held for a Protestant alliance centered on Frederick of the Palatinate. The Rosicrucian Manifestos were anonymous pamphlets that presented the achievements of one Christian Rosenkreutz, an adept of magic, alchemy, and Cabala, and his league of associates who were to bring enlightenment and well-being to the world. No one knows if such a society actually existed at the time of the manifestos, and many think the manifestos were a prank. Nevertheless, people interested in hermetic magic and science, such as Robert Fludd and Michael Maier, were supportive of the manifestos and many attempted to contact the mysterious Rosicrucian society. Later, in the eighteenth century, Rosicrucian societies were actually formed, supposedly along the lines of the manifestos, and such people as German world traveler and revolutionary Georg Forster and his friend, a leading anatomist Samuel Thomas von Soemmerring, were members.

Yates claims that the manifestos were issued in hopes of Frederick and his queen founding a hermetic utopia. She notes that much of the symbolism in the Catholic propaganda broadsheets designed to discredit the Winter King, as he was derisively called, corresponds to symbolism of the Rosicrucians. She points out that Michael Maier's illustrated works (such as *Atlanta Fugiens*) also matches Rosicrucian symbolism and symbolism associated with Frederick and his bride and that Johann Valentin Andreä, who wrote *Christianopolis*[69] and other utopian tracts, was a native of nearby Württemberg present in Tübingen when Duke Frederick I of Württemberg, an alchemist seeking alliance with Elizabeth of England, was elected to the Order of the Garter and elaborate ceremonies, drama, and music were presented for the duke and an English ambassador sent by James I. Frederick of the Palatinate brought his bride to a redesigned palace in Heidelberg, where English plays were performed and geometrical gardens with mechanical marvels were constructed. Frederick also received the Order of the Garter while in England. Much of the propaganda after his defeat played on the fact that the Winter King in his headlong flight from the Catholic troops left his valuable Garter behind.

(The Garter was brought by a looting Walloon soldier to Ferdinand who bought it for one thousand talers.)[70] Yates suggests that the Garter ceremony scenes, actually witnessed by Andreä in Württemberg, were worked into the *Chemical Wedding of Christian Rosenkreutz*[71] in terms of the ceremonies at the Heidelberg castle where the newlyweds had settled.[72] Yates suggests that since Frederick of Bohemia and the Palatinate had received the Order of the Garter, and his earlier favorable propaganda had emphasized this fact, the Rosicrucian wedding symbolized the immanent, hoped-for rule of hermetic monarchs.

Yates claims that the ideas of John Dee, an Elizabethan mathematician-magician, were in part behind the Rosicrucian Manifestos. Dee with his disreputable companion Edward Kelley (who was an alchemist, former counterfeiter, and supposed confident of angels) had visited the court of Rudolf II in Prague, where Bruno, Tycho Brahe, Kepler, and a host of hermetic scientists, magicians, and confidence men had also been hosted. Dee combined an interest in mathematics with construction of mechanical "marvels" and with supposed communications with angels through his assistant Kelley. Dee's scientific side is shown by the fact that he was author of the preface to the first English translation of Euclid's geometry and owned the largest scientific library in England, greatly exceeding those of Oxford and Cambridge. Dee was not only a leading scientist-occultist but was also a major proponent of British imperialism. He is said to have coined the term *British Empire* (a fact that Yates, oddly, does not mention to strengthen her case). He also wrote tracts propagandizing for British naval power as a foundation for empire.[73] Dee's combination of Protestant religion, occultist science, and mathematics and ambitions for the Protestant political rule of Queen Elizabeth of England certainly parallel the goals of the Rosicrucian Manifestos and the hopes for a Protestant alliance placed on Queen Elizabeth of Bohemia. Similarly Yates notes that the printer Jean Théodore de Bry, who had moved to the Palatine, the printer of the tributes to Frederick and Elizabeth was also the printer of works of Maier and Fludd.[74] Several other figures with hermetic and Paracelsian interests as well as patronage of Frederick of the Palatinate figure in the printing of these pro-Rosicrucian alchemical works. One is Oswald Croll, who was patronized by Christian Anhalt, a major strategist and instigator of Frederick's election to the crown of Bohemia.[75] Maier also dedicated one of his books to Anhalt.[76]

Although two leading historians of the seventeenth century, Christopher Hill and H. R. Trevor-Roper, praised Yates's account, a number of reviewers criticized the circumstantial nature of the evidence. The most critical review, that of Brian Vickers, does not deal with the issue of Yates's account of the Winter King and Queen, concentrating instead on Yates's often extreme claims of influences of or sympathies for Rosicrucianism in the works of Bacon, Newton, and other scientific figures among the membership of the Royal Society.[77] Vickers does, however, criticize Yates's use of the anti-Frederick propaganda to draw parallels between the symbolism used by Frederick and that of Rosicrucianism, stating, "Thus a series of rather caustic satires on Frederick of the Palatinate for his presumed associations with the Rosicrucians (pp. 55ff.) are taken in a completely solemn manner, as an actual account of the Fraternity's doctrines. . . . One does not have to be a literary critic to perceive that this is an illicit maneuver. 'A parodies B's extravagant claims: remove the parody and you have an accurate account of B.' But if you remove the satire from a satire, what is left?"[78]

Vickers places this within a list of other cases to argue that Yates takes works of obvious ridicule of the Rosicrucians literally, and several of his criticisms hit home. In this particular case, however, one may ask why the satirists chose to associate Frederick with Rosicrucian symbols rather than symbols of other movements, say, simply English political or religious symbols of extremist Protestant sects. Why not, for instance, associate Frederick with Protestant iconoclasm, of which some of his fanatical assistants in Prague were guilty? For the satire to have some effect, presumably the associations with Rosicrucianism would have to be seen as meaningful. Why was there a concentration on these symbols, and not simply the use of them as a small part of a wide variety of symbols of other movements (political or religious) that the Catholics would see as deserving of opposition or ridicule.

Vickers also argues, with fair plausibility, that Andreä, who later explicitly criticized Rosicrucianism, may himself have been writing a satire on Rosicrucianism in his *Chemical Wedding*, and was upset by the reception of the work as another contribution to Rosicrucianism. But even if this is true, the reception of a work tells us something about the times, and even its content and structure, regardless of the author's actual intentions. For instance, even if the Rosicrucian manifestos were a hoax, the fact that such figures as Fludd and Maier in various parts of Europe eagerly looked upon them as harbingers of the hoped-for age of hermetic rule reveals much about the atmosphere of the times and commitments of the audience of the manifestos. The positive reception of the Rosicrucian writings, even if the writings were hoaxes, shows the expectations and hopes of numerous alchemists and political thinkers.

Descartes was with the Catholic troops that invaded Bohemia, at the Battle of White Mountain, and with the troops that entered and sacked Prague.[79] During his service he had the visions that told him that mathematics was the key to knowledge. Around that time he collaborated with a mathematically talented individual who was one of the first to issue a publication attempting to contact the Rosicrucians. When Descartes returned to Paris, he was able to tell others about the events in Bohemia and was in fact thought by some to be a Rosicrucian himself, both because of the locations of his recent travels and because of his interests in mathematics and collaboration in studies with a seeker for the Rosicrucians, Johann Faulhaber. Descartes was also suspected of being a Rosicrucian because his provisional rules for living (*Discourse on Method*) included following the customs of the country in which one happened to reside which resembled those of the Rosicrucians.[80] Later, one of his most interested correspondents on philosophy was Princess Elizabeth, the daughter of the Winter Queen. According to his biographer, Descartes was invited to the Palatinate, but eventually went to Sweden, to attempt to gain the support of Queen Christina for Elizabeth. Descartes, Leibniz, and other savants in science and philosophy of the later half of the century did not retain their interest in the Rosicrucians, but the hope for a synthesis of science and benevolent political rule aroused hopes that were channeled toward the new science rather than alchemy and toward the benevolent despots rather than the Winter King and Queen.

CHAPTER 8

Gilbert and Early Modern Theories of the Magnet

*I*n 1600 William Gilbert, both a founder of modern experimental science and an animist, published *On the Magnet*, the most thorough experimental investigation of magnetism since Petrus Peregrinus, containing numerous suggested practical applications of magnetism to navigation. Like many of the moderns, Gilbert eloquently denounces the philosophers and speculators who preceded him. He does not embrace the new mechanical philosophy, but presents a purposive account of the forms and forces of magnetism.

Gerrit L. Verschuur, a writer on contemporary astrophysics, presents Gilbert as ending "The Era of Superstition" and "sorting out the wheat from the chaff of fiction about magnetism."[1] It is mistaken to see Gilbert as having been alone in the transition to experimental work on magnetism and to portray him as dispensing with all the medieval "superstition," of the occultism, and the desire to "satiate readers greedy for hidden things."[2] Gilbert certainly criticizes gross superstitions, but he does not banish animism or occultism, and he retains a kind of animism in his later unpublished book, *De Mundo: A New Philosophy of the Sublunary World*.[3] Furthermore, the association of magnetism with mystery, vital forces, and a world organism continues through the eighteenth century into the early nineteenth century in the work of the Romantic natural scientists, such as Ritter and Ørsted, who contributed to the understanding of the interaction of electricity and magnetism.[4]

The Magnet, Experiment, and "Mother Earth"

The parts of Gilbert's work best known to moderns are his simple, clearly presented experiments. Gilbert resolved to "begin with the common stony . . . matter, . . . that we may handle and may perceive with the senses; then to proceed with plain magnetic experiments, and to penetrate to the inner parts of the Earth."[5] He was first to clearly distinguishes between electric and magnetic phenomena,

141

not using the word *electricity* itself, but writing of "electrics" as bodies that attract amber.

Gilbert invented the *versorium* (electroscope), a device consisting of a small needle that when brought close to an electrified body would point toward it. He performed experiments pointing compass needles toward magnets and toward the north, noting the "dip" of the compass needle (tilting of the north pole of the needle toward the ground), which varies with geographical position. Gilbert also experimented on a small magnetic globe (*terrella*) used as a model of the Earth. From experiments with the terrella, Gilbert concluded that the Earth is a great magnet with a North and a South magnetic Pole. Gilbert's translator, Fleury Mottelay, claims that Gilbert's pride in his own experimental discoveries is shown by the fact that he puts an asterisk next to the lucid description of each one. Gilbert's experiments became an established part of all later physical science.

Gilbert's kind of denunciation of the speculations of Scholastic predecessors is found among many of the new philosophers of the Renaissance, but he makes an exception for "God-like Thomas" (Aquinas), whom he praises for his "perspicacious mind."[6] Despite Gilbert's criticisms of the Scholastics, he borrows the Aristotelian distinction between form and matter for his discussion of the difference between magnetism and electricity. Gilbert states that electricity depends on a material cause and is produced by material emission, while magnetism does not and depends on form. Gilbert's form of magnetism differs from the forms of the Scholastics. He rejects the multiplication of species to which Roger Bacon appealed, and posits magnetic form operating as an efficient rather than a formal cause. (An efficient cause can be a physical action, such as the hammering by a sculptor of a statue, while the formal cause can be the purely formal structure, such as the shape of the marble.)

Gilbert appeals to an animistic, magical tradition, earlier and more primitive than that of the Scholastics, which is centered on reverence for the Earth. In his preface he refers to "that great lodestone, our common mother (the earth)."[7] In discussing his method, he refers to "true demonstrations and . . . experiments that appeal plainly to the senses, . . . pointing with the finger, to exhibit to mankind Earth, mother of all."[8] Gilbert appeals to ancient, pre-Christian, pagan worship of the Earth. In *De Mundo*, he associates gravitational attraction with the teleological drive to return to the mother. "Everything terrestrial is reunited to the earth; . . . [with] a propensity toward the body, toward a common source, toward the mother where they were begotten, toward their origin, in which all these parts will be united and preserved, and in which they will remain at rest, safe from every peril."[9]

Carolyn Merchant discusses "The Earth as Female" as an object of respect and the shift to the notion of an earth despised and open to abuse and despoliation with the shift from the pagan and occultist beliefs to the modern, mechanistic beliefs. She does not mention Gilbert and his representation of the Earth as mother, along with his nonmechanistic philosophy of magnetic "coition."[10] Perhaps this is because, while Gilbert is not a mechanist, as an experimentalist he is not completely immersed in the occultist views that mechanism supplanted. He combines aspects of experimental philosophy with a traditional reverence for the Earth that his mechanist successors reject. Contemporary doctrines of Gaia, the Earth goddess (which range from scientific claims about homeostatic equilibrium in the Earth's atmosphere because of the action of the respiration of bacteria to claims that the biogeochemical system of Earth is literally a living organism) find historical antecedent in Gilbert's view that the Earth is our mother.[11]

In a chapter entitled "The Magnetic Force Is Animate, or Imitates a Soul: In Many Respects It Surpasses the Human Soul While That Is United to an Organic Body," Gilbert discusses the soul of the Earth. He criticizes Aristotle for believing both that only the heavens and the stars but not the universe is animate and that the stars are alive but the Earth is not. He says, "Aristotle's world would seem to be a monstrous creation, in which all things are perfect, animate, while the Earth alone, luckless small fraction, is imperfect, dead, inanimate, and subject to decay. On the other hand, Hermes, Zoroaster, Orpheus, recognize a universal soul. As for us, we deem the whole world animate."[12] He further writes: "Pitiable is the state of the stars, abject the lot of Earth, if this high dignity of soul is denied them, while it is granted to the worm, the ant, the roach, to plants and morels."[13] Gilbert agrees with the first Greek philosopher-scientist, Thales, that "the magnet has a soul, for it moves the iron."[14] Gilbert believes that "the Earth's magnetic force and foremate soul or animate form of the globes, that are without senses, . . . exert an unending action, quick definite, constant, directive, motive, imperant, harmonious, through the whole mass of matter;"[15] and that the moon acts "through the joint action or conspiracy of the two bodies and, to explain my thought with the aid of an analogy, through magnetic attraction."[16] This conception of the Earth's soul or spirit operating through magnetic force following laws without errors was the inspiration for Kepler's account of gravitation as a magnetic force in which the gravitational spirits of the planets act like automata. Coition acts "without end,"[17] and does not propagate through a medium. Thus, Gilbert's magnetism will supply a model for action at a distance.

Gilbert holds the traditional view that metals grow in the Earth, and speaks of the Earth as the "mother of metals."[18] As part of his worship of the Earth and of the magnet, Gilbert claims that iron rather than gold is the most noble of metals.[19] He considers the Earth to be fundamentally in nature a perfect sphere.[20] Gilbert is very much a part of the natural magic tradition; although he criticizes and ridicules many of his predecessors, he speaks of Giambattista della Porta, author of *Natural Magick,* as "a philosopher of no ordinary note."[21] Even Marie Boas, who emphasizes the more empirical and "modern scientific aspects" of such figures as Boyle and Newton against the more recent alchemical interpretations of these figures, claims that Gilbert is part of the "natural magic" tradition.[22]

The similarity of Gilbert's magnetic philosophy to the hermetic and alchemical traditions is shown in his distinction between attraction and coition, a terminology he introduces in his distinction between electric and magnetic bodies. He prefers the term *coition* for the interaction of magnetic bodies because they operate on each other mutually. Although Gilbert understands *attraction* to mean that one body unilaterally operates on the other, with the latter as a purely passive partner, he is scientifically mistaken in believing that electrical attraction is not mutual in contrast to magnetic coition; however, the point of interest for his philosophy is that he understands the cosmically important magnetic interaction to be a mutual and equal interaction in contrast to what he sees as his predecessors' one-sided view of electrical attraction. The alchemists and hermeticists envisioned interaction of scientists with nature as an equal sexual intercourse, while Bacon and the proponents of the new mechanical philosophy describe the interaction of the scientists with nature as one of domination and manipulation.[23] Gilbert, the natural magician, sees the fundamental natural interaction as an egalitarian sexual coition.

Even Gilbert's use of the terrella as a model of the great magnet, the Earth, and the extreme trust with which he interprets results on it (even when misleading, such as magnetic dip), suggests microcosm/macrocosm doctrine of the occultists and hermeticists. That is, each part of the universe mirrors the whole, a doctrine found in Chinese correlative cosmology and in Leibniz. For Gilbert each little bit of magnetic Earth is a mirror of the great Earth mother herself. Gilbert interprets (or misinterprets) the experiments on electricity with no conception of the equality of action and reaction, which is not to be found until Newton promulgates his third law. To complete such an analysis, even later theories of charge induction and dielectrics would be needed.[24] Despite the criticisms based on centuries of scientific hindsight, Gilbert's investigations are extraordinarily clearheaded and precise for his day. Galileo evaluated Gilbert quite positively. In the *Dialogue Concerning the Two Chief World Systems*, in which Galileo confronts the new Copernican system with the old Earth-centered one, Galileo writes (through Sagredo, a thinly veiled spokesperson for his own views against the simpleminded Aristotelian Simplicius), "I have the highest praise, admiration and envy for this author, who framed such a stupendous concept concerning an object that innumerable men of splendid intellect had handled without paying any attention to it.[25]

Galileo wishes, however, that Gilbert had "a little more of the mathematician, and especially a thorough grounding in geometry, a discipline that would have rendered him less rash about accepting as rigorous proofs those reasons that he put forward as verae causae for the correct conclusions he himself had observed."[26] Galileo's spokesperson says, "Since we see that the ordinary human mind has so little curiosity and cares so little for rare and gentle things that no desire to learn is stirred within it by seeing and hearing these practiced exquisitely by experts . . . these are concepts and ideas for superhuman souls."[27]

Francis Bacon's negative attitude toward Gilbert shows up the peculiar mix of rigorous experiment and animistic speculation in Gilbert's magnetic philosophy. Many commentators have been surprised that Bacon, who advocates careful inductive experimental method, criticizes Gilbert. Gilbert's *On the Magnet* would seem to be the best experimental work at the time that Bacon wrote. One explanation is that Bacon was primarily referring to Gilbert's later unpublished work on meteorology and cosmology, *De Mundo*, which extends the animistic and cosmological speculations at the end of Gilbert's *On the Magnet*. *De Mundo* is more thoroughly occultist and animistic than Gilbert's much more famous experimental work on the magnet. Bacon possessed an unpublished copy of this work; however, since it was not actually printed for another four decades, it would seem foolish for Bacon to spend much time criticizing a work that his readers had never seen and were unlikely to see.

Another simplistic theory holds that Bacon was simply jealous of the results of Gilbert's experiments, and wished to play them down. A third theory, more plausible, is that Bacon was so narrow in his empiricism that he rejected Gilbert's development and testing of broad theories. This view is similar to the one Hesse presents following Popper's view of science as theoretical conjectures and refutations. Hesse rejects the simplistic view of Bacon as a totally atheoretical fact gatherer, but notes that Gilbert's use of experiments to eliminate theories is closer to the spirit of later science than Bacon's construction of tables of presence and absence of qualities.[28] Certainly Bacon is justified in criticizing some of Gilbert's more animistic and empirically unfounded speculations. Bacon was critical of astrology and parts of alchemy; however, he was also critical of astronomy in general, and disliked the somewhat arbitrary mathematical constructions of some astronomers to fit the available data.

The best-known criticism of Gilbert by Bacon appears in *New Organon*: "The race of chemists again out of a few experiments of the furnace have built up a fantastic philosophy, framed with reference to a few things; and Gilbert also, after he had employed himself most laboriously in the study and observation of the lodestone, proceeded at once to construct an entire system in accordance with his favorite subject."[29]

Part of the criticism is directed against Gilbert's use of his experiments to generalize about the cosmos. At the end of *On the Magnet*, Gilbert defends the motion of the Earth, and gives a favorable presentation of the Copernican system of the Sun-centered universe.[30] Galileo and Kepler as well as later scientists and historians of science hold this to Gilbert's credit; however, Bacon was very skeptical concerning astronomical theories in general and speculations concerning the physical arrangements and causes of the motions of the planets in particular. He wished to limit astronomy to purely mathematical descriptions of the apparent motions. Thus, for Bacon, the defense of Copernicanism was an unscientific part of Gilbert's philosophy and was to his discredit. There are also elements in Gilbert's later astronomical speculations of a highly teleological, or purposive, nature that were probably offensive to Bacon. For instance, he notes in an aside that the tilt of the Earth's axis is "for the everlasting good of man."[31]

Much of Gilbert's defense of the motion of the Earth is relatively weak. He gives a series of old arguments and uses a great deal of pure rhetorical assertion. Some of his rhetoric is similar to that of Copernicus, and distinguishes between the complex mathematical curve fitting of the mathematicians and the need for simple causes on the part of "philosophers." He claims that "we must pardon slips in mathematicians, for one may be permitted in the case of movements difficult to account for to offer any hypotheses whatever in order to establish a law and to bring in a rule that will make the facts agree. But the philosopher never can admit such enormous and monstrous celestial constructions."[32] The "monster" rhetoric is from Copernicus himself.[33]

146 Renaissance Occultism

Peter Urbach, in his reevaluation and defense of Bacon's philosophy of science, defends Bacon's evaluation of Gilbert.[34] Urbach denies that Bacon was a narrow empiricist of the "strict Baconian method" to which later scientists, such as Darwin, claimed to adhere. To make his case, however, Urbach defends almost all of Bacon's shortsighted evaluations of contemporary science. Urbach can trade on the general modern rejection of the nonmechanistic traditions (hermetic, occultist, and animist), so since Gilbert uses purposive explanations, references to souls or spirits—even if of a relatively automatonlike sort—Bacon is correct to reject Gilbert's magnetic philosophy. Urbach's special pleading neglects Bacon's overall tone of rejection of what, after all, was the best example of experimental science and application of empirical method in Bacon's day.

Bacon's extreme distrust of speculation led him to reject even Gilbert's most brilliant hypothesis of geophysics—that the whole Earth is a magnet—and was skeptical about Gilbert's notion (now believed correct) that the core, not minor surface features, are what accounts for Earth's primary magnetism. It is now thought that the rotation of the metallic core of the Earth is what sets up the Earth's magnetic field, and that the reason that the Earth, but not neighboring planets, has a strong magnetic field is the size and speed of the rotation of the Earth's magnetic core. Mars is too small to have a powerful, magnetism-generating core, while Venus rotates too slowly. The outer planets lack the metallic core of the inner planets, but their huge size and the dissociation of atomic particles allow them strong magnetic fields. Urbach defends Bacon by noting that the latter accepted the idea that the Earth was magnetic, while basing this notion on the magnetism of features of the outer surface of the Earth—but such an admission is hardly a piece of astute science. It was precisely the notion that certain isolated features of the surface of the Earth accounted for its magnetism that misled all theorists of terrestrial magnetism before Gilbert. The notion that there must be a large magnetic mountain at the North Pole that accounted for the behavior of the compass, or the legends of magnetic islands that pulled the nails from passing ships, were precisely the sort of ideas generated by the hypothesis of surface magnetism that Bacon grants and Urbach defends. Gilbert's distinguishing the nonmaterial nature of magnetic coition from the material effluvial nature of electric attraction had virtues that were realized only later. Kepler took inspiration from Gilbert's magnetical theory of the solar system. In Gilbert's posthumous and unfinished work, *De Mundo*, he claims that the magnetic sphere of influence of the Earth extends as far as the Moon.[35] The acceptance of empty space combined with the rejection of material effluvia paved the way for the action-at-a-distance conception of gravitation in Newton.

Although Gilbert's magnetic philosophy was associated with animist and nonmechanist conceptions of "virtues," his instrumentation and technology were associated with the tradition of British instrument makers and technicians. Gilbert had built on *The New Attractive* of Robert Norman, who had discovered the dip of the compass needle in the course of affixing needles to compass cards, finding the inconvenient result that the needle sometimes tilted downward so sharply at its north end that it would touch and rub against the compass card. Gilbert's experimental work and knowledge was much indebted to the artisans, as Zilsel emphasized.[36]

Followers of Gilbert and the "Magnetical Philosophy"

Stephen Pumfrey has examined the social interests of the partisans of the "magnetical philosophy" after Gilbert, as well as the unique historical conjuncture that led to the rapid acceptance of the "observation" of magnetic variation.[37] Gilbert believed that the geographical North and South Poles coincided with the magnetic North and South Poles. This was a fundamental part of Gilbert's magnetic philosophy because he believed the magnetic nature of the Earth accounted for its rotation. Thus, Gilbert's Copernicanism was strongly tied to his view that the magnetic pole could not be different from the geographical pole. Advocates of Gilbert's magnetic philosophy who were also Copernicans linked their own modernity to Gilbert's position concerning the poles. In contrast, on the continent of Europe, the view that dominated concerning magnetic variation, or declination, was that the phenomenon was accounted for by a difference in location between the magnetic and geographical poles of the Earth. John Dee, a magician and mathematician, had held the same view in the previous century; however, the majority of English practitioners strongly rejected the European position and adhered to Gilbert's theory of the Earth.

Another factor besides Gilbert's magnetic theory of the Earth's rotation that partially accounted for the difference between the theories of the English and of the continental Europeans was that magnetic declination was much stronger in arctic regions than in tropical ones. (This is because the difference between meridians leading to the geographical North Pole and the meridians leading to the magnetical North Pole is much more noticeable in the Arctic Circle than at the Equator.) The Spanish and Portuguese dominated the southern sea routes to Latin America and the East Indies, and magnetic variation was not so much as a problem for them as it was for those exploring the far north. The English were forced to explore the north in the futile search for a Northwest Passage to the Indies. British explorers, such as Henry Hudson and William Baffin, were concerned with accurate observations of magnetic declination, and Baffin himself contributed data to the dispute concerning Gilbert's theory.

Gilbert's own account of the magnetic declination attributed it to irregularities in the distribution of the land masses of the Earth. Although Gilbert had ridiculed earlier Renaissance thinkers who attributed the magnetic pole itself to a magnetic island or mountain in the north, he made an appeal similar to this in accounting for the variation of compass direction from geographical north. Gilbert claimed that it was the mountains and ocean depths that made Earth's "essentially" perfect sphere irregular and accounted for the observable magnetic declination. According to Gilbert, the distribution of magnetic declination or the divergence between the north compass point and geographical north, should have remained constant within recent historical time.

Gilbert did allow for regular geological catastrophes throughout history (such as the sinking of Atlantis), which could vary the pattern of magnetic declination in *De Mundo*. There is some resemblance between Gilbert's contrast of a perfectly spherical Earth governed by the Earth's soul and the irregularities produced by geological catastrophes with Thomas Burnet's *Sacred Theory of the Earth*, where a spherical, biblical Earth underwent geological degeneration because of the Fall of humans.[38]

One of the difficulties in evaluating the nature of magnetic declination was the uncertainty of the observations. Sailors' compasses were not of high quality. Many modified their compasses to correct for the local magnetic declination by bending the compass needle or turning the compass card, and did not take declination seriously. European theorists treated magnetic declination as simply a product of observational error. English translations of French and Portuguese tracts claiming this were persuasive. Pumfrey has argued that the anomalous observations allowed for the entry of social influences into interpretation. The sailors' ambiguous data could be interpreted as random error from observations of the identical north geographical and magnetic direction. Other theorists believed that there was a hidden, underlying pattern. A concentration on compass observations taken in the far north led theorists to believe that variation was not simply a matter of observational error.

There was a strong technological and economic interest in determining geographical longitude. It was hoped that the discovery of systematic patterns of magnetic declination would help sailors to determine longitude and sail more efficiently. Edward Wright, who wrote fulsome praise of Gilbert in the preface to *On the Magnet*, was strongly involved in the effort to determine latitude by magnetic declination. Pumfrey has argued that Wright actually wrote the chapter in Gilbert's book on the data of variation.[39] The British figures involved in the study of geographical magnetism had close personal and institutional ties to Gresham College at Oxford and were central to the development of the Gilbertian account of declination. Their activities were closely tied with the craft tradition of such figures as Robert Norman and Gilbert himself.

Pumfrey argues that the acceptance of the supposed discovery of secular magnetic variation in 1634 by Gresham College professor Henry Gellibrand was overdetermined by the acceptance of Gilbert's magnetic philosophy by the relevant scholars, the interest in improving British navigation, the system of shared craft practices that the Gilbertians inherited, and the institutional and community ties linking the relevant practitioners. Gellibrand actually discovered the time variation of magnetic declination, or the variation of the variation. He was not the first to observe variation, a topic of mariners of the 1400s. The Chinese knew about it at least four centuries before that; however, Gellibrand's discovery was eagerly embraced by a tightly knit community of investigators. Had it occurred at either an earlier or a later period, it would probably have been dismissed as simply experimental error. Shortly after Gellibrand's discovery, the British magnetic theory began to disintegrate.

The mechanists had to rid themselves of the magnetical philosophy of Gilbert along with other animist, occultist, or alchemical theories. The magnet had been, since the Middle Ages, a prime example of natural magical powers. It was imperative that magnetism be explained in a mechanistic manner in order to remove it from the clutches of the alchemists and Scholastics. The Jesuits, in their rear guard defense of Aristotelian Scholasticism, combining it with reinterpreted results of the new science, made frequent reference to the magnet.

The mechanists treated magnetism as a material effluvium, or emanation. Gilbert held that electricity was such a material effluvium but of the nature of soul or spirit. The mechanist attack on the magnetic philosophy attempted to show that magnetism was affected by mechanical interventions. The effect of heat on magnetism was well known—Gilbert had talked about it in demagnetizing iron—however, he described it in terms of confusion of the form of the iron by heat. The mechanists had long suggested that heat was a form of motion; hence, the effect of heat on magnetism would fit well with the mechanical philosophy. Walter Charleton, a follower of atomist Pierre Gassendi, claimed that the spiritual quality of the magnet was disproved by the effect of fire upon it.

Henry Power had incorporated a book on magnetical philosophy into his work, *Experimental Philosophy*.[40] Power noted that after one had heated a magnet and let it cool, striking it with a hammer would effect the strengthening of its magnetic power. This experiment (which was later taken as clear evidence of mechanical effect on magnetism, showing the mechanical nature of magnetism itself) was, when first communicated to the Royal Society, barely noticed. This was, according to Pumfrey's social account, because the Magnetics Committee of the society was at that time embroiled in the difficulties of measurement of magnetic variation.[41] John Wilkins, who had been involved in the disputes between mechanists and occultists during the Civil War, was instrumental in bringing Power's experiments to the attention of important physicists of the time, such as Robert Boyle and Robert Hooke.

Besides experiments that involved the mechanical effects of heat and percussion on the magnet, there was another program pursued primarily by Boyle that attempted to show that magnetism was exuded from the lodestone like a kind of atmosphere.[42] Boyle had developed a vacuum pump to remove the atmosphere for the purposes of chemical experimentation. He thought that if the air around the magnet was removed, the magnetic particles would also be sucked out and the magnet would lose its power. Other experiments of this sort involved placing magnets under water so that the emissions of their magnetic particles would be deflected or slowed. These experiments were failures.

Nevertheless, the effect of heat and hammering on magnets was successful enough to be the basis for an assault on the magnetic philosophy. Robert Hooke (1635–1703) was a central figure in this attack. He was apparently the first person to attempt to determine a law of magnetic force, but Newton was the first to succeed.[43] Hooke relished controversy, as shown by his disagreement with Newton over the theory of light, which caused Newton to shun publication for a long time thereafter. Hooke had a straw man to attack in the form of Martin Lister, a physician who had integrated magnetism into a theory of chemical mineralogy.[44]

Lister was follower of Gilbert's magnetic philosophy, as was Hooke though he was critical of medical applications of magnets.

Lister claimed that fire alone did not create or destroy magnetism, that magnetism alone had such a power. Lister asserted, following Gilbert, that air could condense into showers of iron. He also suggested that rainstorms were magnetic phenomena, and even that hammers and drills affected magnetism only because they were themselves magnets. Hooke and his allies had a field day against poor Lister, signaling the death knell of the magnetic philosophy.

Temporary Replacement of the Magnetical Philosophy with Anti-Copernican Hermeticism cum Mechanism

The triumph of the mechanical atmosphere theory of magnetism had a paradoxical effect. Now that magnetism was no longer distinguished from other mechanical processes, such as those of gases and electricity, the old occultist superstitions about magnetism could regain scientific respectability because now virtually any mysterious power of the magnet could be attributed to a simple mechanical emission. An ironic consequence of this new shift in thinking was that Gilbert's own theory of the whole Earth as a great magnet was rejected by the mechanist geophysicists and was replaced by an alternative theory that was propounded by extreme occultist and early China scholar Athanasius Kircher.

Among Kircher's numerous, elaborately illustrated works were ones on magnetism.[45] The center of his museum at the Vatican, full of Egyptian obelisks, speech machines, and other imposing objects, was an unimposing little magnet, for the magnet was the paradigm of occult power.[46] Kircher also wrote an extensive and widely read work on the interior of the Earth.[47] He had a complex, well-staffed alchemical laboratory at the Collegio Romano, and thought alchemical processes and apparatus were simply imitations of volcanoes, underground caves, and subterranean hot springs.[48] Kircher's alchemical investigations were closely tied to his investigations of metallurgy. By tying alchemy to metallurgy and pharmacy (spagyric alchemy), Kircher could distinguish his own practical alchemy from the literally diabolical alchemy of transmutation. For Kircher "the geocosmos was the archetype of alchemical transformation."[49] He claimed that the "magnetism" of the Earth was caused by long subterranean fibers. The variation with time of magnetic declination was now claimed to be due to the shifting of these strings, or fibers, within the bowels of the Earth. Kircher also used his own theory of the mechanical magnetism of the Earth to argue against Gilbert's theory of the Earth as a great magnet, which Kepler used to explain planetary motion, and thus to argue against the Copernican system and for the immobility of the Earth. Kircher had become a star at the papal court after the condemnation of Galileo.[50] He argued that if the Earth were truly a great magnet, mountains would collapse and iron tools could not be used to dig and lift Earth.[51] Ironically, the theories of an extreme hermeticist now mechanically reinterpreted were able, for some (mainly but not solely Continental Catholic) investigators, to displace Gilbert's more moderate (and accurate) theory of the Earth as a great magnet.

Robert Boyle and Henry Oldenburg of the Royal Society in London were negative about Kircher's theories, but attended to Kircher's work, given his connections with numerous princes and kings. Oldenburg, industrious correspondent for the society, whose members had initially approached Kircher's work on the Earth with interest, and some of whose English contacts had been very favorably impressed by Kircher and his museum, was somewhat doubtful about the direction of Kircher's theorizing in the *Subterranean World*. Oldenburg criticized it as based more on collections of the known than upon novel observation or experiment, and noted that attempted replication of the first experiment based on Kircher's book (concerning the formation of fossils) had failed.[52]

For a few decades between its publication and Newton's hypothesis of universal gravitation and action at a distance, Gilbert's magnetic philosophy went through a period of popularity, followed by a precipitous decline. His experimental results and clear explanations were universally accepted, but the theory and cosmology that went along with them were violently rejected due to the rise of mechanistic philosophy and its popularity in the period of the Restoration after the Civil War of the 1640s. The alchemical, Paracelsian, and hermetic occultist philosophies were associated with the radical reformers of the Puritan Revolution. Such figures as John Webster linked hermetic and alchemical worldviews with the notion of a science for the people in which agricultural and medical applications would guide the development of theoretical science. Both the notion of science in the direct service of the people (as opposed to the sovereign) and the occultist and alchemical doctrines associated with it came under suspicion at the end of the Civil War with the Restoration of the monarchy. The criterion of demarcation between science and political, religious, and philosophical or ethical issues that was drawn by the Royal Society was not just an academic matter. It was an act of self-protection against accusations that the new science was associated with freethinking or religious heresy, with a higher status for women in the professions, with radical political reforms, or even with revolution.

While Gilbert's magnetical philosophy was rejected in Restoration England, Kircher's outdated and bizarre conglomeration of mechanistic occultism displaced both Copernicus's and Gilbert's approaches in Counterreformation Europe for several decades. Johannes Kepler, though, had already combined elements of geometric mysticism, astrology, and Gilbert's magnetical philosophy to defend Copernicus and find laws of planetary motion that, in turn, would be explained by a theory whose acceptance had been prepared for by the Gilbertians—Newton's gravitational attraction.

CHAPTER 9

Kepler and the Magnetic Force Model of the Solar System

———

$\mathcal{J}$ohannes Kepler (1571–1630) developed the modern model of the solar system with its elliptical orbits and correct estimates of the speeds of the planets. His laws were later to be the basis of Newton's laws of motion; however, Kepler was also heavily immersed in the Renaissance universe of Pythagorean and Neoplatonic mysticism and astrology. He truly stands between the occultist universe of the Renaissance magician and the modern universe of laws. Kepler was guided and motivated in his discovery of the laws of the solar system both by a mystical search for Pythagorean harmonies—geometrical and musical—and by an erroneous physics of celestial magnetic forces, but this was the first time that astronomical theories were built to conform with a genuinely physical explanation.

Kepler's Early Life: The Mystery of the Universe

Kepler led an exciting and tumultuous life for a scientist. He was buffeted by vicious religious and political conflicts, beginning with the Reformation and ending with the disastrous Thirty Years' War. His mother was tried for witchcraft, and the "aunt" who raised his mother was executed as a witch. He led a life of wandering like that of his reprobate father, a soldier of fortune who disgraced his family by joining the forces of the opposing religion and barely escaping the gallows. One of Kepler's brothers was an epileptic, psychopath, and army camp follower.[1]

Kepler himself, although miserable and sickly as a youth, was recognized as bright enough to pursue advanced studies. At age thirteen, he entered a theological seminary and later attended the university of Tübingen. While teaching a class, the young Kepler had a flash of insight that the distances of the planets from the Sun could be explained by the five regular solids: the cube, the tetrahedron, the octahedron, the dodecahedron, and the icosahedron. This was the basis of his

Pythagorean and Platonic work, *Mystery of the Universe* (translated as *Secret of the Universe*).[2]

These solids are the only possible three-dimensional figures all of whose sides, faces, and angles are equal. For instance, the cube has twelve equal edges, six equal faces, and twenty-four equal planar angles. The tetrahedron has four sides, each of which is an equilateral triangle. The faces of the octahedron are eight triangles, of the dodecahedron twelve pentagons, and the icosahedron twenty triangles.

Modern crystallography, chemistry, and atomic physics find regular solids in nature. For instance, the lattice of table salt (NaCl) is cubical, while carbon atoms in the diamond are tetrahedral. Kepler wrote *New Year's Gift; or, The Six-Cornered Snowflake*, which deals with the regular hexagonal pattern of the snowflake, and is an early work crystallography.[3] In twentieth-century development of mathematics, these regular solids reappear in unexpected places.[4]

Regular solids have been objects of fascination since ancient times. The cube was, of course, long well known, and dice in the form of dodecahedrons were found among the ancient Etruscans. In *Mystery of the Universe*, Kepler claimed that a cube inscribed within the sphere of Saturn would contain a sphere the size of the orbit of Jupiter. A modern commentator on Kepler's *Mystery* notes that Kepler made the orb of the earth as thick as the eccentricity of the orbit. If one does this for each planetary orb, one gets Kepler's results. A tetrahedron within the sphere of Jupiter's orbit would exactly contain a sphere the size of the orbit of Mars with a thickness equal to the eccentricity. A dodecahedron within the sphere of Mars would contain the orbit of the Earth. If an icosahedron is placed on the inner surface of the Earth's orbit, then its inscribed sphere is the orbit of Venus. And finally, if an octahedron is inscribed on the orbit of Venus, its inscribed sphere is the orbit of Mercury. One science historian notes, "Until quite recently twentieth century cosmologists would have been very pleased if their theories had fitted observations as well as Kepler's do."[5] With the difficulties during the 1980s concerning "missing mass" in the universe, and the 1994–1995 claim that some stars seem to be older than the universe, this is an understatement, though our measurements on the solar system are vastly more accurate than Kepler's and undermine his theory. In his last work, *The Harmonies of the World*, Kepler added musical chord ratios to correct for the elliptical orbits he later discovered.

This model shows Kepler's total commitment to the hypothesis that the Sun is at the center of the solar system. Kepler was the first major astronomer to commit himself openly to Copernicanism. Galileo, who is famous for his later defense of Copernicanism, and who suffered trial and house arrest for it, had initially remained silent. For many years, Galileo taught the Earth-centered, Ptolemaic solar system and in class presented the standard arguments against the moving Earth. Even Galileo's *Dialogue Concerning the Two Chief World Systems*, for which he got in trouble with the Church, was superficially presented as a balanced debate between partisans of the old system and the new. During the period when positions on astronomy would not have brought persecution by the Church, Galileo kept silent about the Copernican system, apparently fearing ridicule from academic colleagues. In contrast, Kepler's commitment to Copernicanism was open from the beginning. Luther, the founder of Kepler's church, condemned Copernicanism in his table talk, in terms far more ignorant and crude than any used later by the Catholic Church.

Copernicanism led Kepler to a historical discovery. He could not believe that Copernicus could have written his great work, with its mathematical labors, without actually believing in the reality of the system that appears within it; yet, the preface to Copernicus's *Revolutions* makes the disclaimer that the theory presented is a mere calculating devise, not meant to be literal. Kepler suspected, and asserted for the first time in print, that the preface was a forgery. Andreas Osiander, Copernicus's pious assistant, had added the preface to the manuscript at the last minute, and the dying Copernicus, publishing as he perished,[6] signed off on the now partially forged manuscript. Other Copernicans had been aware of this. In Rheticus's own copies, Osiander's preface is crossed out with red pencil or crayon. Rheticus had originally arranged for the publication of the *Revolutions*, but was absent during its preparation by Osiander.[7] He proceeded against the printer to have a new preface distributed.[8]

Despite the importance of Kepler's commitment to Copernicanism, the beautiful mathematical ratios of distances of the planets from the Sun turn out to be a delusion. Kepler himself later recognized this, but he continued his pursuit of Platonic mathematical harmonies, later embodied in the form of musical theory, to the end of his life. The young Kepler, nevertheless, had introduced his discovery with a characteristic burst of enthusiasm. The discovery "loosed a flood of tears."[9] "I believe it was by divine ordinance,"[10] and "it will never be possible for me to describe with words the enjoyment I have drawn from my discovery."[11] A quarter century later he wrote, "It would be mistaken to regard it as a pure invention of my mind. For as if a heavenly oracle had dictated it to me, the published booklet was . . . immediately recognized as . . . true throughout (as it is the rule with obvious acts of God)."[12]

Kepler as Astrologer

Kepler was the last major astronomer who took astrology seriously. Astrology led him to accept the notion that gravity connected the Earth and Moon. Kepler had sufficient belief in astrology to consider it worthwhile to devise his own horoscopes; however, he viewed the astrology of his day with considerable skepticism. Within his own, highly modified version he was cautious about making specific predictions for the political or military careers of individuals. He despised the astrologers who used their power over superstitious minds to gain wealth by flattery.[13]

For Kepler, astrology was part of his science. Unwilling to admit this, many historians of science have attempted to excuse Kepler's astrological writings, saying that they were produced simply as part of his duties to royalty or in order to make money. This was true of Kepler's calendars and prognostications. He wrote, "To defend my annual salary, my title, and my position I must humor uneducated curiosity," and he considered such works a better source of money than begging.[14]

Kepler rejected natal astrology, in which the configuration of the stars at the time of a person's birth was seen to determine personal behavior. He believed that stellar influence could be overruled by the environment and the individual's will. It was the attack on astrology in defense of human freedom and dignity by Giovanni Pico della Mirandola that swayed Kepler to reject natal astrology.[15] Surprisingly, Galileo, who rejected Kepler's theory of the Moon causing the tides as occult astrology, did take an interest in natal astrology. Galileo drew up horoscopes for eminent people, most notably for Cosimo de' Medici,[16] and also for himself, friends, and relatives.[17]

Kepler, on the other hand, was very cautious about making astrological predictions, even when his royal patrons demanded them. He drew up a horoscope for Albrecht von Wallenstein, the swashbuckling leader in the Thirty Years' War, but stopped his own evaluations of Wallenstein's prospects in 1634, prophesying "horrible disorder" for March 1634. Wallenstein was murdered at the end of February.[18] Kepler foresaw turmoil and misfortune for Rudolf II of the Holy Roman Empire (who supported Kepler along with numerous astrologers, alchemists, and magicians in Prague); however, he advised Rudolf II's courtiers to discourage him from concentrating on astrology, suggesting that if Rudolf believed the astrological predictions, he would not bother to use political means to protect himself. Kepler encouraged Rudolf to put aside concerns about past losses and concentrate on present tasks.[19] He recommended that "astrology . . . should be kept completely out of the emperor's sight."[20]

Despite this, Kepler wrote works on astrology not demanded by his position at court or needed to make money.[21] He rejected the doctrines of the signs of the zodiac and astrological houses (regions of the sky associated with such human issues as family, money, and travel), but did believe that celestial bodies influenced human life. He advised, "Do not throw the baby out with the bath water."[22]

Given Earth-centered astronomy, astrology was based on a reasonable, inductive inference. The Sun influences night and day, seasons and crops. Kepler and other astrologers (rightly) believed that the Moon influenced the tides. It was only a step to generalize from the two "planets," the Sun and Moon, to other planets. One leading historian of astronomy notes that ancient astrology was more scientific in its search for regularities than was the competing religious appeal to the arbitrary whims of gods.[23] Kepler no longer considered the Sun a planet, but thought that celestial bodies influenced the Earth. For Kepler, astrology was the lowly applied science while astronomy was the divine spiritual science dealing with God's plan of the universe.[24] The opposite evaluation is made by science-minded people today, for whom astrology smacks of spiritualism, while astronomy is a useful science. Kepler stated several times that astrology is the foolish daughter of astronomy, who by her often foolish predictions supports her impoverished mother.[25]

For Kepler the planets influence earthly events by purely geometrical means. Both astronomy and astrology deal with geometrical harmonies: astronomy in terms of God's view of the universe, astrology from the earthly viewpoint. As heavenly geometrical patterns affect earthly creatures through perception, psychology plays a role in astrology. Since there are astrological effects on the weather and agriculture, Kepler needed perceivers to mediate these effects. Kepler attributed a soul to the Earth to perceive geometrical arrangements of the planets. He justified this by appealing to the generation of minerals and the exhalation of vapors producing rivers and sulfur springs.[26] "Flexible portions in the interior of the earth . . . might play the part of lungs or bronchi."[27] Kepler in the *New Astronomy* considered planetary minds an alternative hypothesis to that of magnetic force in explaining the planetary motions.

Twentieth-century physicist Wolfgang Pauli has said that astrology is part of Kepler's general physics. By this he meant that astrology deals with the physical effects that the heavenly bodies have on the Earth. For Kepler, physical effects of the stars and planets on the Earth made much more sense than they did for an Aristotelian. Aristotle's followers believed that the stars and planets were made of a substance very different in nature from the four basic earthly elements. Kepler believed that the planets were made of the same stuff as the Earth. For Kepler, and for Newton later, there is no basic distinction in physical processes between the realm beneath the Moon and the realm above.

Kepler believed planets affected Earth through light and drew conclusions about the effects of different planets in terms of the colors of light they emit. His astrology is not supernatural. In comparison, the geometry that is the basis of structure in both astrology and astronomy is identified with God. "Geometry is coeternal with the mind of God . . . ; it is God himself."[28]

Kepler's greatest innovation in the nature of explanation in astrology was his demand of a *causal* explanation for any astrological effect on a person or on the weather. Giving to various constellations names of mythological figures associated with water or fire was for him a matter of human convention and had no explanatory force. He ridiculed the notion that a star appearing in the so-called fiery trigon, should lead to conflagrations, droughts, and wars. Kepler was measuring astrology by the new causal criteria rather than those of symbolic meaning. This is obviously part of the major transition from the medieval and Renaissance outlook to that of modern science. Kepler's "cure" for astrology "would soon kill the patient."[29]

What finally undermined traditional medieval astrology was not the shift to the Copernican Sun-centered worldview. Kepler's case shows that this shift need not have been fatal; however, the infinite universe (which Kepler rejected) undermined astrology more significantly with respect to the influence of the fixed stars. Kepler accepted a finitely distant sphere of the heavens as location of the stars. In contrast, in an infinite universe, with stars distributed in the depths of space, the apparent configurations and distances between the stars on the sphere of the heavens was only an accident of perspective from the Earth. Some of the stars that appear near to each other in terms of our visual lines are really at different depths in space, and may be much further from one another than one of them is from another star at the same distance from Earth in another region of the sky.[30]

Kepler's culmination of reformed astrology appears in book 4 of *The Harmonies of the World*, containing also books on mathematics, music, and astronomy as parts of a single structure. For Kepler the harmonies of astrology are an effect of angles between celestial bodies *as seen from Earth*, based on constructable regular polygons, as are the harmonies of astronomy. Because the angles of astrology are valid only from an earthly viewpoint, they are less fundamental than astronomical ones for the whole universe.

Kepler and Tycho Brahe: Uneasy Collaboration

Kepler was able to gain access to a large store of the best astronomical observations of his day by working for the Danish astronomer Tycho Brahe. Tycho was a nobleman given lordship over an island and a large income in order to collect data for astrology. Tycho's lifestyle was hardly that of a twentieth-century scientist. His nose had been cut during a duel and was said to have been replaced by a gold and silver replica.[31] He presided over a large court, including a fool, and had riotous banquets and drinking parties. When he traveled, he took his entourage as well as his astronomical instruments—a veritable "traveling circus."[32] He would casually mention various astronomical observations to Kepler in the midst of wide-ranging dinner conversations with his numerous guests and hangers-on. He threw bits of data to Kepler the way he threw bits of food to his dogs and witticisms to his fool. Tycho called his early observations "childish" and his better observations "virile," manifesting the seventeenth-century association of science with manliness found later in the Royal Society. He was vain and tyrannical, while Kepler was sensitive and irritable. Despite their disagreements, they worked to avoid a major break for each knew he needed the other. Tycho had the data Kepler needed but was working within his compromise astronomical system in which the Sun traveled around the Earth but the other planets traveled around the Sun (the Tychonic system, which appealed to astronomers concerned both to improve their theory and to retain the centrality of the Earth). Tycho knew that Kepler was a far better mathematician than he, while Kepler knew that only Tycho had the observations that he needed. Kepler wrote that Tycho had riches (the data) that he did not know how to use, and for this reason the data must be pried from him, in order to put it to better use.[33]

Under Tycho's sway, Kepler began calculations of the orbit of Mars. He had bet that he could finish the problem in eight days, but the project took eight years. After Tycho's sudden death from a burst bladder at one of his feasts, Kepler found his next major patron in Rudolf II, the emperor in Prague. Here Kepler gained the exalted title of Imperial Mathematician. In some ways, the court of Rudolf was much stranger even than the entourage of Tycho Brahe.[34] Rudolf II was a patron of the fine as well as the occult arts. Rudolf, who was eager to hear all secrets and marvels, had large collections of strange objects from all over the world arranged in "Wonder Cabinets."[35] Rudolf's court sponsored and received visits from many astrologers, alchemists, and magicians, including such major figures in the hermetic tradition as Giordano Bruno, a partisan of a revived Egyptian religion from Italy, and John Dee, a mathematician, book collector, and magus from England. John Dee's questionable companion and "skryer" of visions of angels,

Edward Kelley, a suspected counterfeiter, stayed on after Dee left and gained importance at court through his supposed alchemical abilities. When Kelley was unable to produce more of his magical powder, he was jailed. He was released (with a promise to make the philosophers' stone) but while under house arrest he killed a guard. He was imprisoned again and died attempting to escape.[36] Among such a mixture of charlatans and brilliant thinkers Kepler pursued his mathematical "war with Mars."

In Prague, Kepler was able to finish his *New Astronomy*, which contains his first and second laws. The first law maintains that planets travel around the Sun, not in circles, but in ellipses with the Sun at one focus. The second law holds that planets move at changing speeds such that the line from them to the Sun sweeps out equal areas in equal times. It was in this work that Kepler developed his magnetic theory of gravity.

While in Prague, Kepler first heard of Galileo's telescopic discoveries concerning the planets when the news was shouted from a coach by his friend Wacker von Wackenfels. Wacker was advisor to Rudolf and a figure in the political intrigues of the court, a humanist involved with Neoplatonic and Paracelsian writings, author of a now lost utopian tract, and a patron of Bruno. Kepler dedicated the science fiction work *The Dream* to Wacker.[37] Although Kepler was not political like Wacker—let alone a religious innovator and possible spy like Bruno—his friendship with Wacker suggests the political atmosphere in which the astronomer lived.

Galileo had heard of the Dutch manufacture of the telescope, and with it discovered the mountains on the moon, phases of Venus, and satellites of Jupiter. (Kepler was the first to use the term *satellite*.) Galileo reported his observations in a brief but eloquent pamphlet, *The Starry Messenger*. He named the four moons of Jupiter after the Medicis, wealthy patrons of the arts and philosophy in Florence.

Galileo's report rocked the intellectual world. It showed that the heavenly bodies had features similar to the Earth, breaking down the traditional Aristotelian distinction between the matter of the Earth and the matter of Heaven. Galileo's discovery evoked strong reaction. He exhibited his telescope at a house party but was unable to convince the powerful guests that his observations were correct. Two philosophy professors refused to look into the telescope, the others were unable to make out the blurry images through the crude and shaky instrument.

Kepler jumped into the fray with his own *Conversation with the Starry Messenger*, although he was Galileo's only defender. Despite this, Galileo evaded lending Kepler a telescope. When Kepler was finally able to borrow one from a local duke, he made the observations (and in a few weeks developed a theory of the telescope) that he presented in his *Dioptrics*. Galileo did not have a theory of the telescope with which to justify the claim that what was seen in the telescope was really out there in space and not merely an artifact of the lenses. Kepler supplied this crucial justification for Galileo's claims and made a major contribution to optics as well. Galileo never thanked Kepler for his public defense, and seems never to have bothered to learn Kepler's optical theory, despite the fact that it, for the first time, rigorously justified the veracity of Galileo's observations.

The following year yielded multiple misfortunes for Kepler. His mother was accused of and tried for witchcraft. His patron Rudolf II lapsed into melancholia and paranoia, abdicated, and died. Kepler lost his major supporter, although not his position as Imperial Mathematician. Kepler's wife became ill, epileptic, and deranged; his favorite son died of pox. Soon thereafter his wife died, and later his daughter. When Kepler moved to his new location, in the backwater of Linz, he ran into religious trouble. He took his Protestantism seriously and refused to convert to Catholicism when it was convenient, but he did not adhere to all tenets of Lutheranism. His minister refused him Communion. His mother was accused of poisoning several townspeople with witch's potions and of giving the evil eye to children. In Kepler's birthplace, a neighboring town of some thousand adults, thirty-eight witches had been burned in the previous fifteen years. Kepler rushed home to defend her, and had to pay for and assist in the legal defense, as well as pay for her imprisonment. She remained in prison for over a year, then was unable to return home because of threats of lynching. Kepler could feel guilty as well as distracted by the legal defense and his mother's tragedy. In Kepler's own utopian work, *The Dream*, the first science fiction account of a journey to the Moon[38] he portrays his mother as a dispenser of hallucinogenic herbs who could call up the powers of demons, and in the story he uses his mother's spells as the means to travel beyond the Earth.[39]

The Music of the Spheres

During the miseries of this period, Kepler escaped into contemplation of the cosmic harmonies and produced his *Harmonies of the World*, in which he carried on the search for Pythagorean and Platonic harmonies that he had begun in his *Mystery of the Universe*. This time, however, he sought the harmonies within music as well as in geometry. The ancient Pythagoreans and Plato had held the doctrine of the harmony of the spheres. Each planet was said to emit a note, and the set of planets was claimed to produce a chord, which we cannot hear because we have accommodated to it.

Just as the early Pythagoreans had found symmetry and harmony in the regular solids as manifested in gems and crystals, so they found evidence of the numerical structure of the world in the discovery of the simple ratios that account for musical chords. Pythagoras (or his early followers) discovered that the ratios of lengths of vibrating strings or the ratios of lengths of tubes of wind instruments accounted for the musical chords. The ratio of two to one accounted for the octave, the ratio of two to three accounted for the fifth, and the ratio of four to five accounted for the third. There was a long tradition of investigation of these musical harmonies in relation to astronomy. Ptolemy, who wrote an astrology as well as his now better-known *Almagest* on astronomy, gave numbers to the orbits of the planets.

Kepler, using the Sun-centered astronomy with its elliptical orbits, searched for features of the planetary orbits that he could match with his musical notes. He tried a number of different features, such as the size of the orbits, the maximum speed of the planets, and the time taken to cover a given distance. None of these gave him his relationships, but he found that the ratios of the angular velocity of the planets—the angles swept out by a line from the Sun, for each planet's slowest and fastest speed—gave the desired musical ratios. Saturn gave a major third, Jupiter a minor third, Mars a quint. The different planets' extreme velocities also held musical ratios to one other.

In a book that contains the third law, Kepler made another enthusiastic outburst, while deceiving himself once again concerning the validity of his ad hoc reasoning of the harmonies of the planets: "Yes, I give myself up to holy raving. I mockingly defy all mortals with this open confession. I have stolen the golden vessels of the Egyptians to make out of them a tabernacle for my God, far from the frontiers of Egypt. If you forgive me, I shall rejoice. If you are angry, I shall bear it. . . . I am writing a book either for my contemporaries, or for posterity. It is all the same to me. It may wait a hundred years for a reader, since God has also waited six thousand years for a witness."[40]

Kepler's "holy raving" was not wholly delusory. His musical ratios were the end of the tradition of the harmony of the spheres within serious science, but book 5, which this outburst introduced, also contained Kepler's third law, that the squares of the time of revolution of the planets about their orbit vary as the cubes of their average distance from the Sun. Kepler did find the sort of reader he was waiting for in Isaac Newton;[41] however, Kepler is nowhere mentioned in book 1 of Newton's great work, in which he deduces the elliptical orbit from the laws of motion.[42]

In *The Mystery of the Universe*, Kepler had crudely attempted to associate various human qualities with the faces of the regular polygons that he claimed explained the distances between the planets. Five years later, Kepler sketched in letters to friends a system involving geometry and music that would account for astrological influences.[43] The initial tie between music and astrology comes from Ptolemy's astrological work, which associated the orbits of the planets with musical notes. Kepler, in a purely geometrical account, replaced Ptolemy's Earth-centered system with a Sun-centered one that made many of Ptolemy's explanations unusable.[44]

Kepler treats the ratios of numbers that are associated with musical harmonies by dealing with the relationships between sides of polygons. Very much in the spirit of his early work in which the five regular solids were used to explain the distances between the planets, he uses regular polygons to explain the numerical ratios that result in harmonies in music. The early Pythagoreans recognized that ratios of lengths of strings or of pipes accounted for harmonies of notes produced by those strings or pipes, basing their music on ratios of the numbers 1, 2, 3, and 4. During the Renaissance, practical musicians demanded that music theory take into account more harmonies than had the Pythagorean theory. Giuseppe Zarlino introduced the numbers 5 and 6 into the ratios. Galileo's father, Vincenzo, was a music theorist with a system very similar to Zarlino's despite all the polemics between the two. Kepler followed Zarlino,[45] though he also read the elder Galilei's book. One musicologist lamented that Paul Hindemith, in his opera *The Harmony of the World*, had to reintroduce twentieth-century music theorists to Kepler, who was well known in previous centuries.[46]

Kepler gave a reason that such ratios as 1:2, 3:4, and 4:5 were harmonious while ratios such as 6:7 or 12:13 were not. It was that three-, four-, five-, and six-sided polygons could be constructed with straight edge and compass, but polygons with seven, eleven, or thirteen sides could not. The polygons constructable by straight edge and compass had measurements that were commensurable, expressible by fractions. The other polygons had measurements that were incommensurable. These ratios were called *irrational* by the ancient Greek Pythagoreans, who meant the term in a literal sense, and the terminology (minus the literal connotations) remains in mathematics today. Kepler in the second book of *The Harmonies of the World* investigates polygons that fit together, having what he calls "sociability,"[47] and that form tilings, or tessellations, covering a surface. Kepler was the first to work out the mathematics of such tessellation, and ideas were still being mined from his work by mathematicians in the 1970s.[48] For Kepler geometrical congruence and musical harmony both mean fitting together and are tightly linked.

Kepler applied his polygons and ratios to music theory, astrology, and astronomy in the last two books of *The Harmonies*. He set up ratios between the slowest and fastest speeds of each planets, that is, the speeds at the point most distant from and closest to the Sun. He found these ratios to be "harmonic," corresponding to the ratios that formed musical harmonies. He also made comparisons of the slowest speed of an outer planet to the fastest speed of the one next inside as well as ratios of the fastest speed of an outer planet to the slowest speed of the next planet within. All three of these series of ratios among the planets closely approximated harmonic ratios.[49] For Kepler, this showed in strict mathematical terms the harmony of the spheres, of which Pythagoras and Ptolemy had spoken, presented in precise mathematical form.

Modern astronomers today talk of "resonances" in the orbits of planets that give rise to the stability of the orbits. These correspond in many ways to Kepler's ratios.[50] Resonance holds when "the ratio of the angular velocities of these motions are rational numbers."[51] A twentieth-century compendium of celestial mechanics states, "The solar system is full of resonances. . . . The satellite systems of Jupiter and Saturn show them, the ring/moon system of Saturn has them; the asteroid belt exhibits them with respect to Jupiter; Jupiter/Saturn, Neptune/Pluto and so on exhibit them; . . . none of these are really explained. Interpreted in terms of mean motion commensurability, yes, explained no."[52] Interestingly, the same term, *resonance*, that expresses in modern form the Keplerian concern with ratios of small integers for orbital attributes is also used to translate a key notion in the Taoist *Huai-nan-tzu*.[53]

Kepler's Last Years

The later years of Kepler's life, after he lost most of his family, his major patron, and his ties with the Church, are filled with wanderings. During this period he published a major textbook of his astronomy, the *Epitome of Copernican Astronomy*. After a long, unpleasant struggle with Tycho's possessive but ignorant heirs (ending with disagreements over the dedications), Kepler also finally published his major record of astronomical data based on but extending Tycho's observations. After many misfortunes and difficulties, financial and physical, including a year of begging the three cities now responsible for the emperor's delayed payments to him, troops billeted in his printing shop, a peasant revolt that burned what had been printed so far at that shop, attempted extortions and threatened lawsuits with another printer, Kepler finally published the *Rudolphine Tables*, named after the emperor.[54]

Also in his later years, he engaged in a controversy with Robert Fludd, an extreme representative of the earlier hermetic and alchemical worldview. Fludd defended numerology, or number mysticism, while Kepler denied the validity of Fludd's mystical use of mathematics and defended his own mathematics. This debate is often presented as mathematical physics versus number mysticism,[55] but we have seen that Kepler's own Pythagoreanism, while more rigorous and systematic than Fludd's mysticism of individual numbers, was hardly contemporary empirical science. Kepler's mathematical physics was strongly tied to doctrines and philosophies that some have called mystical.

Fludd says that Kepler is concerned merely with the superficial aspects of things and not with their essences, and that he piles up dry axioms and theorems rather than expressing the real nature of things. Kepler says that Fludd's mathematics is "enigmatic, emblematic, and Hermetic," while Fludd says that Kepler is the worst sort of mathematician who "concerns himself with quantitative shadows."[56] Fludd claims that he himself holds nature by the head in its intellectual principles, while Kepler grasps nature by the tail. Kepler replies, "I hold the tail, but I hold it in my hand; you may grasp the head mentally, though, only, I fear, in your dreams."[57] Similarly, Kepler claims, "I play, indeed, with symbols, . . . But when I play, I never forget that I *am* playing. For nothing is proved by symbols alone" unless it is "an account of the connections between things and their causes."[58]

It is significant that a leading twentieth-century physicist, Wolfgang Pauli, whom Werner Heisenberg called a "night bird,"[59] should have written on the Kepler-Fludd controversy in a book coauthored with psychoanalyst Carl Jung. Pauli is most famous for his exclusion principle of the holistic correlation of the spins of particles in a system. The exclusion principle suggested the scientific respectability of acausal synchrony to the psychoanalyst in the latter's conception of what Kepler's biographer Arthur Koestler called "the roots of coincidence."[60] Pauli devoted his one lengthy historical-philosophical work to the Kepler-Fludd debate, and was struggling with the two sides of his own nature in writing this piece; indeed, he underwent psychoanalysis with Jung. (Pauli was the anonymous source of the diagrams, mandalas based on his own dreams, in Jung's *Psychoanalysis and Alchemy*.)[61]

Kepler's Struggle to Discover the Mathematical Structure of the Solar System

Central to Kepler's vast innovation in astronomy was his concern with the actual physical forces that accounted for the orbits of the planets. A great aid to Kepler's groping conceptualization of these forces was the analogy of gravity to magnetic force. The route along which Kepler struggled, however, from various elements of the ancient and medieval view of the world toward a modern one, involved a gradual transition from concepts of spirits and Aristotelian species to a notion of physical forces.

Astronomy before Kepler was concerned with the purely geometrical accounting for the successive positions of the planets. Any geometrical device that could account for these positions, so long as it was constructed of circles, was acceptable regardless of its consistency with any conceivable account of physical causes. This lack of concern with the fit between mathematical curve plotting and physical causes was evident throughout the period from shortly after Plato up through the work of Copernicus.

Within this nonphysical curve fitting, there were some considerable mathematical restraints, in the directive to account for the motions of the planets by perfectly uniform motion along perfectly circular orbits. This program was traditionally called "Plato's problem for saving the phenomena." Actually, this specific formulation traces back not to Plato but to a remark by Sosigenes, a Greek astronomer living in Roman Egypt during the second century C.E. Sosigenes claimed that Plato laid down the principle that all celestial motions must be accounted for in terms of uniform motion on perfect circles.[62] Later astronomers throughout late antiquity up through Copernicus, and even Galileo and Descartes, used only circles.

As motions became more complex, more circles were added to the system. These included circles rotating around a point on another circle: epicycles. There also were circles centered not on the Earth, but at a point slightly away from the Earth: the eccentric. Finally, there were motions that were uniform in velocity, not around the center of the circle, but with respect to a different point within from the circle: the equant.

The original model of the universe had the Earth at the exact center and the planets traveling in circles around the Earth. This was far too simple, and soon a model in which the Earth was still at the center but in which the planets traveled on spheres that rotated on different axes attached to the sphere just outside of them was introduced. This model was the system of homocentric spheres invented by Greek mathematician Eudoxus, who studied with the Egyptian priests and was a contemporary of Plato. The Eudoxian model was that upon which Aristotle based his physical cosmology. Curves, such as a figure-eight, can be generated by it; however, the mathematics of the Eudoxian model was soon found inadequate to account for the complex, observed motions of the planets. The succeeding model (using epicycles, eccentrics, and finally the equant) came to be universally accepted, through Ptolemy's great work known in the West as the *Almagest*.

The Ptolemaic model did a reasonably good job of saving the phenomena. Ptolemy (if the work concerning the theory of knowledge attributed to him is genuine) was himself philosophically a skeptic. He is credited with a treatise (not of great quality) on the uncertainty of all knowledge.[63] This fits well with Ptolemy's apparent attitude toward the role of astronomical knowledge, as not directly describing ultimate reality, but only saving the appearances. There were discrepancies between what Ptolemy's astronomy actually described and the standard, physical picture of the universe from Aristotle, which went along with it. In the Aristotelian physical universe, the Earth is at the dead center, and solid physical spheres rotate around the Earth. Yet, in the Ptolemaic model the epicycles carry the planets on smaller circles around and through the supposedly physical spheres, if they exist. This is physically unacceptable. Furthermore, the central notion of the whole world system—that the Earth is at the dead center of the universe—is spoiled by the alternative method of allowing some of the great circles of the planets to be centered at points away from the Earth. Thus, there came to be a discrepancy between the simply physical model of spheres traveling around the Earth and the complex mathematical model of circles mounted on circles or of circles not centered at the Earth.

In the Middle Ages, there were approaches now called mathematical astronomy and physical astronomy.[64] The mathematical astronomy did not treat its complicated wheels upon wheels as representing the actual physical structure of the universe, but treated them as convenient calculating devices that predicted the positions of the planets at different times. On the other hand, the physical astronomy gave a neat picture of perfect crystalline spheres uniformly rotating about the Earth. This picture was the basis for the philosophical and literary portrayal of the universe that we find in the theology of Saint Thomas Aquinas, and in Dante's epic poem *The Divine Comedy*. It was grossly inaccurate, however, as a description or prediction of the detailed motions and positions of the planets.

The modification in Ptolemaic astronomy that most disturbed Copernicus was, not as tradition has it, the progressive complication of added epicycles. (Girolamo Fracastoro, using the old Eudoxian model, had the largest number of circles, seventy-nine. Fracastoro's model, outdated at its creation, led to the myth that Ptolemaic astronomers were using eighty epicycles when Copernicus arrived. Fracastoro's circles were homocentric spheres not Ptolemaic epicycles, and no Ptolemaic astronomer had anywhere near that many circles.)[65] What bothered Copernicus most was the introduction of the equant. The equant destroyed the whole point of Plato's program to save the phenomena by using perfectly uniform motions. The equant made the motions with respect to the Earth *non*uniform. The motions were uniform only with respect to an arbitrary point away from the Earth.

Copernicus revolutionized our view of the universe by placing the Sun, rather than the Earth, at its center. In a brief passage, he relates this to the hermetic idea of the Sun as a "visible god."[66] Copernicus is very much part of the traditional astronomical program, and shifted the center to the Sun to preserve uniform circular motion. Another myth is that Copernicus eliminated the epicycles. His model is approximately of the same complexity as Ptolemy's.[67]

Copernicus did not much concern himself with the physical forces that accounted for the motions of the planets, but Copernicus's new model of the universe did destroy the traditional Aristotelian account for gravity. For Aristotle, bodies had "natural places." The natural place of earthly bodies was the center of the universe. The natural place of fire, for instance, was above the atmosphere of air. Thus, where unconstrained, rocks fall down and fire reaches up. In Aristotle's universe, the center corresponds with the center of the Earth. In the Copernican universe, however, the center is now the Sun. Thus, Aristotle's doctrine of natural place cannot be used to explain why rocks fall toward Earth. The Copernicans had to reject the doctrine of natural place, and they introduced the tendency of physical bodies to approach one another and cohere. Copernicus wrote: "Gravity is . . . a . . . natural appetition given to the parts of the earth by divine providence of the Architect of the Universe in order that they may be restored to their unity."[68]

The Copernicans could account for gravity on the individual planets and celestial bodies, but did not need an explanation for the circular motion of the planets themselves. Even Galileo appealed to the "natural" circular motion of the heavenly bodies to account for their behavior. Galileo never accepted Kepler's physical account of the noncircular orbits of the planets, and criticizes talk about a force of gravitation that causes the tides.[69]

It was not the early Copernicans who attempted to develop a physical model for gravitational attraction, but the astrologers. The Copernicans kept as much of the old Aristotelian system as possible. They considered gravity to be a tendency to move within each heavy body, not an attractive force. The astrologers, on the other hand, with their conception of the affinities between the heavenly bodies and earthly happenings, defended notions of natural attractions between the celestial bodies and Earth.[70]

On the issue of the tides, it was not the Copernicans but rather the occultists and astrologers who presented the doctrine that the tides depended on the attraction of the Moon. On doctrines that resemble correlative cosmologies, the watery nature of the Moon attracted the waters of the Earth.[71]

Galileo, as a good Copernican, argued that the ebb and flow of the tides was brought about purely by the Earth's rotation and rejected the notion that the Moon accounted for the tides. He argued this despite the fact that the rotation of the Earth does match the timing of the tides. For Galileo, gravity belonged to the realm of magic and astrology, as indeed it had.

Magnetism supplied a natural model for those who wished to describe gravitation as an attractive force. The magnet was easily observable and had long been an example of magical power. In the Middle Ages, the magnet and the *Echeneis* fish that supposedly could stop ships were traditional examples of magical powers.[72] Medieval Scholastic writers, such as the Averroës and Thomas Aquinas, described the action of the magnet as based on a "magnetic virtue" that is gained as iron ores alter their forms in the region of a magnet.[73] This explanation is similar to the Copernican account of gravity; however, the astrologers and physicians appealed to a sort of sympathy between like substances to account for the magnet. William Gilbert used this doctrine of sympathies in his account of magnetic "coition" over the more violent "attraction." Kepler eventually found in Gilbert the source for a model for gravitation.

Kepler's interest in the changing velocities of the planets was closely tied to his ultimately unsuccessful search for a correct force law. His notions of magnetic force in the end did not play a role in the actual mathematical formulations of what came to be known as his three laws; however, his general concern with the actual physical causes of the velocities and distances of the planets guided his concerns in the search for mathematical laws of planetary motion. It is ironic that the earliest detailed historical study of Kepler's discovery procedures totally leaves out the theory of magnetic forces that move the planets.[74] Devoted as the author was to chronicling Kepler's steps, this historian thought that the force hypotheses were irrelevant to the mathematical calculations (implicitly holding that equations and predictions are all that matter), thereby missing one of Kepler's great innovations—the introduction of physical forces into astronomy.

The theory of the magnet played a central role in Kepler's physical explanations of the solar system. The novelty of his work was precisely in giving physical, not purely mathematical, accounts of the motion of the planets.[75] Kepler shows he is aware of the novelty of his demand, as he subtitles his *New Astronomy* "An Astronomy Based on Causal Considerations." Ptolemaic mathematical astronomy did not concern itself with the real causes of motion, and physical hypotheses about the heavens were primarily about the crystalline spheres that supposedly carried the planets. When Ptolemaic astronomers added circles on the circles of the orbits of the planets, this led to complications for any physical hypothesis about physical spheres. It was possible to model an epicycle as a small sphere with the planet embedded, rolling around on the inside of a bigger sphere. It was also possible to model an eccentric as a sphere about the center of the universe with a circular tunnel centered on another point, but it was impossible to model an equant. Copernicus was able to use a device (probably borrowed surreptitiously from the Eastern astronomers of Maragha) that constructed an equivalent epicyclical model. Mathematical astronomers using the model of epicycles with epicycles crossing through other spheres simply ignored physical considerations.

Tycho, on the basis of his finding that comets were not atmospheric phenomena but bodies that traveled from deep space across the orbits of the planets and back again, rejected the existence of the crystalline spheres. Kepler jokingly writes that "the hypothesis appears, so to speak, to collapse of its own weight."[76] The obvious next question was, If there were no crystalline spheres to hold the planets, what held them in position? Tycho, despite his monumental step of discarding the spheres, was not bothered by this question, but Kepler was.

Besides the question of what held the planets in place there was the question of what made the planets move. The early mythological explanation of the Babylonians and others was the gods—or that planets themselves were gods. Aristotle had a less personalized explanation: the unmoved mover, an impersonal god, of "thought thinking itself" was the object of desire of the planets that strive for the unmoved mover but are constrained in their spheres, thus traveling in circles. Some medieval explanations appealed to individual planetary movers. Kepler alternated between explanations in terms of planetary minds and explanation in terms of physical forces, but he leaned toward the latter.

Kepler wanted accurate predictions, but he wanted more. He wanted physical explanation of what most previous astronomers took for granted. Why were there six planets? What explained the distances of the planets from the sun? The early hypothesis concerning the regular solids that so excited Kepler was a delusion, but it shows that Kepler from the start wanted rigorous explanations for the geometrical structure of the solar system; he did not want merely predictive calculating devices.

Neither Ptolemy nor Copernicus, for all the world-shattering dispute about whether the Earth or Sun was center of the universe, really was concerned with physical explanation. Copernicus sought a mathematical description that was as simple and elegant as possible, which led him to place the Sun at the center. It is Kepler, not Copernicus, who makes a transition to a physics of the heavens. "The Copernican disturbance" contrasts with the "Keplerian revolution."[77] Historians have come to realize that Copernicus is very much in the tradition of ancient astronomy. It is Kepler who, hesitatingly and confusedly, gropes forward to modern *physical* astronomy, despite his being immersed in the worldview of Pythagorean mathematics worship, astrology and speculation about the worldsoul—indeed, precisely *because* he is immersed in that worldview.

Kepler's Use of Souls: From Spirits to Computing Devices

Kepler's thought spans two worlds. It harks back, much more than that of Galileo or Descartes, to medieval concepts, such as astrology and Neoplatonic harmonies, yet leads toward physical ideas more modern than Galileo and Descartes with respect to elliptical orbits and physical forces. Koestler's term for Kepler's life—"the watershed"— is very apt.

In his youth Kepler carefully studied the work of Julius Scaliger, who followed Arabic philosopher Averroës in having intelligences move the heavenly bodies, a doctrine that traces back to the ancient Stoics.[78] Even in his juvenile *Mystery of the Universe*, in which Kepler had souls moving the planets, he was interested in the causes for the speeds of the planets, and wished to quantify the activity of those souls. Kepler formulated the hypotheses that "either the souls which move the planets are the less active the farther the planet is removed from the sun, or there exists only one moving soul in the center of all the orbits, that is the sun which drives the planet the more vigorously the closer the planet is."[79]

A footnote added to the second edition (1623) of *The Mystery of the Universe* encapsulates the transition from souls to forces in a single sentence: "If the word soul (*anima*) be replaced by force (*vis*), we have the very principle on which the celestial physics [of *The New Astronomy*] is based." The transition from animism to mechanism, from the living universe to the mechanical universe, is here in a nutshell.[80]

Kepler's path from souls to forces was not a straight and easy one. Even in his mature work, *The New Astronomy*, Kepler supplements his physics of magnetic force with an account in terms of planetary minds. Now these souls have no free will. They are purely cognitive beings who think in purely mathematical terms, making mathematical calculations based on their positions and observations concerning the position and diameter of the Sun.[81] They are more like cybernetic devices than like traditional celestial souls. Kepler formulates precisely what calculations they would need to perform and what celestial features it would be possible for them to perceive. He does not make the usual hand-waving explanation that "mind does it"; he works out the details of exactly what such a planetary mind would have to do.

Kepler notes that a planetary perceiver could not use the empty point at the center of an epicycle or the eccentric point to guide the planet, as there would be no material body there to perceive. The planetary mind has to perform all its calculations based on the perceived size of the Sun. This principle, that the planetary mind must function solely in terms of the Sun and not in terms of abstract geometrical points it could not perceptually locate, eliminates several accounts of the path of the planet.

Kepler argues that the epicyclical model does not make physical sense, either in terms of physical forces or in terms of minds.[82] It does not make physical sense because force could not issue from an immaterial, empty point. It makes no psychological sense because the body cannot use geometrical imagination to guide movements about the unoccupied point at the center of the epicycle. We humans would need paper and pen, or at least memory of such guidance.[83]

Of course one might object that ordinary considerations of embodied human minds or of forces exerted by spirits might not be governed by ordinary restrictions of the sort Kepler considers. But he replies: "Those sublime considerations of the . . . operations of the blessed angels . . . are irrelevant. For we are arguing about natural things that are far inferior in dignity, about powers not endowed with a will to choose how to vary their action, and about minds that are . . . yoked and bound to the celestial bodies that they are to bear."[84]

Later, considering the difficulties of the magnetic-force hypothesis, Kepler considers and compares the mental hypothesis. He uses the consideration of what physical attributes a planetary intelligence could perceive to guide its mathematical calculations, and notes that the increase of the Sun's diameter could be perceived.

The planetary mind senses the tension between direction of the striving of the planet and the natural attraction in the direction of the Sun. The force to be overcome by the planetary striving would vary as the sine of the angle. The mind would sense this force and seek to regulate the perceived size of the Sun by varying its distance from the Sun.[85]

Kepler then makes a systematic comparison of the planetary-mind hypothesis with the force hypothesis. Kepler writes that his reader may not have so firm belief as does he in "this perceptive cognition . . . , which I so easily accept, and bestow upon the planetary mind."[86] The purely magnetic hypothesis has the difficulty that the planet must be oriented in a constant direction throughout its travels. The Earth's axis of rotation remains constant but does not point in the direction that his theory needs. Nevertheless, his arguments support the force hypothesis on grounds of simplicity, since the mental hypothesis must also make use of physical forces as a basis for its estimates.[87]

Kepler's Magnetic Model

Kepler, borrowing from Gilbert, also developed force models based on magnetism. In a letter Kepler claimed that his program was to explain the universe purely mechanically. "My aim in this is to show the celestial machine not to be divine organism but rather to be a clockwork . . . , all the . . . movements are carried out by a single, . . . magnetic force, as in . . . a clockwork all motions by a single weight."[88] In *The New Astronomy* Kepler claimed that the motions of the planets were accounted for by a force like magnetism, but for the Moon developed a theory of gravitational attraction. Kepler claimed that the Earth and Moon exerted mutual attraction. This accounts for the tides and for the motions of the Moon. The attraction depends on size, and the Earth contributes a great deal more than the Moon, using the model of two magnets of unequal sizes.

When Kepler discussed the motion of the planets in relation to the Sun, he dropped the mutual-attraction model he used for the relationship between Earth and Moon. In the case of the Sun and the planets, the Sun emits a rotating ray of gravitation, or magnetism, that sweeps the planets around as would a paddle. The Sun's beam of magnetism rotates like the beam of a lighthouse (in terms of twentieth-century cosmology, like the beam of a pulsar). The beam weakens with distance that accounts for the slower movements of the outer planets relative to the inner planets.

Kepler introduced several ways of discussing the nature of the magnetic rays and their effect that are not consistent with one another. In one conceptualization, they function like elastic bands. He also described them as producing a circular vortex motion; the planets, because of their inertia, slow down because of vortical motion.

This can be considered one of the sources of the later vortex theory of planetary motion, developed by Descartes and Leibniz and their followers, a theory in which gravitational force was replaced by the whirlpool motions of the fine material particles of the ether. This material-vortex theory became the major competitor to Newton's theory of action at a distance. Kepler's analogy between gravity and magnetism pointed more in the direction of the force theory than toward later vortex theory.

Kepler correctly located the motive power of the solar system in the Sun, now in the center.[89] He used the analogy of an orator who turns around, glancing at everyone in the audience surrounding him.[90] The Sun rotates, and with it rotate the magnetic filaments, which are purely magnetic-force-like, but are imagined on the pattern of iron filings about a magnet. (In this respect they are similar to Faraday's lines of force, and this aspect of Kepler's theory of the moving forces of the solar system is more like a field theory than like Newton's action at a distance.)

Kepler believed that the magnetlike force that drives the planets involves a rotating Sun. Force here does not attract but pushes. Lines of circulation of the force circle the Sun in parallel with its equator. Kepler believed that this explains why the planets are in the same plane (of the zodiac or ecliptic). There is no force pushing outward from the poles of the Sun, as the magnetic filaments rotate with the Sun.[91] Rotation is necessary for the carrying of the lesser bodies in their orbits around the central body. Earth, Jupiter, and Saturn must rotate because they have satellites. (Saturn's blurredly observed rings were treated by Kepler as satellites.)[92] Since the planes of the circles of the magnetic filaments are perpendicular to the axis of rotation of the Sun, Kepler argued, the planets must all rotate in the single plane.

One of Kepler's magnetic physical-force models explains why a planet has a noncircular orbit that is closest to the Sun at one point (the perihelion) and furthest away at another (the aphelion). It is claimed that the planet has a constant direction of magnetic orientation that remains parallel to previous orientations as the planet travels around its orbit. The planet has two poles, one of which is attracted toward the Sun and the other of which is repelled. As the planet travels one side of the path with its pole attracted to the Sun it nears the Sun, and as it travels on the side of its orbit with its pole repelled by the Sun the planet retreats.

Kepler later added *inertia* (a term he coined) of the planet as another factor. He presented a figure with an oar in a swirling river. The sailor rotates the oar once as the planet travels around twice. As the river flows counterclockwise, the oar is pushed closer on the left side of the orbit and further away on the right side.[93] This model is a vortex model of motion, with its flows.[94]

Kepler was aware of the inverse-square law in his work on optics and realized that the intensity of light varies as the inverse square of the distance from its source. The whole tradition of the Neoplatonic metaphysics of light that can be traced down through the medieval, Augustinian, and Franciscan Scholastics supplies a powerful source of imagery for the understanding of forces acting on distant objects.[95] Kepler was aware, however, that light itself could not be the source or bearer of the Sun's force. If gravitation acted like light and was blocked by objects, leaving a shadow, a planet eclipsed by another planet between it and the Sun would stop, or at least slow down.

Kepler claimed that light is a material thing, or at least "substantial." At one point he described light as operating through the effect of place or location. Light "cannot be regarded as something that expands into the space between its source and the movable body, but as something the movable body received out of the space that it occupies."[96]

Even though Kepler initially toyed with an inverse-square-force law, he dropped it in favor of a simple, inverse-distance law. This erroneous move was due both to his innovative interest in the causes of speed of the planets and his inadequate understanding of inertia. Kepler had an Aristotelian conception of inertia as a kind of laziness of bodies, a tendency to slow down. Galileo and Descartes had a different conception of inertia, as a tendency to stay in the same state of motion whether at rest or at a constant speed (later Newton's first law), but neither Galileo nor Descartes ever accepted Kepler's notion of elliptical orbits.

Kepler recognized early that the planets slowed down in those portions of their orbits further from the Sun. He also believed that force was proportional to velocity rather than to acceleration (here closer to Descartes than to Newton). From this, Kepler inferred that the force was inversely proportional to distance, rather than to the square of the distance.

A puzzle of Kepler's theorizing is that the theory of gravity in the Introduction to *The New Astronomy* seems more advanced and sophisticated than the magnetic theory used later for detailed calculations concerning planetary motion. The theory of the Introduction sounds like Newton's gravitational theory, but the theory presented later uses the analogy with light and involves magnetic force decreasing as the inverse of the distance rather than the inverse square.

One explanation of this discrepancy is that Kepler's qualitative speculations were sounder than the detailed hypotheses constrained by other considerations in his calculations.[97] One reason for Kepler's inverse law was that it fit with his wrong distance law; but in combination with wrong mathematical techniques, it led him to the correct elliptical shape of the planetary orbits.

Another explanation of the difference between the Introduction and the body of the book is that the former discusses the attraction between Earth and the Moon, or between Earth and bodies made of "earthy" stuff.[98] The theory later in the book deals with the relation between the Sun and a planet, two bodies made of different sorts of material. (Kepler denied fire was an element, considering it an active principle.)[99] Thus "kindred" corporeal substances showed a sort of mutual attraction not shown between Sun and planet.[100] Kepler claimed that the attraction of gravitation is proportional to the bulk of the attracting body. The Earth attracts the Moon more strongly than the Moon attracts the Earth—like two different-size magnets.[101] Thus, "if the earth should cease to attract its waters to itself, all the sea water would be lifted up, and would flow onto the body of the moon."[102]

When Kepler developed the physical-force account of planetary motion later in the work, he did not use the notion of gravitational attraction, but still depended on William Gilbert's notions of magnetism. Here the magnetic force is not attractive, but pushes the planets around their orbits. This shows the tremendous power of the circular metaphor over Kepler, even after he rejected circles for ellipses. Kepler could not conceive that the natural motion of the planets might not be along closed curves but in straight lines.[103]

The explanation to the puzzle of why Kepler's gravity in the Introduction seems so modern while his account in the body of the work is so different is that Kepler did not consider moving forces to be gravity. His theory of moving forces is often called theory of gravitation because those forces do the explanatory work that was later done by Newton's gravitation. But Kepler considered *his* gravity to explain why the Earth stuck together, now that the Earth was not in the center of the universe.[104] What Kepler called gravity did not play a role in the account of planets' revolutions. The latter was explained by a different force, also analogous to one found in the magnet, but not involving attraction. Kepler followed Gilbert in distinguishing *two* spheres of action of the magnet: a smaller sphere (that of *coition*) in which the magnet attracts bodies, and a much larger sphere (of "direction") in which the magnet orients compass needles or iron filings.[105] Gilbert experimentally observed the attraction of magnets to each other at close range. He was unaware of the very weak magnetic attraction at great distances but was aware that compass needles pointed north all over the Earth (or at least along the north-south line, as the Chinese considered the south end of the needle to be the pointer), and so considered this directive capacity to be unlimited in range. Following Gilbert, Kepler restricted gravitational attraction to the near vicinity of the Earth, pulling stones to Earth and attracting the Moon. The larger sphere produces circular lines of force that orient or carry other bodies in a closed curve around the central body. This is the means for the Sun to carry the planets in their orbits. Later, in *The Dream*, he speculates that the Sun's attractive capacity may reach all the way to the Earth;[106] however, only with respect to minor tidal effects and not with respect to the motion of the Earth.

In *The Epitome of Copernican Astronomy*, a late summary of his discoveries, Kepler describes his astronomy as made of three components: Copernicus's astronomical theory, Tycho's observational data, and William Gilbert's philosophy of magnetism.[107] Elsewhere (in letters) he describes his mathematics "of computing not from circles but from natural faculties and from magnet properties,"[108] and "the hypothesis is physical because it uses the example of the magnet."[109]

Kepler's physical hypotheses, based on magnetic analogies, are often crude and mutually inconsistent, yet they are what guide him in overthrowing the ancient astronomical tradition still followed by Copernicus.

Accustomed as we are to Newton's gravitation, we tend to dismiss Kepler's force models as simply wrong; however, it was Kepler's laws that supplied a test for Newton of the power of his inverse-square law. Kepler's use of analogies with experimentally accessible magnets built the bridge to later theories of gravitation. The followers of Gilbert's magnetic philosophy in England incorporated Kepler's speculations concerning magnetlike forces in the cosmos in their theories, which paved the way for Newton's gravitation.[110]

The Discovery of Kepler's Laws

Kepler is highly unusual among scientists of previous centuries in that he left extensive accounts of the path to his discoveries in his texts. He described each step he made, whether correct or incorrect, as well as his feelings of triumph, discouragement, and self-disparagement.

The ancient Greek geometers left only their finished proofs but no account of how they found them, leading later writers to believe that the Greeks had a secret method of discovery. Newton presented his works in terms of laws and their deductive consequences, and covered the tracks of his discoveries. Indeed, Newton presented his major work on mechanics, *Principia*, using older geometrical techniques and without divulging the calculus he had actually used. From Newton's voluminous, obsessive notes we can discover what he read, but we do not have any account of his actual process of discovery.

Modern mathematical and scientific research papers have followed the model of the Greeks and Newton. They present their finished conclusions and results without any explanation of the discovery process. This has led one scientist to call contemporary research papers "fraudulent,"[111] because they "not merely conceal but actively misrepresent the reasoning that goes into the work they describe."[112] It has only been within the last few decades that scientists themselves have begun to become more willing to discuss their processes of creation and discovery. Perhaps Kepler's own intuitive, concrete approach to science is congenial to an emphasis on the method of discovery as opposed to the finished formal deduction. His chronicles of his thought processes make Kepler a favorite subject for philosophers of science concerned with discovery.[113]

Kepler started his assault on the motion of Mars with a recognition that the motion of the planet was nonuniform. He still accepted the circular orbit, but recognized that the speed of the planet varied. He struggled to reconstruct the details of Mars's motion based on the circular orbit and to fit Tycho's data. He first used an eccentric circle and got results extremely accurate for his time, much better than those which satisfied Ptolemy and Copernicus. He recognized a discrepancy of eight minutes of arc, less than one-eighth of a degree, and his demand for precision was so great that this discrepancy led him to reject his theory.

Kepler's attitude is characteristic of that of modern science, and contrasts with that of Ptolemy. Ptolemy often corrected his observations and reported observations to degrees of accuracy that would have been impossible to obtain with his crude instruments. Ptolemy's nonchalant attitude toward observational data led one modern writer to call him a fraud,[114] a highly misleading accusation, since it overlooks the different attitude toward observational data of ancient astronomy in contrast to modern.[115] Some claim that the ancients did not really "test" their theories against observational data or perform experiments in the modern sense.[116]

Kepler was driven by the relatively minor discrepancies initially to question the method of areas that he was using, and they finally led him to give up the circular orbit. Kepler's willingness to throw out the result of years of arduous calculations, and then renounce the assumption of two millennia of astronomical tradition, was based on his concern with physical causes. The concept of circularity had been as tied to the notion of planet as tangibility is to the notion of physical object.[117] The reverberations of the shift initiated by Kepler extended into popular thought. Marjorie Nicolson in *The Breaking of the Circle*[118] has traced the influence of this change on the poetry of the seventeenth century, most famously in John Donne's.

And new philosophy calls all in doubt,
The element of fire is quite put out;
The Sun is lost, and the earth, and no man's wit
Can well direct him where to look for it. . . . .
'Tis all in pieces, all coherence gone.[119]

Art historians have contrasted Baroque and Renaissance art in terms of the ellipse versus the circle.[120]

Kepler's path from the circle to the ellipse, however, was not direct. The first noncircular curve that Kepler used for the planetary orbits was the ovoid, or egg-shaped, curve rather than the ellipse. Kepler actually used ellipses in his mathematical calculations as approximations to the ovoid, but did not yet think of actually using elliptical orbits.

One reason for this that has been suggested is that the ovoid curve has only one focus, or center, like the circle and unlike the ellipse, which has two foci, with the Sun located at one.[121] Another more metaphysical reason may have been involved. Often associated with Neo-Pythagoreanism and the Greek mystery religions, and with the musical tradition in which Kepler was intensely interested, was Orphism, the Greek cult of Orpheus with his lyre, who traveled to the underworld and returned. In Orphism the universe was a cosmic egg. (Orphic accounts of the origin of the universe as a sort of bubble in a cosmic fluid that then rapidly inflated have a resemblance to recent scenarios for the origin of the universe out of a quantum fluctuation in the vacuum.) It is possible that the egg and its Orphic symbolism were more aesthetically and religiously satisfying to Kepler than the ellipse.[122]

Kepler struggled to find the area of the ovoid curve. (The formula for exact area was not discovered until 1960).[123] He attempted to approximate it with the ellipse, stumbling upon a numerical correspondence. The width of the sickle-shaped areas that lay between the circle and the fattened oval of the orbit was .00429 of the radius of the circle. He found that the result of "optical equation" (the secant of the angle between the Sun and the orbit's center) was also .00429. The identity of these two numbers gave him the key of how the planet's distance from the Sun varied with its position. He quoted Virgil on nature as a lascivious girl unwilling to deliver herself to him.[124]

> Galatea seeks me mischievously, the lusty wench,
> She flees to the willows, but hopes I'll see her first.[125]

He attempted to develop a curve that fitted his equation. One he constructed, based on an error he made, was bulgy, or "cheeky," and was not an ellipse. He next attempted to construct an elliptical curve, not knowing that his original curve was an ellipse. He battled to understand why his orbit of Mars did not have the librations that the geometrical construction needed.

After many struggles using ellipses to approximate the egg-shaped orbit, and using epicyclical librations to correct the circle, it finally struck Kepler that the ellipse itself could be taken as the *actual* orbit. In a famous passage showing his endearing psychological honesty, Kepler writes: "What need is there for many words? The very truth, and the nature of things, though repudiated, . . . sneaked in again through the back door, . . . I rejected the reciprocation [hypothesis] . . . and began by recalling the ellipses, . . . although they coincide exactly. . . . Despite my . . . searching about almost to the point of insanity, I could not discover why the planet, . . . would rather follow an elliptical path. . . . O ridiculous me!"[126]

Was Kepler a Mystic?

Marie Boas writes, "Of all the astronomers of the post-Copernican period, the most difficult to appraise and appreciate is Johannes Kepler.... Mystic and rational, mathematical and quasi-empirical, he constantly transformed apparently metaphysical nonsense into astronomical relations of the utmost importance and originality."[127] Kepler says, "Whoever wants to nourish his mind on the mystical philosophy... will not find in my book what he is looking for."[128] Those who wish to deny that Kepler is a mystic can take comfort from this quotation.

Many historians call Kepler a number mystic or mathematical mystic.[129] Others, who were concerned to defend Kepler as a scientist, have denied that he was a mystic. Edward Strong began his investigation of early modern science expecting to find Platonism the framework that led to the origins of science, but was so horrified by the sort of magical, Neoplatonism believed in by Renaissance writers that he then denied that Platonism was influential on science and argued that technical procedures led to the rise of science.[130]

Bruce Stephenson, a physicist-historian who studies the detailed mathematical arguments of Kepler's physical cosmology, several times contrasts his account with those of unnamed antagonists who call Kepler a "numerically inclined mystic."[131] He nowhere characterizes the notion of "mystic" except in opposition to reason although granting that "mystical" is "a much-abused word." He thinks that showing that Kepler's theories, including the astrological ones, were rational shows that Kepler was not a mystic. He associates mysticism with mystical experience and points out that Kepler's accounts of dreams and visions, in *The Dream* and *The Harmonies of the World*, are conscious fictions, thinking this absolves Kepler of accusations of mysticism.[132] Others wish to split Kepler in two: the empirical scientist and the mystic, but modern historians realize that this presentation of two Keplers does not do justice to the unity of Kepler's thought.[133]

In attempting to assess whether Kepler was a mystic, we should examine the meanings of the word *mystical*. One of the core meanings has to do with mystical experience, generally one of unity with God, the Universe, or Nothingness. This experience is often characterized as ineffable and not literally describable. Another dictionary meaning is "obscure or unintelligible." This comes from modern criticism and debunking of mysticism and of religion in general, leading some mistakenly to associate *mysticism* with *misty*.

Insofar as he attempted to produce clear and understandable mathematical demonstrations and to fit his results to observed data, one cannot claim that Kepler was a mystic in the derogatory sense of pursuing incomprehensibility. Kepler believed that statements of the harmonies are mathematical expressions understandable by any rational mind. Kepler never claimed to be communicating to a special group of initiates in an esoteric language. His results were meant to be public, and in this respect he is part of the shift from the esoteric nature of occultism to the public nature of modern science.

There is a third characterization of mysticism as based on intuition. In this sense, some of Kepler's discoveries can be called mystical. Kepler himself often celebrates his flashes of intuitive insight, especially in *The New Astronomy*, where he discusses his method of discovery, his false steps, and his triumphant successes. More than any modern scientist, and in strongest contrast to Newton, Kepler reveals his thought processes. Insofar as scientists have such flashes of intuition, it is misleading to call all such creative minds mystical; however, insofar as Kepler, more than any other great scientist, celebrates these instances of insight, he can be considered a mystic. That is, if the emphasis in a theory of knowledge is on acts of intuitive insight, such a theory is often called mystical, especially if the acts of intuition are connected to God and religion. Kepler often wrote about how his insights are gifts of God.[134] Unlike some devotees of intuitive insight and imagination, however, Kepler tested such insights against observation and rejected them when they fail.

The content of his theories is geometrical, and Kepler claimed that geometry is a part of God. He certainly believed that ultimate reality is mathematical. This has led many to call him a number mystic. There are two senses, however, in which this is misleading. First, for Kepler abstract numbers were of no great significance. He believed that genuine numbers are magnitudes and have physical dimensions. Such numbers are measurements tied to geometrical figures. For this reason, J. V. Field denies that Kepler is a number mystic.[135] Kepler is still a mathematical mystic, however, a geometry mystic if you will, because geometry rather than pure numbers is the ultimate basis of intuitions of physical structure. To deny that Kepler is a number mystic in this sense does not ally him with antimystical empiricists or skeptics.

Second is the contrast between mathematical mysticism and mathematical science.[136] It is certainly true that Kepler was concerned with genuine mathematical computation and argument, not simply with the contemplation of numbers of geometrical figures in isolation. He also wished to compare and correct his mathematical conceptions with nature. In this sense, Kepler certainly can be strongly contrasted with those who simply meditate on the mystical significance of individual numbers or shapes and associate them with particular qualities.

The incident that shows up the contrast between Kepler and the number mystics of a more traditional sort who were influential in the Renaissance is his polemical exchange with Robert Fludd. Kepler dryly lists the differences between the two. Fludd appeals to numbers for their symbolic and personal associations; he emphasizes pure numbers, such as two in itself or three in itself. Kepler treats numbers as physical dimensions or measurements. Fludd does not engage in technical, mathematical calculation, nor is he at all seriously concerned with comparing his numerical speculations with empirical observations. He simply *asserts* that the traditional Ptolemaic astrology is superior to the Copernican system in accuracy.

Certainly one can portray this confrontation as one between modern empirical science and occult obscurantism; however, Kepler clearly held beliefs close to those of the mathematical mystics despite the fact that he engaged in careful computations and empirical refutations. He certainly believed that aesthetic contemplation of the mathematical harmonies in geometrical form gives a kind of deep insight into nature. To him such geometrical contemplation was akin to the contemplation of God since geometry is in the mind of God. "To contemplate these Axioms . . . is sublime and Platonic and resembles Christian faith concerning Meta-physics and the doctrine of the soul . . . Geometry, coeternal with God and shin-ing in the divine Mind."[137]

Kepler certainly had ecstatic experiences of insight and celebrated them, as in his famous claim to have stolen the golden vessels from the temple of the Egyp-tians, for which the Creator has waited six thousand years for recognition. Simi-larly, "I certainly know that I owe it [the Copernican theory] this duty, that as I have attested it as true in my deepest soul with incredible and ravishing delight,"[138] and, speaking of his regular polyhedron hypothesis concerning the planetary dis-tances, "The intense pleasure I found in this discovery can never be put into words."[139] Such psychological "peak experiences" (to use Abraham Maslow's term) of insight and discovery might not have counted as mystical in Kepler's day, but do today to many personality psychologists.[140]

CHAPTER 10

Newton: Alchemy and Active Principles

$\mathcal{N}$ewton was deeply immersed in the study and practice of alchemy during the period of his creative scientific work. Newton's most noteworthy work on alchemy was precisely during the period of his devising, interpretation, and attempted extension of the program of mechanics in his greatest work, the *Principia*. Newton ceased to study alchemy and also stopped producing any truly novel and creative scientific work shortly after his mental breakdown and his move to London. The concepts of active spirits in nature described in alchemy are the source of Newton's concepts of force and of other active principles in the universe. It is primarily these concepts which distinguish Newton's physics and natural philosophy from that of his predecessors, such as Hobbes and Descartes and their followers.

Newton and the Magi

Newton wrote at least 650,000 and perhaps as many as 1,200,000 words on alchemy.[1] The sheer volume of this production as well as the amount of time Newton spent on chemical experiments related to the alchemical works that he read shows that Newton's alchemy cannot be dismissed as a minor foible of old age or a hobby, as some once attempted to do. The existence of this huge mass of alchemical writings by Newton was long an embarrassment to those who, for at least two centuries, wished to present Newton as the ideal of the modern mechanical and empirical scientist. In 1727, when Thomas Pellet reviewed Newton's papers for the family after Newton's death, he scrawled "not fit to be printed" on a package of Newton's alchemical manuscripts.[2] David Brewster, who wrote the major, worshipful nineteenth-century biography of Newton, admitted the existence of this material, but bemoaned it, writing "We cannot understand how a mind of such power . . . could stoop to be even a copyist of the most contemptible alchemical poetry and the annotator of a work, the obvious production of a fool and a knave."[3]

The economist John Maynard Keynes, after the dispersal of Newton's manuscripts at auction, reassembled about half and wrote a brief lecture in 1946 entitled "Newton the Man," which discussed the strange involvement of Newton with alchemy. Keynes, who questioned the eighteenth-century idea of the harmonious, self-regulating market, also undermined the Enlightenment view of Newton the physicist. "Newton was not the first of the age of reason. He was the last of the magicians, the last of the Babylonians and Sumerians, the last great mind which looked out on the visible and intellectual world with the same eyes as those who began to build our intellectual inheritance rather less than 10,000 years ago."[4] Although Keynes called Newton "a magician," he did not regard Newton's alchemy as part of his science, calling it "wholly magical and wholly devoid of scientific value."[5]

Despite Keynes's having let the alchemical cat out of the physicist's bag, orthodox historians of science continued to downplay or dismiss Newton's alchemy through the early 1960s. In 1958, A. Rupert Hall and Marie Boas wrote an article on Newton's alchemy in which they claimed that Newton's only interest in alchemical literature was to extract purely scientific chemical and metallurgical principles.[6] This was in part motivated by his interest in telescope mirrors, but when Newton describes the star-shaped crystal called "the star Regulus" by alchemists, and notes that it might be useful in the making of telescope mirrors, one suspects correlative cosmology is an influencing factor.

During the late 1960s and early 1970s Piyo Rattansi and Richard Westfall, among others, made claims as to the importance of Newton's alchemical work and incorporation of ideas from it on his theory of forces. In the 1970s Betty Jo Teeter Dobbs[7] and Karin Figala[8] initiated major, detailed research on Newton's alchemical writings. It is perhaps significant especially with regard to Keller's claims about the alternative presentation of gender relations in alchemy that both are women.

Some 10 percent of Newton's library consisted of magical works and some 20 percent scientific works. Over half of the scientific works were chemical and alchemical in nature. Newton tackled alchemy with the same intellectual concentration and systematic organization with which he tackled both science and theology. He made himself a dictionary of chemical and alchemical terms, and it contained around five thousand references to some one hundred fifty alchemical books. Dobbs notes that never before or since was the traditional alchemical literature studied with such thoroughness.[9] Newton brought to it an experimental accuracy that none of the traditional alchemists possessed. He also brought to the alchemical literature a combination of sympathetic acceptance and scientific seriousness that none of the more recent historical scholars of alchemy has had. Most of the later historians sympathetic to the alchemical worldview treat it psychologically or humanistically and are unable or unwilling to check scientifically its chemical result the way Newton did.[10]

Newton also applied his extraordinary powers of mental concentration and physical energy to the experimental replication and testing of alchemical claims. He worked in his laboratory for weeks on end, alternating all night vigils with his lone assistant and sleeping only on alternate nights. While Newton certainly attempted to extract experimental results from the alchemical treatises, it would be misleading to claim that he was merely extracting the empirical scientific gold from an otherwise incoherent alchemical dross. As a master of experimental science he concentrated on the chemical experiments that could be extracted from the metaphorical writings. Beyond this he also took the traditional alchemical symbolism and doctrines more seriously than did many of his scientific contemporaries.

His notebooks involve the translation of the rich alchemical imagery into a chemical language, but he also seems to take quite seriously the connotations of the traditional alchemical symbolism and uses it to describe his own experiments and their significance. Terms such as the *green lion*, the *oak*, *Mars*, *Venus*, *Saturn*, the *net*, and the *star Regulus* are translated into talk about chemical substances and physical structure. He was fascinated by Vulcan's golden net in which he caught Venus and Mars. Newton found a metallic, netlike crystal that captured the chemicals corresponding to Venus and Mars. Similarity, he grew star-shaped crystals that alchemists had identified with various stars in the heavens.

Newton believed in the alchemical claims to secret knowledge and the need to keep secret their deeper doctrines. For instance, he wrote to chemist Robert Boyle to remind him not to make public some of the more profound of the alchemical secrets. Boyle, rather than Newton, is a good example of someone who was concerned with alchemy only to the extent to which he could translate it into empirical chemistry and was less interested with the symbolism and worldview of traditional alchemy. Despite this, Boyle, Newton, and Locke were collaborators in alchemical investigations.

In contrast to Boyle, Newton practiced alchemy in relative secrecy. His laboratory was in a shed, probably against the side of a chapel at Cambridge University. Newton never openly published an alchemical treatise; there are only scattered remarks in his published writings in which we can discern alchemical ideas and terminology. When he moved to London to become a public servant (Warden of the Mint), he ceased his alchemical activities, which would have been even more suspect there than at the university.[11] It is perhaps significant that in his role at the mint, Newton acted as a detective to hunt down, punish, and often have executed various counterfeiters, who were often former alchemists or their associates.[12] Frank Manuel has speculated in a general psychoanalytical form on motives for Newton's aggressive action against counterfeiters in terms of psychic revenge on his stepfather,[13] but the alchemist-counterfeiter connection may offer more specificity regarding concealment of a suspect past and put these motives in an institutional context.

Newton granted the claims of the alchemists to ancient wisdom. His view of history was that of a continuous tradition of wisdom, stretching back to the ancient Hebrews, Greeks, and Phoenicians, which could be recovered, in part through empirical research. Newton accepted the Neoplatonists' view that Moses was to be identified with atomist Moschus the Phoenician.[14] He also believed that Pythagoras had known the law of gravitation, and that alchemy stemmed from Hermes, or Thoth, that the works of more recent alchemists carried down in secret language this ancient tradition, and that this profound knowledge should be kept secret from the vulgar and transmitted only to initiates. In this respect, and probably reinforced by his own somewhat secret and paranoid personality, Newton accepted the esoteric and secretive approach to knowledge of the alchemists that stood in strong contrast to the notions of publicity and publication that were fostered by the Royal Society of which he became the head.

Newton and Cambridge Platonism

Besides alchemy, the other major contender for status as the source of Newton's rejection of contact-action mechanism and Descartes's philosophy were the Cambridge Platonists, a school of philosophers and divines at Cambridge University in the mid and late 1600s. Henry More, one of the most important, was personally acquainted with Newton and was the tutor of the apothecary who had taught Newton in his youth. Newton had read one of More's books before he became an undergraduate and absorbed some of More's philosophical views as well as the Cambridge Platonist religious position, latitudinarianism.

Among the more philosophically oriented of the Cambridge Platonists was Ralph Cudworth, who was opposed to the austere predestinarian religion of Calvinism that dominated the college and opposed the materialist philosophy of Thomas Hobbes. He rejected the theology of John Calvin and the philosophy of Hobbes for being extremely deterministic. For different reasons, they denied an ultimately humanly comprehensible spiritual direction to the universe. In his *True Intellectual System of the Universe*, Cudworth attacks the atomistic materialism of Hobbes and what he conceives of as the hylozoistic materialism of the soon-to-be published *Ethics* of Spinoza. Both *Hobbism* and *Spinozism* were the scare words for politically radical materialism during the succeeding century, the way *Marxism* or *Communism* have been in the twentieth century.

Cudworth criticizes materialism by drawing out the implications of atomic explanation and attempting to show that it cannot account for life or mind. He replaces the dualism of matter and mind with a polarity of active and passive. He not only rejects Descartes's dualism but denies the identification of spiritual activity with consciousness. Cudworth appeals to "plastic natures" as a kind of vital spirit that shapes matter and accounts for living things. His theory was used by naturalists and biologists, including John Ray, who was a later acquaintance of Newton, and by vitalistic materialists, such as Denis Diderot in the next century. It also is a source for nineteenth-century concept of the unconscious mind in psychoanalysis via the French psychologist Pierre Janet.[15] Cudworth's work was the primary reference source for discussions of the Greek pre-Socratic philosophers up to the late nineteenth century. He believed in the "Ancient Theology," wherein the earliest peoples possessed all wisdom, which was transferred to the present in an incomplete and simplified form. Cudworth, following Ficino and the Neoplatonists of Florence, claimed that Moschus the Phoenician, and another figure, Moschu, were the same as Moses of the Bible. Supposedly Moses-Moschus, in earliest times held the doctrine of atomism. Newton took over this doctrine that the earliest ancients (such as Moses, Hermes, and Pythagoras) had all the wisdom of modern science in a deeper form.[16]

Henry More also rejected the form of mind/body dualism held by Descartes. Although More was initially very excited by Descartes's philosophy and exchanged letters with him, he gradually came to be an opponent of Descartes and the mechanical philosophy. (In a letter "on the automatism of brutes," Descartes claims that swallows migrate in spring and fall the way clocks go off, due to a literally mechanical procedure.)[17] More saw the work of Hobbes and Spinoza as displaying the evil consequences of the mechanical philosophy and of materialism.

More did defend a dichotomy of spirit and matter, but claimed in opposition to Descartes that spirits as well as matter could be extended in space. Spirits are indivisible while matter is divisible, and spirits can interpenetrate with matter while pieces of matter are mutually impenetrable. This doctrine probably played a role in Newton's striking metaphysical hypothesis in his early work *De Gravitatione* (On Gravity) that God created matter simply by endowing certain regions of space with impenetrability. Indeed, More goes so far as to claim that atoms can be held together only by means of spirit. This may be one of the sources of Newton's speculations on active principles in relation to interatomic forces.

More's impact on Newton is probably much greater with respect to the concept of absolute space. He was receptive initially to the Cabala, the body of medieval Jewish mystical writings read by (or at least alluded to by) many of the hermeticists and other Renaissance occultists. More claimed that behind ordinary material extension, which is finite and divisible, there is a metaphysical extension infinite and indivisible. He went so far as to say that space is a kind of obscure representation of the divinity of God. Newton's doctrine that space is God's sensorium, and his doctrine of an absolute space behind the "vulgar" relative space of physical measurement, is evidently in part developed from More's doctrine of space. Newton opposed Descartes's identification of matter with extension with the doctrine of absolute space as distinct from matter.

In his doctrine of space More made use of cabalistic ideas that can be traced back in part to the Old Testament.[18] The doctrine of one God led to an emphasis on the presence in all places: God is everywhere. In Hebrew the term *place* was used for *God* in the first centuries of Judaism. In Arabic the term *holy place* means tomb, or place of a saint. As local gods were unified to one God, not of a particular place but of all places, the use of *place* as a term for God became more abstract and general without any connotation of particular location.[19]

The notion of God being present in all parts of space is eloquently expressed in Psalm 139: "If I take the wings of the morning, / And dwell in the uttermost parts on the sea; / Even there shall Thy hand lead me, / and Thy right hand shall hold me." In the Mishnah Torah, and in the Greek translation of Exodus in the Septuagint, one finds the use of the term *place* for God. The cabalistic literature of medieval Judaism (such as the book of Zohar and the Cabala of Luria) continues this identification of God with place and expands it to the identification of God with the space of the universe; in the latter God is claimed to contract himself, leaving a spiritual realm of space. The Cabala was appropriated for Christian purposes during the Renaissance by Pico della Mirandola, whose famous oration *On the Dignity of Man*, often cited as a classic humanistic work, was imbued with cabalistic doctrines.[20]

Agrippa, said to be a source of the Faust legend, pursued cabalistic doctrines. So did Michael Maier, whose alchemical writings were intensively studied by Newton, and the Paracelsian and Rosicrucian Robert Fludd, who along with Henry More was one of the leading scholars of Hebrew literature of his time. More translated and commented on the Cabala in numerous writings; he refers to the cabalistic idea that space is a quality of God, as presented by Agrippa in his *Occult Philosophy*. Given Newton's strong interest in biblical history and in the prophesies of the Old Testament, it is highly appropriate that he borrowed from More doctrine that is traceable back to the earliest Hebrew doctrine, despite Newton's own repudiation of the Cabala.

Two women played an important role in the development and propagation of Cambridge Platonism. Ralph Cudworth's daughter, Lady Marsham, was educated by Cudworth himself and was a friend and patron of John Locke and an acquaintance of Lord Shaftesbury. Lady Marsham wrote a book of her own and propagated her father's ideas in England and in Europe. Philosophically more original was Anne Conway, the center of an intellectual circle that included More and Cudworth, a collaborator of the Platonists and writer on witchcraft, Joseph Glanvill, and the young Mercurius van Helmont, whose father, Jan, was an important figure in Paracelsian alchemy and medicine. Anne Conway, like More, made use of ideas from the Cabala, though unlike More she did not deny cabalistic influence.

Conway was very much a part of the Cambridge Platonist movement and engaged in a wide-ranging correspondence with leading intellectuals. Among those with whom she corresponded were such Quakers as William Penn. Conway wrote a major work, *The Principles of Most Ancient and Modern Philosophy*;[21] unfortunately, she was only recently discovered to be its author. The book was edited posthumously by Mercurius van Helmont, and her name did not appear on the title page. Thus, her ideas, including her use of the term *monad*, which Leibniz adopted, were mistakenly credited to van Helmont.[22]

Conway differed from More and Cudworth in totally rejecting dualism in any form. Cudworth had retained a dualism of active and passive, and More retained a dualism of mind and matter similar to Descartes's, but unlike Descartes allowed mind to be extended and indivisible. Conway denied that there was any purely dead matter. She also held a kind of evolutionary doctrine that lower forms transform into higher forms as well as a doctrine of reincarnation and theodicy. She claimed, in a manner similar to Leibniz, that each creature contains within it an infinity of smaller creatures.[23]

The Cambridge Platonists have often been cited as the major source of Newton's doctrine of active principles in matter. This is probably because they appear to be a more respectable source for Newton's philosophy than the alchemists and the magical tradition. Certainly it is true that Newton learned from Henry More, both in his doctrine of space and in aspects of his religious politics.[24] Nonetheless, for every page that Newton wrote concerning the Cambridge Platonists, he wrote hundreds of pages on alchemy.[25] Some scholars have attempted to attribute Newton's doctrines of active principles in relation to force to his readings in Cambridge Platonism;[26] however, given the time sequence and the verbal associations that one finds in some of Newton's passages concerning force and active principles, it seems reasonable to weight the impact of his alchemical studies more heavily.

Newton's Doctrine of Force and Active Principles

Newton's work is Janus-faced: in sharp contrast to Kepler, the public face of science and the private thoughts of the scientists are most sharply distinct in Newton.[27] Newton's great work of mathematical physics, the *Principia*, is primarily a work of mathematical deduction with certain important short sections on methodology and philosophy of nature. Newton claimed in the *Principia* that he did not feign or frame hypotheses, in particular with respect to the true nature or cause of gravitation (*hypotheses non fingo*).[28]

The claim not to be making hypotheses became the basis for positivist, empiricist British science in the succeeding two centuries. During the Enlightenment and the nineteenth century, Newton was seen as a cautious inductivist and a positivist. By the time of his burial in Westminster Abbey, alongside major political figures, Newton had become a pillar of the British establishment. He was the ideal scientist, in terms of the predictive success of his mechanics and the cautiousness of his public writings on the cause of gravitation. In his alchemy, theology, and metaphysics, he would not have been a proper mechanist and positivist by Enlightenment or British nineteenth-century standards. The public image of Newton as the ideal scientist and as a figure of national pride was maintained at all costs.

Thus, when Newton's alchemical manuscripts were discovered among his literary remains, it was necessary to suppress them. We know now that the man who claimed not to frame hypotheses struggled for decades with numerous hypotheses—scientific and metaphysical—concerning the ultimate nature of force. Some of Newton's biblical chronology had been published (against Newton's wishes) during his lifetime and shortly thereafter.[29] Jean Baptiste Biot, a French student of the optical activity of crystals, made the first critical study of Newton's lifetime activity in 1820. Biot explained away the theological and biblical history writings by Newton's nervous breakdown.[30] David Brewster, a British optical physicis, who wrote the first full-scale biography of Newton (1831, 1855), was the first to study seriously the unpublished papers of Newton. Brewster defended Newton against all criticism, but did report the existence of the alchemical writings despite being embarrassed by them.[31] The alchemical writings attracted little attention from idolizers of Newton during the next century, however.

In the 1920s, Edwin Burtt emphasized the extent to which Newton and many other great figures of early modern science held metaphysical doctrines.[32] Burtt tended to underestimate and mistakenly disparage the quality of Newton's philosophy of nature.[33] He wished to show that Newton, the patron saint of modern science, was a worse philosopher than the other early moderns, but did not emphasize the alchemical and occult inspirations for Newton's natural philosophy. Since

Keynes's essay, writers of recent decades have examined the alchemical roots of many of Newton's speculations on force, and have noted that it was primarily alchemy that contributed to Newton's philosophy in its opposition to the mechanical philosophy of causation purely by contact action.

Newton was strongly akin to the alchemists and the hermeticists in his claim that forces and active principles are at work in the world; however, he distanced himself from them insofar as he did not wish to attribute activity or vital force to matter itself. Newton's doctrines were thus a compromise between the pure mechanistic materialism of Hobbes's philosophy or Descartes's physics and the doctrine of an active and perhaps living matter of the alchemists and the magicians.

Newton strongly distinguished matter from spirit and degraded the status of matter in contrast to God or soul. For Newton, God's infinity and eternity are manifest in the infinite and eternal existence of absolute space and time that are the sensoria of God, not characteristics of the purely physical universe. Into this space and time, Newton's God creates as little matter as possible for his creation, given the degraded and lowly status of matter. Psychoanalytic writers and biographers have suggested associating matter with feces.[34] In Newton's alchemical writings, one frequently finds talk of the separation of waste as "feces" from chemically more purified forms of matter.[35]

One can also discern Newton's rejection and degradation of the female principle in his downgrading of matter. He was angered that his mother had remarried after the death of his father, and rants against the "whore" of the Catholic Church in his theological writings. In his alchemical writings, one also finds reference to substances such as the "pure milky virgin-like Nature drawn from ye menstruum of o^r sordid whore."[36] More frequently than other alchemists, he discussed and used "the net," based on the legend in which Vulcan traps Mars and Venus *in flagrante delicto*.[37] Alchemists tended to treat the relation of the investigator to nature as a relationship of equals, with the investigator as male and the investigated nature as female. The relation of mind to nature was a coition of equal sexual partners. In the writings of Francis Bacon and of Glanvill, Oldenburg and other boosters of the Royal Society, the relation to female nature was one of domination, and the scientist was to purge the female in himself. Part of the changed portrayal of the relation of male mind to female nature included the downgrading of the status of matter in seventeenth-century mechanistic thought in contrast to Renaissance and medieval thought about nature. Matter is divested of its powers and portrayed as barren, an object for manipulation.[38] Thus Newton, although taking seriously the ideas of the alchemical tradition, and absorbing the notion of powers or influences into his own metaphysics of forces, breaks with one aspect of the alchemical tradition implicitly in his rejection of the equality of male investigator and female object of investigation.

Because of the doctrine of the barrenness and passivity of matter and its complete separation from spirit that we find in Newton, matter must be completely separated from the active principles (largely borrowed from alchemical notions) that account for force, particularly gravitation. Tracing Newton's doctrines on this issue from his published writings alone is difficult. Newton was cautious in presenting his true private thoughts on the ultimate nature of matter and force, for a number of reasons.

First, Newton's personality was very secretive and distrustful, verging on paranoid, though he raised his system to an impersonality that became the scientific and Western worldview.[39] Newton saw himself as playing a special role in the universe. Despite humble statements about his "standing on the shoulders of giants" and being a child picking up pebbles on the shore of the vast ocean of truth, he conceived of himself as having special access to the divine mind.[40]

Second, Newton held Unitarian doctrines rejecting the Trinity. (Leibniz compared Newton's God to the lazy God of the heretical Unitarian Socinians, "who lives only from day to day.")[41] Newton had to be very cautious in keeping secret religious beliefs that could lead to loss of his university position. He was also aware that his view of space and time as divine attributes resembled pantheism, to which he was dangerously close. (Again Leibniz focuses on space and time as the sensorium of God for Newton.)[42]

Third, he shunned public debate. He was disturbed by an early controversy with Robert Hooke concerning optics. Later Newton preferred to guide his side of the controversy from backstage, but not to present his views in public. In the case of the debate with Leibniz over the nature of space, time, and God, Samuel Clarke presented Newton's views, with Newton coaching Clarke. In the case of the priority dispute over invention of the calculus, Newton secretly chose a biased panel of judges and wrote their verdict claiming that Leibniz had plagiarized. Newton hid behind a similar panel in his dispute with John Flamsteed over access to the latter's telescopic data.[43] Given all this, we can hardly expect Newton to be publicly presenting his views on theological and metaphysical issues.

Newton's views on the status of forces and active principles in relation to matter are also difficult to discern because he struggled for decades attempting to give an account of the ultimate nature of force in relation to matter. He shifted back and forth between various hypotheses and emphases in his attempt to account for gravitation and other less well known forces of matter. At one level, we can find the "positivist" Newton who denies that he is making hypotheses and claims that he is dealing with principles "deduced from phenomena." Newton asserts that his scientific claims are all inferred from experiments and observable phenomena, and that he has no interest in ultimate metaphysical factors that cannot be connected with experiment and observation. Nevertheless, he is aware that the purely "mechanical" system that attempts to explain everything by contact action is at fault. He realizes that the concept of force cannot be completely unpacked in purely descriptive, observational terms.[44]

Ernan McMullin summarizes four versions of Newton's explanatory account. Newton appealed at various times to the ether, the peculiar qualities of light, active principles, and the power of God.[45]

Newton started out reading Descartes and other early modern physicists. He rejected the debilitated Aristotelian physics he was taught at the university for the mechanistic physics of Descartes, but he very rapidly recognized the inadequacies of Descartes's physics, which he criticizes in his early work, *On Gravity*.

Descartes himself showed indirectly that he realized the peculiar characteristics of heaviness (later to be Newton's mass and gravity) in his letter to Princess Elizabeth of Bohemia on the mind/body problem. He suggests that she use the notion of heaviness as a means for understanding how the mind can act on the body (replacing one mystery with another).[46] In fact, Descartes's followers thought that the mind acted in matter by changing the direction of motion without changing its velocity, and claimed that this did not disobey his principles of the conservation of velocity. Newton realized that the conservation of momentum involved direction as well as straight-line velocity, thus the conservation of a directed quantity, one with both magnitude and direction.

Newton recognized early that the ether was an inadequate explanation for gravity. He knew that the physics of swirling vortices of fine material particles making up the ether would be inadequate to account for the motions of the planets. Some sort of force different from the material vortices of the ether was needed. Nevertheless, the ether was a tempting, convenient, apparently materialistic hypothesis, and Newton in his later life returned to versions of it.

The young Newton thought that perhaps light with its peculiar characteristics could yield insight into the nature of gravitation and force. There is the long, medieval tradition of light metaphysics, stemming from Neoplatonism, in which divine activity is described in terms of the radiation of light. Newton presented a paper "The Hypothesis of Light" that has often been described as a prime example of mechanical explanation; however, it contains language that suggests a very different background, where Newton speaks of nature as a "perpetual worker" and in which alchemical terminology creeps into his discussions. Earlier readers did not notice the alchemical terms and images, but modern historians, aware of Newton's alchemical writings, have gone so far as to call it "an alchemical cosmology."[47]

Newton believed that light was a good candidate as an entity more active than ordinary, passive matter. At various times, and especially in Query 30 of the *Optics*, he hypothesized that light could be part of the composition of material bodies. (The "Queries" appear at the end of Newton's *Optics*. This was the main location in his published works in which Newton allowed himself to speculate publicly concerning the nature of matter.) In a paragraph found only in a Latin version of the *Optics*, Newton raised the speculation that "gross bodies and light interconvert and bodies receive much of their activity from the particles of light in their composition."[48] One of the difficulties shared by Newton's speculations about light and his later speculations about ether is that when light is considered as composed of particles, then it appears that particles of light are material. Material particles cannot be sources of activity, according to Newton's doctrine of the passivity of matter, however. In Query 22, in a passage very similar to the one in Query 30 cited above, Newton asks, "Do not bodies and light mutually change into one another?" This would suggest that light is something different from matter, although they can transform into one another.[49] This transformability of an apparently nonmaterial substance into matter could allow Newton to equivocate on the status of light as he later did in discussing the ether. In a late draft of the

Optics he suggests that anything that can be transmuted into an ordinary body or matter should itself be called matter. In this way, Newton could call the ether material at the same time that he suggested it was something different from matter. Perhaps he was considering a similar move with respect to light. The alchemical component in Newton's speculations concerning the interactions of light and matter is suggested by his late claims that sulfur attracts light more strongly than any other chemical element.[50] This special role for sulfur comes from the alchemical role that it traditionally played as the most active of elements.

In his later work Newton appealed to the electrical spirit in an attempt to find a source for attractive forces. In 1707, Francis Hauksbee demonstrated a number of electrical phenomena to the Royal Society, showing Newton that there was an electrical force that seemed to be treatable in the manner of gravitation. On this basis, he speculated about the existence of an electric spirit, which could fill space and mediate between light and mechanical motion. Newton suggested in the 1713 edition of the *Principia* that this "most subtle spirit" could account for forces of attraction and repulsion between the atoms of bodies. The electric spirit was presented as something that, unlike the ether, was nonmaterial. Despite this fact, he wrote relatively little about it in his published works and dropped the draft of the Queries in which he had discussed the topic.

Strangely enough, Newton in his old age returned to the ether that he had rejected in his mechanics as a young man. This seems extremely odd, and is somewhat similar to Kant's return to the attempted deduction of the ether a century later in the *Opus Posthumum* after Kant had totally replaced the notion of matter with that of force.[51] Newton's new ether, however, was not the same as the old. In one version, it is treated as not an ordinary material entity, but more in line with the electric spirit. In another version, however, the ether is material, but contains forces of repulsion between its particles. Here Newton seems simply to have pushed back the issue one step. The original ether of Descartes and his followers was conceived of as a purely mechanical particulate explanation for gravity. Newton rejected this sort of explanation in favor of action at a distance. Newton's new ether requires action at a distance in the form of repulsive force.

This seems to be the very sort of thing that was to be explained by the ether. Perhaps Newton thought that it was safer to allow direct activity by the ether than by ordinary atoms of matter, since the ether's status was more mysterious. In fact, he cannot call the ether truly material if it is to be the source of forces. At times he seems to be attempting to classify the ether somewhere between ordinary matter and spirit.[52] This notion of entities intermediate in status between matter and spirit is evidently another derivative of Newton's study with the Cambridge Platonists.

Like Newton after him, More had believed that nonmechanical phenomena, such as cohesion, electricity, and gravity, could only be explained by a nonmaterial entity. Spirits and forces relate directly to material objects in a way that pure mind did not. Aspects of More's concept of spirit were transformed into Newton's notion of force. In addition, Newton's own concept of "spirit" was closely related to that of More. The electrical spirit was apparently material but its qualities were intermediate between ordinary passive matter and spirit.[53] Newton's ether, in its later version, did contain material particles; however, they were vanishingly small. In addition, the activity of a repulsive force, which is extremely powerful between the tiny particles, accounted for the behavior of the ether. Insofar as Newton attributed to the ether material particles, an ether explanation of forces seems to lead to an infinite regress. If the ether explains forces of cohesion, electricity, and so forth, then the repulsive forces within the ether are themselves the explanations of these forces.

Newton was aware that appealing to a purely mechanical ether led into an infinite regress. He knew that forces were needed, and to explain these forces purely by material particles and their interaction would lead to the postulation of further forces, and these forces would then be explained by further particles, and so forth. Newton seemed to think, however, that something was gained by postulating an ether of tiny particles related by repulsive forces. Perhaps short-distance action at a distance would seem a little less mysterious and nonmechanical than the long-distance action at a distance in an infinite universe.[54] Part of his motivation for this later postulation of the ether may have been Leibniz's criticisms of the concept of gravitation as "occult." Perhaps Newton's ether would look more material to the Cartesians and others swayed by the arguments of Leibniz. Apparently Newton was also pleased by the fact that the vanishingly small ether particles would support his general doctrine regarding the paucity of matter in the universe and the infinite space in which it is contained.

The last sort of explanation for gravity and other active principles that Newton occasionally tried is that of the direct action of God. Newton and Halley were said to have laughed together at the "theory" of Newton's young protégé, Fatio, that gravity was simply the direct action of God.[55] Nevertheless, at one point Newton did toy with such a solution, which resembled the "occasionalism" of Malebranche and Leibniz.

This was a view that Newton, in dispute with Leibniz, would elsewhere reject; however, Newton did wish to preserve room for the direct intervention of God in the universe, though limiting it to God's tinkering with the motions of the planets as the solar system ran down. Leibniz, in correspondence with Clarke, ridiculed Newton for portraying God as such a bad workman that he had built a mechanism he had to reset from time to time.[56] Nevertheless, Newton believed that this preserved the physical universe from being portrayed able to run independently of God. Newton thought that the latter alternative would later lead to atheism, as indeed it did. The marquis de Laplace, on professing to prove the stability of the solar system, also is alleged to have claimed to Napoleon to have no need of the hypothesis of God.

Newton, like Descartes, apparently thought that not only God's will but human will (and for Newton also the activity of animals) could add new motion into the universe, and because conservation of momentum involved conservation of direction as well as magnitude, such activities could not be sneaked in without disobeying conservation by changing direction alone, as in Descartes's scheme. This new motion could disobey conservation laws; it was not clear whether it was compensated for by the motion lost in collisions that were not elastic.[57]

Politics of the Newtonians

With the rise of economic individualism in early modern Europe, one finds a massive revival of atomistic philosophies. Descartes, Gassendi, and Boyle revived scientific atomism. Thomas Hobbes developed an atomistic materialism as well as a most ruthlessly selfish individualist philosophy of human nature. Physical atomism was the natural philosophy for an individualistic society, though the atheistic implications of physical atomism conflicted with the religious and more conservative political views of the time. One of the concerns of Cambridge Platonists was to retain the new science while denying its materialistic implications. Descartes had done this by supplementing a purely mechanical physical world with nonmaterial souls and God. More, Cudworth, and Newton thought that this was not sufficient, since the nonhuman world was given over entirely to material substance and mechanism.

Newton saw the need to supplement "stupid matter" with nonmaterial active principles so that even the physical world was not totally materialistic and mechanistic. His views also had strong political implications. Newton himself wrote: "I received also much light in this search [of the Scripture] by the analogy between the world natural and the world politic. For the mystical language was founded in this analogy, and will be best understood by considering its original. . . . The whole world natural consisting of heaven and earth signifies the whole world politic consisting of thrones and people."[58]

Despite his heretical Unitarian views and his secretive activities in alchemy and theology, Newton became an important political influence in England in his later years. In the 1680s King James II took the throne and was a supporter of Roman Catholicism. Newton led the opposition within Cambridge to James's Catholicism just prior to the Glorious Revolution of 1688. Newton's own political views were tied to latitudinarianism. He opposed the radical political currents of lower-class revolt, the materialism of religious freethinkers, and the Roman Catholicism of supporters of King James II.

Newton's friend Robert Boyle was associated with a circle, the "Invisible College," which was involved with the colonization of Ireland as well as with the founding of the Royal Society.[59] There were numerous strands in the entanglement of the founding of modern English natural science with the development of the British Empire and trade. John Dee was involved with the Muscovy Company in seeking a northeast passage to Asia, north of Russia, and is credited with coining the term *British Empire*.[60] Several authors in the Royal Society explicitly linked the conquest of nature through knowledge with the imperial conquest of nations. Latitudinarian Joseph Glanvill linked the conquest of the Americas with the conquest of knowledge. Bishop Thomas Sprat tied the spread of science with the civilizing of savages, and John Dryden poetically linked the expansion of commerce with the expansion of knowledge.[61]

This linkage of the expansion of science and colonial expansion was not merely metaphorical, but existed in the careers of a number of early modern scientists and philosophers. Atomist Thomas Hariot was involved with the Virginia Company and was Walter Raleigh's spokesperson while Raleigh was away sailing the oceans.[62] Hariot astounded the Native Americans with his magnetic compasses and lodestone.[63] John Locke received much of his income from the slave trade, through involvement with the Royal Africa Company.[64] William Harvey's brother was involved in the Baltic trade, and wrote on the circulation of gold in the economy, as his brother discovered the circulation of the blood in the body.[65]

Even the famous story about Edmund Halley's visit to Newton, and Halley's almost forcing Newton to write up his physics in the *Principia*, may have a political aspect. Halley was concerned to gain the favor of James II for the Royal Society. Although James was a Catholic, his court contained many libertines who subscribed in a dilettantish way to Epicurean philosophy. These fashionable courtiers criticized the Royal Society from the viewpoint of Epicurean atomism. Halley wrote an ode to Newton that also fawned upon the king, made reference to Epicurean themes, and used the Roman Epicurean poet Lucretius as its model. It has been surmised that Halley was attempting to persuade court Epicureans of the value of Newtonian science and the Royal Society.[66]

Newton secretly labored on biblical history and the interpretation of the Apocalypse. In so doing, he was a part of conservative millenarian speculation was pursued by the latitudinarians. We tend to think of speculations about the end of the world during this period as being primarily pursued by the radical reformers and egalitarians; however, more conservative theorists also engaged in such speculation.

Thomas Burnet, author of *The Sacred Theory of the Earth*, developed a science of earth history to correspond exactly to the biblical story. Burnet dealt with the Creation, the Flood, and the Last Judgment. Interestingly, in Burnet's millennium, the wealthy among the saved were to retain their wealth and property, and social distinctions were to be maintained.[67] Newton engaged in millenarian speculations, as did his star pupil, William Whiston, who speculated that the Creation, the Flood of Noah, and the end of the world were all caused by comets. Burnet, far from being praised for his complete reconciliation of geology and the Scriptures, was distrusted for his scientific reinterpretations and ultimately lost his job as Royal Confessor due to his allegorical treatment of the Creation.[68]

William Whiston made the mistake of being too outspoken about his beliefs based on Newton's private doctrines and got in trouble on a number of religious issues. Newton very cautiously and coldly refused to support Whiston in his religious and academic difficulties, despite the fact that Whiston was Newton's leading protégé at the time.[69]

With the success of the Glorious Revolution, the Newtonians and the Anglican clergy associated with them achieved political triumph. Newton, within five years, shifted from being a reclusive scholar known only to experts to being England's most renowned scientist and Master of the Royal Mint. A number of Newton's friends and disciples developed a unified philosophy of science, religion, and politics, presented through the medium of the Boyle Lectures, endowed by Boyle. The lecturers were all Anglican clergy who were disciples or associates of Newton. Richard Bentley, the bishop to whom Newton explained the *Principia* and wrote important letters concerning the role of God in the maintenance of the solar system, inaugurated the lecture series. Clarke, who disputed with Leibniz under Newton's guidance, also delivered lectures. William Derham delivered the last of this major series and presented an elaborate theology of God's purposes in nature that set the stage for later natural theologies, such as that of William Paley.

These lectures were the means by which Newtonianism was presented to the general public and became a worldview. Newton's technical writings would have been incomprehensible to all but a tiny minority of scientific experts. Even John Locke, who returned from exile in Holland after the Glorious Revolution and soon published both the political justification for the revolution and the philosophical justification for Newton's science, had to appeal to Dutch physicist Christian Huygens for an evaluation of Newton's scientific work. Clarke was probably the only one of the Boyle lecturers who grasped much of Newton's technical thought. Even Bentley, who was apparently backed by Newton as the first of the Boyle lecturers, was probably ignorant of much of Newton's work.

The Boyle lecturers, however, were able to convey a general religious and political view to the public based on the broad outlines of Newtonian science. They were concerned with refuting the radical enthusiasts and reformers who believed in life forces within matter and magical and alchemical principles as the basis for a science that served the common people. They were also concerned with refuting the doctrines of Hobbes concerning a purely material universe and a ruthlessly selfish human nature. Although there were relatively few people who openly supported the doctrines of Hobbes, and they were quickly suppressed,[70] the Newtonians realized the intellectual power and consistency of Hobbes's political philosophy and associated Hobbism with the selfish and opportunistic merchants of their day.

The Newtonians were also concerned with refuting freethinkers who leaned toward pantheism. The word *pantheism* was in fact coined by John Toland,[71] a notorious materialistic freethinker and a sympathetic defender of Spinoza's religious toleration. Toland also read, incorporated the ideas, and published hermetic works of Giordano Bruno. Clearly the monistic Bruno-Spinoza tradition was the enemy of the political Newtonians.

The Newtonians defended the political status quo after the Glorious Revolution. One Boyle lecturer (whose sermons were not delivered because of ill health) claimed that if "there is no God nor religion . . . all men are equal."[72] Richard Bentley argued against materialistic atomism by saying that it was absurd that material atoms could "transact all public and private affairs by sea and by land, in houses of parliament and closets of princes."[73] The Newtonian synthesis combined successful physical science with antimaterialism. The cosmic order was supposed to show the naturalness of the social order.

John Theophilus Desaguliers, author of the major scientific text that communicated Newton's physics to the majority of scientists (oddly in a set of notes and corrections to the textbook of Cartesian physics by Jacques Rohault), wrote a poem in which he claimed that the limited monarchy of George and Caroline, the new rulers brought to England by Parliament, was modeled on or justified by the structure of the solar system. In the dedication the author writes, "The *limited mon-*

archy, whereby our liberties, rights, and privileges are so well secured, to us, as to make us happier of our system [the Newtonian]; and the happiness that we enjoy under his present majesty's government makes us sensible, that *attraction* is now as universal in the political, as the philosophical world."[74]

Desaguliers continues with his poem, in which he rejects various systems of astronomy as leading to bad forms of government and being scientifically inaccurate, and concludes his work with a comparison of the royal government to the solar system:

> When majesty diffusive rays imparts,
> and kindles zeal in all the British hearts, . . .
> When ministers within their orbits move,
> Honour their king, and they each other love: . . .
> Comets from afar, now gladly would return,
> And, pardon'd, with more faithful ardour burn.
> *Attraction* now in all the realm is seen
> To bless the reign of George and Caroline.[75]

This poem seems extraordinarily naive, but its author, in fact, has the prestige of being the major scientific educational purveyor of Newton's mathematical physics.

Of course, Newtonian physics as a model for society was very powerful in later periods, especially in economics. Adam Smith wrote a history of astronomy and compared the equalization of prices by the market to water seeking its own level under the influence of gravity. This belongs to the history of social thought, but it is often not realized that the original Newtonians themselves were a political as well as a scientific movement.

CHAPTER 11

Interlude: Leibniz and China

$\mathcal{G}$ottfried Wilhelm Leibniz (1646–1716) is one of the leading logicians and mathematicians of all time as well as a major philosopher. His defense of the principle of continuity and the physical continuum against atomism and his doctrine of the relational nature of space and time are important sources of ideas in field theory.[1] Leibniz links the three holistic traditions that contributed to field theory. He saw a strong affinity between his metaphysics and Chinese, Neo-Confucian cosmology. More than any of the other early modern philosopher-scientists, he incorporated Renaissance notions of the reflection of the whole universe in each of its parts into his understanding of mathematical physics. Finally, his emphasis on force as the essence of matter was developed by the Romantics in the "dynamical" interpretation of matter that led to field theory.

Leibniz spent part of his career as a diplomat, and his universal philosophical goals included his attempts to reconcile various religions and cultures of the world, such as his youthful project of reconciling Protestantism and Catholicism through a series of "Catholic Demonstrations." This project lay behind some of his fundamental metaphysical principles.[2]

One of Leibniz's earliest forays as a diplomat was to write a proposal to King Louis XIV, requesting that he divert his efforts from wars in Europe to a conquest of Egypt as a way of distracting him from an imminent invasion of the Netherlands.[3] Leibniz wrote several more drafts, including lengthy proofs in the manner of geometry, and traveled to Paris in a vain attempt to gain an audience with the king. Leibniz's visit instead led to contacts with physicists and mathematicians that stimulated his later invention of the calculus.

Leibniz successfully proposed to Peter the Great of Russia (1672–1725) that the latter found an academy of sciences, and that Peter maintain diplomatic and intellectual contacts with the Chinese. When the czar visited Germany, Leibniz attempted to meet and convince him of the advantages of Russia's becoming the link between Europe and China.[4] Leibniz early on had plans for the conversion of China to Christianity via Russia. He first suggested a mission to China in 1687.[5] He was fascinated by the size and age of the Chinese empire, and considered China a kind of "anti-Europe," "an Oriental Europe."[6] Leibniz was keenly interested in China from philosophical, religious, and political viewpoints. He devoured what was available about China and defended the accord, or "accommodation," of Chinese philosophy and culture with Western Christianity. Leibniz's interests in China also appear in his accounts of binary arithmetic. He hoped that China, whose society he considered ethically and organizationally superior, could be converted to Christianity by showing the correspondences between ancient Chinese thought and his own metaphysics. He wrote, "We need missionaries from the Chinese who might teach us the use and practice of natural religion, just as we have sent them teachers of revealed theology."[7] A single emperor, K'ang Hsi, having great intellectual curiosity and a critical, skeptical mind, ruled China throughout the whole of Leibniz's adult life. K'ang had Euclid translated and wrote knowledgeably of the medical theories of his day. It is no wonder that Leibniz held K'ang Hsi up as the ideal of a benevolent despot, and hoped for philosophical and spiritual reconciliation of Europe with a China ruled by such a philosopher-king.[8]

Leibniz was as well informed on China as anyone in Europe who had not themselves visited there.[9] His attempts to draw out the similarities of Chinese philosophy of nature and the trigrams of the *I Ching* with his own philosophy and mathematics were part of his efforts to harmonize and synthesize all cultures. These efforts included the study of Jewish Cabala, Islamic philosophy, and Chinese thought. Leibniz's philosophical views were sufficiently similar to those of Chinese natural philosophy that Needham proposed that Leibniz was influenced by Neo-Confucian thought.[10] More likely Leibniz's thought was already sufficiently similar to certain Chinese views that he could find support and sustenance from them.[11]

The Jesuit Mission to China

Leibniz's knowledge of China was made possible by the Jesuit missionaries. The early history of modern European contact with China involved the Portuguese trade, which followed the capture of Malacca in 1511. They founded the colony of Macao around 1555, which was not officially recognized by China for over three centuries.[12] Once Jesuit missionaries were established in Macao, news concerning China filtered back to Europe and was devoured by Leibniz. In the mid-1500s the Jesuits planned a massive spiritual takeover of China. Matteo Ricci, the major figure in initiating the Jesuit penetration of China, spent two decades in China after his arrival in Macao.[13] The Jesuits had a policy of attempting to accommodate Christianity to the local cultural context, which involved far more than a merely tactical presentation of aspects of Christian doctrine. The Jesuits were greatly impressed by the civilization and the high culture of China, as had been Marco Polo centuries earlier and as later would be the eighteenth-century advocates of enlightened despotism who made China a model for Europe.[14]

The Jesuits initially attempted to move about China in the robes of Buddhist monks. This had been successful in Japan and was thought to enable Jesuits to be recognized by the Chinese as playing familiar role. Buddhist-like wandering Jesuit monks soon found that they were treated badly. Ricci concluded that Buddhism was a religion of the very poor, low-status people of China. Jesuit monks resembling Buddhist monks were at a disadvantage in comparison to the more aristocratic Confucians in the emperor's court, so Jesuits should ally with Confucianism rather than Buddhism. Whether as a political move or a genuine misunderstanding, Ricci in later writings unfairly disparaged and attacked Buddhism as superstitious.

The Jesuits recognized that Confucianism was the religion associated with high culture, and that the Confucians were the court intellectuals. The Jesuits began to dress in the robes of Confucian scholars. Ricci saw that the way to ingratiate himself with the Chinese emperor and court was to trade on the power of Western scientific knowledge, particularly up-to-date Western astronomy, which was important in the court for calendars and astrology.[15]

The Jesuits found that their competitors were Islamic astronomers, who had introduced ossified forms of Islamic astronomy into China. These late Islamic astronomers were not creative, and had become repeaters by rote of simplified doctrines. In addition, they were purely empirical observers. The Jesuits were able to displace the competition by means of superior tables, more accurate mathematical predictions, and better Western astronomical equipment.

Chinese astronomy, previously quite advanced, had deteriorated in recent centuries because Chinese skeptical philosophical doctrines (similar to the Western skepticism of the Renaissance) had undermined scientific interest and motivation.[16] Europe had been moving in the direction of humanistic skepticism during the same period but suffered a failure of nerve in the course of the violent religious wars of the Reformation and the early seventeenth century and retreated into a new metaphysical and political dogmatism reinforced by mathematical, mechanistic science.[17] China retained the skeptical attitude toward its own philosophical cosmology and science, but paid the price in scientific decline.

Ironically, the astronomy that the Jesuits introduced to displace traditional Chinese astronomy was itself becoming scientifically outdated in Europe. The Jesuits introduced medieval, Earth-centered Ptolemaic astronomy into China just as the Copernican revolution was taking place in Europe. The Jesuits taught the Chinese a medieval cosmology based on Aristotle and Ptolemy in which the stars were made of a matter (the quintessence, or fifth element) different from that of things on Earth, and the universe was relatively small. Furthermore, this cosmology claimed that the physics of the heavens was different from the physics of the Earth, and that the stars and planets were eternal and unchanging.[18]

The Chinese traditional cosmology portrayed a universe in which the stars were born and evolved and were part of a unified cosmos with the Earth, and in one version the Earth floated in empty space in an unbounded universe. In Europe, the Jesuits initially fought the idea that new stars appeared in the heavens because medieval European astronomers had recorded no new stars for the last thousand years. Meanwhile, Chinese astronomers had accurately recorded supernovas, such as the one that formed the Crab nebula in 1054.[19] Another changing phenomenon in the heavens that the Chinese had extensively observed for centuries but that was initially rejected by the Jesuits in Europe was the existence, and changing position and numbers, of sunspots.[20] Galileo and Kepler were engaged in controversies concerning the existence of sunspots and the "New Star" of 1604. Meanwhile, the Chinese were being taught that the phenomena they had earlier recorded could not exist.

Ricci and the Jesuits supplied the Chinese with outdated yet computationally impressive European astronomy while they themselves were progressively transformed by the profundity and sophistication of Confucian philosophy. They increasingly made more frequent accommodations with Confucian doctrine, such as using traditional Chinese words for "God," or "heaven" *(shang-ti)*, and introducing ceremonies, such as ancestor worship, into the Catholic Mass.

This led to the rites controversy, in which competing nations and religious orders criticized the Jesuit accommodation of Confucian rites in their Catholicism. The politics of the conflict were complex, including competition of the French and Portuguese as well as various religious orders for influence in the Chinese court and commerce. Jesuit versions of the Catholic religious rites that incorporated Chinese ceremonial forms and terminology were eventually discontinued by papal bulls of 1715 and 1742, which declared that the Chinese were pagans and did not have a religion that could be harmonized with Catholicism. In 1724, the Chinese emperor revoked his own edict of toleration and condemned Christianity. In the late eighteenth century, the Jesuits were expelled from France and generally suppressed, virtually ending the whole Catholic missionary effort in China, and with it one of history's greatest attempts at multicultural synthesis.[21]

Leibniz, the *Book of Changes*, and Binary Arithmetic

Leibniz learned about China from Athanasius Kircher, who produced elaborate and fantastic illustrated books deeply imbued by hermeticism and belief in visual symbols. One of Kircher's books illustrated Chinese ideograms, which along with Egyptian hieroglyphs fascinated Kircher and then Leibniz in their possibility to directly represent the essence of things.[22] From Kircher's and the Jesuits' accounts Leibniz came to believe that the ancient Chinese sages and the earliest Western religious figures were presenting a single doctrine.

In the early seventeenth century, Kepler had been asked to supply astronomical data to the Jesuit mission in China. Along with the astronomical tables, Kepler wrote a letter in which he presented the "Noachite" theory of the earliest Chinese sages, according to which Noah's four sons traveled to the four corners of the Earth. Shem was the eponymous founder of the Semites. Another son of Noah traveled to northeast Asia and became the first emperor of China. According to the more speculative theory of some Jesuits, Hermes, Moses, and the earliest Chinese sages all presented a single doctrine that had since degenerated. This doctrine of a unitary ancient wisdom is similar to that of Isaac Newton with respect to Hermes, Moses, and Pythagoras.

The most influential missionary figure for Leibniz was Joachim Bouvet, who introduced Leibniz to the diagrams of the *I Ching*, presenting Leibniz with a chart of the sixty-four hexagram (six-line) patterns that allegedly stemmed from the ancient Chinese sage Fu Hsi.

Bouvet claimed that all mathematical knowledge and ancient Chinese science was contained in the diagrams. He proposed that Fu Hsi was the same as the ancient Egyptian Hermes and also Zoroaster, the ancient Persian sage. Bouvet pointed out correctly that Fu Hsi's name means "Dog Man," and that Hermes was represented by the ancient Egyptians as a man with the head of a dog.[23] Leibniz was sympathetic to this tracing back of all wisdom to the patterns in the *Book of Changes*. Bouvet noted that the sequence of patterns read from top to bottom corresponded to Leibniz's own recently discovered binary arithmetic. (Many accounts erroneously claim that Leibniz first discovered this.)[24]

Leibniz realized that the Hindu and Arabic decimal system was only one of many possible bases for arithmetic, and that instead of the base ten of our Arabic number system, one could use a base twelve or even a base sixty, such as the Babylonians used. (The Babylonians related their base-sixty system to the supposed 360 days of the year, and it is the source of our 360 degrees in the angle of a circle. This base-sixty system was revived in the 1980s by philosopher Saul Kripke to simplify Godel's theorem in logic.)

Leibniz was the first to understand that the simplest number system was binary, which using only the digits 0 and 1 built up all the numbers. This is used in modern digital computers, and 1 and 0, representing *yes* and *no* or *true* and *false*, are used in truth tables for the analysis of logic. Leibniz was astounded to find in what he thought was one of the earliest Chinese sages this fundamental mathematical theory, which he thought he had just invented. This made Leibniz sympathetic to the most extreme claims of Bouvet. After all, the foundations of all of mathematics were to be found in the diagrams in the *Book of Changes*. Leibniz could elaborate on Bouvet's claims by writing that the Creation of the world was basically the transition from 0 to 1, and with that the generation of all the other numbers and structures of the universe was possible. Leibniz's doctrine of Creation was thus consistent with the doctrine in Lao Tzu that "one made two, two made three, and three made all the [infinity of] things."[25] Leibniz claimed that the six days of Creation, followed by the seventh day, were tied to the binary number system. In the twentieth century the profundity of the hexagrams has been similarly claimed by noting that the DNA code is one of four "letters" that can correspond to the four possible pairs of the broken and connected lines of the *I Ching*, forming sixty-four three-letter "words," like the sixty-four hexagrams made of the four different pairs of lines. The "letters" are the bases of DNA, and the "words" are the sixty-four codons for the amino acids that make up the proteins that catalyze the biochemical reactions that generate us.

Bouvet believed that the Fu Hsi diagram of the hexagrams of the *I Ching* contained the principles of both Chinese and Greek music, which had been lost, as well as all the mathematical and physical knowledge that could be rationally derived from these mathematical principles.[26]

Bouvet's model fit very well with Leibniz's rationalistic conception of knowledge. For Leibniz, the inventor of propositional symbolic, or mathematical, logic two hundred years before it was reinvented and much expanded by Frege, Russell, Whitehead, and Peirce, the universe could be understood as a logically deductive system, in which for God all truths were true by logical definition alone. For us, there is a difference between necessary truths, which can be found to be true by definition by finite analysis, and contingent truths, brute facts about the world which cannot be so analyzed. God, with an infinite mind, can intuitively grasp what for us would be the result of an infinite analysis. For such a mind all truths are true by definition, and the universe can be looked upon as a deductive system. Leibniz added the extra principle of God's goodness, and this would distinguish our world from all the logically possible ones in terms of the maximization of goodness. Leibniz's followers, such as the Wolffians, however, emphasized even more the purely logico-deductive nature of the universe (although they lacked Leibniz's extraordinary mathematical abilities and creative insight). Leibniz invented not only symbolic logic, but codiscovered with Newton the differential and integral calculus that is the basis of most of today's physics, and also discovered topology (what he called *analysis situs*), which was truly developed only in the twentieth century. The thesis that arithmetic can in principle be deduced from logic, which was developed by Whitehead and Russell in *Principia Mathematica* in the early twentieth century, is also Leibniz's, and Russell wrote a book on Leibniz before embarking on it. John A. Wheeler, among other physicists, embarks on theories that attempt to generate space and time out of purely logical principles that are very much in the spirit of Leibniz.[27]

Leibniz envisioned a "universal characteristic" (*characteristica universalis*) that was a universal mathematical language in which all subjects could be formulated. (The "ideal language" approach to philosophy in terms of an ideal language of symbolic logic, in Russell and the logical positivists, likewise stems from Leibniz.) Leibniz thought of this language as one in which all knowledge could be unified and in which all disputes could be resolved. "Come let us calculate" would be the invitation given to resolve any intellectual dispute.

Leibniz's universal mathematical language would not be one of purely logical definition and deduction. It would also be one in which the symbols directly and literally represented the fundamental things that exist. This would be the "alphabet of thought." Leibniz's ideal was that the true natures or essences of things would each be represented by a single symbol, which would directly reveal that nature to the reader. This was where Chinese ideograms came in. Leibniz had read early European students of the Chinese language, such as Andreas Müller and Kircher, who claimed that Chinese was a purely pictographic language, in contrast to phonetic languages. Kircher and Bouvet thought that Egyptian hieroglyphics and Chinese ideograms were originally closely related or identical, and represented an earlier, uncorrupted, direct insight into things. This has resonances of the doctrine of the so-called Adamic language, the language spoken by Adam before the Fall, the language in which Adam "named the animals" in the Book of Genesis.[28] This language directly represented the true natures of things, but it was lost in the Fall and further destroyed in the plurality of languages that were created when God said in the wake of the Tower of Babel, "Between us, let's descend, baffle their tongue until each is a scatterbrain to his friend."[29] Francis Bacon also believed that Chinese ideographs were "real characters" that represented things and notions.[30] Leibniz was taken with the figurative conception of the Chinese language as purely pictorial and thought that this might be a model for "alphabet of human thought."[31] When the symbols of the *I Ching* were found to correspond to the most basic foundation of arithmetic, Leibniz was understandably impressed.[32]

Leibniz's Defense of the Chinese Natural Theology

Besides claiming that the trigrams of the *Book of Changes* were an ancient discovery of binary arithmetic, Leibniz claimed that the Confucian philosophy of China was harmonious with Christianity and close to his own system of philosophy.[33] Leibniz follows Ricci in this accommodationist approach and opposes two later scholars—the Jesuit Nicholas Longobardi and the Franciscan Antoine de Sainte-Marie—who argued that Chinese religion was atheistic and materialistic and was incompatible with Christianity. Longobardi, in particular, emphasized the materialism of the Chinese. Leibniz was sent tracts by these two scholars, and his discourse was a reply to their claims. Leibniz argued that the Confucian philosophy held the seeds of a spiritualism compatible with Christianity. He attributed this spiritual wisdom, just as he and Bouvet had attributed the binary arithmetic, to the earliest Chinese sages. Just as the binary arithmetic attributed to Fu Hsi really first appears in a work in the eleventh century C.E., however, so the views that Leibniz attributes to the earliest pre-Confucian sages really belonged, for the most part, to Neo-Confucian scholastics of the twelfth and thirteenth centuries C.E. and later.

Following Ricci, Leibniz was contemptuous of the Chinese Buddhists and Taoists as superstitious idolaters. He also was negative toward contemporary Confucian intellectuals, correctly described by the Europeans as secular and atheistic; however, Leibniz believed that these modern Confucians held an atheistic degeneration of the ancient doctrine, rather like contemporary European skeptics and materialists. When even Confucius deviates from what Leibniz thinks of as the true philosophy, then Confucius must have strayed from the true doctrines of the ancient Sage Kings. Ironically, this doctrine attributed by Leibniz to the most ancient mythical sages is really a doctrine of the much later Neo-Confucians.

The philosophy of nature of the Neo-Confucians was actually highly dependent on the Taoist philosophy that Ricci and Leibniz dismissed as the superstition of the common people. Leibniz really does succeed in finding a true affinity between his own philosophy and the dominant natural philosophy of China, however. His own philosophy emphasized the mirroring of the whole universe in each individual (the microcosm/macrocosm doctrine), stressed the harmonic coordinating of independent events across the universe (the preestablished harmony), and understood the universe as a hierarchy of organisms (subordinate/dominant monads). Insofar as his philosophy contained these three elements, it was far more similar to the Chinese view of the world than were competing European mechanistic philosophies.[34]

Leibniz emphasized the terms *li* and *ch'i*, fundamental in the Neo-Confucian scholasticism of a major theorist, Chu Hsi. *Li* is akin to principle or law or form, especially in Leibniz's account. *Ch'i* refers to a kind of matter or force. In Neo-Confucianism, these two things account for the universe. Longobardi identified *li* with scholastic Aristotelian "prime matter," which is pure potentiality that takes the various forms that make particular kinds of objects. Thus, Longobardi can claim that matter is at the basis of the system, and that the Neo-Confucians are materialists.

Leibniz, on the other hand, claimed that *li* cannot be prime matter, but must be something akin to form, and it must have the characteristics of spirit in order for it to be active and for it to inform the universe. *Ch'i*, in contrast, is clearly much more like matter. Leibniz referred to it as a "primal air" and recognizes it as matter. The Chinese notion of *ch'i* has the connotations that force or energy have in Western thought and thus is less material than the early modern Western conception of matter. This association of *ch'i* with force and dynamics would naturally appeal to Leibniz, since he himself denied the ultimate existence of passive matter and identifies real things with souls on the basis of their force and activity. Leibniz here followed Plato in defining the soul as the self-moved. Recall that in seventeenth-century philosophy, unlike twentieth-century physics, matter is by definition totally passive, and anything that is active is soul.

Leibniz was unsympathetic to Longobardi's notion that *li* is totally passive, stating that if *li* must create *ch'i*, *li* must itself be active. Leibniz ended up identifying *li* with spirit, and claimed that *li* as general principle of the whole universe can be identified with God, even if the Chinese do not explicitly do so. He also liked the idea that the Chinese call *li* the natural order of Heaven, and spoke of Heaven as acting consciously when it really is not so. This fits well with Leibniz's notion of the preestablished harmony, that the activities of the various substances in the universe are coordinated by God without truly interacting with one another.

In many ways, Leibniz's own organic philosophy is very congenial with the Chinese philosophy, but it also differs. His philosophy is highly hierarchical, like the social philosophy of the Confucians; however, Chinese natural philosophy, incorporating Taoism, emphasizes the spontaneous coordination and harmony of elements. This spontaneous harmony is much like Leibniz's preestablished harmony, but emerges "from below." There is a point to Longobardi's identification of *li* with prime matter. In Chinese philosophy, the form is not imposed upon the matter from above, as in Aristotle's philosophy, but rather the form emerges spontaneously from the matter below. This aspect of Chinese philosophy of nature more resembles that of Giordano Bruno, the hermetic philosophers, and those later radical freethinkers who transformed Leibniz's philosophy of spirit into a philosophy of active matter.

Leibniz, Chu Hsi, and Neo-Confucianism

Leibniz's later metaphysics most resembles Chinese Neo-Confucianism, although some components were developed before Leibniz was seriously acquainted with Chinese thought. Leibniz's formal logical approach to substance in terms of containment of properties or qualities, corresponding to the containment of meaning in the subject of a definition or truth by definition, shifts to emphasis of substance as force, pregnant with the future, and as containing vital principle or entelechy and from the substance understood through the "complete concept" (paralleling the fact that the substance contains its properties or qualities) to the notion of substance as "spontaneous principle of action."[35] One factor in this shift is Leibniz's interest in physical dynamics and laws of force, but another, later, reinforcing one may be his interest in Chinese thought. It is significant that Leibniz, in his attempt to relate his own philosophy to those of other cultures, focused on China, rather than, say, Persia or India.[36] Leibniz, in his later works, in which the universe is presented as a system of organisms as well as a system of entities mirroring each other and the universe, develops a philosophy that greatly resembles the Chinese philosophy of nature. The major Chinese philosopher whose doctrines appear to be used by Leibniz is the Neo-Confucian Chu Hsi.

Daniel Cook and Henry Rosemont make a strong case against Needham's claim that Leibniz's later metaphysics was inspired or influenced by the thought of Chu Hsi. They grant that Leibniz was interested in things Chinese from his early years, and claimed the superiority of Chinese medicine to Western medicine;[37] however, Leibniz's earliest requests for information to the mannerist polymath Kircher are simply requests for a variety of miscellaneous information. Leibniz hoped that Chinese ideographs could function as the symbols in his "alphabet of human thought" but soon became disillusioned with this. Despite his interest in the Chinese language, in Chinese society as a model for a rational despotism superior to the European, and in the missionary efforts in China, Leibniz was not exposed to accounts of sophisticated Chinese philosophy—Neo-Confucian scholasticism—until toward the end of his life, and that even then he was acquainted with it only through the secondary accounts of missionaries. Leibniz repeated all the errors of fact in Longobardi's and Sainte-Marie's accounts of ancient Chinese thought in attempting to rebut Longobardi's hostile account. Likewise he used only passages that occur in the Jesuit treatises that he is debating, and seems to have been unaware when different transliterations are referring to the same Chinese term.[38]

Nevertheless, it is notable that Cook and Rosemont entitle their criticism of claims of influence "The Pre-Established Harmony between Leibniz and Chinese Thought."[39] Although denying influence, they note the striking parallels and correspondences between the thought of Leibniz and Chu Hsi. Just as Leibniz's monads contain images of each other without being in genuine causal relations, so Leibnizian philosophy and Chinese philosophy mirror one another without genuine historical influence. It is extraordinary how well Leibniz was able to reconstruct the thought of Chu Hsi, even if his account was based on inaccurate and even hostile secondary sources.[40] This, combined with Leibniz's lifelong sympathetic interests in a variety of topics concerning China, suggests a strong affinity of both the conceptual orientation and sympathies of Leibniz for Chinese thought. Many contemporary historians are critical of the very notion of "influence." This vague term, they claim, should be replaced by more specific references to what texts (or other sources) the author encountered, and what the author made of them, whether (by our lights) correctly or incorrectly. Much recent scholarship has focused on the active reception of ideas, including their modification and transformation, rather than influence. Leibniz absorbed and gained support from Chinese philosophy, and systematically drew the parallels of Neo-Confucianism with his own thought.[41] It is notable that Leibniz, who is the major source of the relational theory of space and time of the Continental continuum theorists and Kant (whose work led to field theory from mathematical and philosophical approaches, respectively), was philosophically favorable toward Chinese thought. This thought was expressed in a language that made a holistic and continuum view of the world more easily formulated and natural, and had centuries before, unknown to him, led to the compass and magnetic declination.

Among the major early modern philosophers, Leibniz was the only one who attempted to retain and reconcile organic Aristotelian and Scholastic elements with modern science. It is perhaps appropriate that Chu Hsi, the Chinese philosopher whose system is closest to Leibniz's, played a role much like that of the great Western Scholastic Thomas Aquinas.

Chu Hsi uses the metaphor of the reflection of the light of the Moon on a multitude of different objects.[42] This image is borrowed from earlier Buddhist thought and has a strong resemblance to the metaphor of Indra's net, which the Buddhists had borrowed from Hindu thought.[43] In Indra's net each node contains a jewel that reflects the whole net as well as all the reflections of the individual jewels in the net. This image obviously fits with the microcosm/macrocosm as well as with Leibniz's doctrine that every individual reflects the whole universe.

It is not surprising that Leibniz found many affinities to Chu Hsi. Chu Hsi's emphasis on principle, or *li*, is similar to Leibniz's emphasis on form or essence and to his logico-mathematical conception of the cosmos. For Leibniz, each individual substance has a complete idea or form that contains within it logically all the behavior of the individual as well as all its relations to other individuals represented as qualities of the individual in question. Thus, each monad is completely programmed. It has a substantial form somewhat analogous to the DNA of an individual organism, which contains the information for its structure and activity.

Leibniz was a preformist in biology, believing that all the features of an organism were contained in its original germ cell, which in Leibniz's sexist theory was the sperm cell. Leibniz's preformism is similar to the notion held by those who overemphasize the role of DNA and underemphasize the role of the environment in the development of the organism. Leibniz also believed in a kind of logical preformism at the metaphysical level, that is, the complete concept of the individual or the substantial form of the individual is an idea rather than a physical structure. The complete concept of the individual conceptually contains all the characteristics of the individual including the time order in which those characteristics are manifested.

Chu Hsi's emphasis on the logical priority of form to matter, despite the actual copresence of the two, fits well with Leibniz's logico-mathematical conception of the individual. Both in turn have resemblance to Plato's notion of form, but both differ from usual Platonism in that the forms are forms of individuals. Chu Hsi distinguished between *li* taken as the forms of classes of individuals and *li* understood as the forms of particular individuals.

Similarly, Chu Hsi's notion of the Great Ultimate had many resemblances to Leibniz's notion of God. Leibniz comments upon the diagram of the Great Ultimate that Chu Hsi borrows from Chou Tun-yi. The Great Ultimate is both a value concept and a logical concept. In this regard, it is similar to the Platonic or Neoplatonic notion of the One, and in this has some similarity to Leibniz's Platonism. The Great Ultimate contains the forms of all the individual things within itself, resembling Leibniz's notion of the mind of God, which contains all the ideas of particular things. In some respects, Chu Hsi's Great Ultimate resembles Malebranche's conception of things as ideas in the mind of God. Leibniz's contemporary Malebranche also wrote a dialogue on Chinese philosophy, but compared to Leibniz's work it is relatively superficial. Malebranche probably would have been surprised to find to what extent the *li* of Chu Hsi resembled his own philosophy.[44] Another similarity of Chu Hsi's philosophy to Leibniz's is that each individual entity contains or manifests the whole Great Ultimate and thus also mirrors the whole system of all other entities. This is similar to Leibniz's and Malebranche's preestablished harmony.

One aspect of traditional Chinese philosophy that would have appealed to Leibniz is the use of harmony as a principle. Twentieth-century works about Leibniz (and many of Leibniz's own accounts) present the preestablished harmony in terms of God's activity in thinking-creating (fulgurating) it, but Leibniz also presents harmony simply as a principle in itself, as in "The Secrets of the Sublime," where particular minds are said to exist because it is harmonious that they reflect the varieties of things, like mirrors.[45]

Another way in which Chu Hsi's universe resembles that of Leibniz is in the notion of material force, or *ch'i*. *Ch'i*, in its crudest and most materialistic interpretation, is a continuous quantity, like a fluid or gas. Some of the early translations of Chu Hsi translated *ch'i* as "air." The resemblances of Chu Hsi's cosmology, in which the universe is generated by the whirl of *ch'i* in which the thicker material condenses at the center as the Earth and the clearer material is whirled off as the heavens, resembles the early Greek philosophy of Anaximenes.

Leibniz's physical universe is also a universe of the continuum. Matter is infinitely divisible and all space is filled with matter. So-called empty space is actually filled with the ether, and the conception of vortices of whirling ether in Descartes and Leibniz has some resemblance to the Chu Hsi universe. At a deeper level, *ch'i*, as we have noted, is of the nature of force. Furthermore, in Leibniz and Kant, the real nature of matter reduces to that of forces.

Chinese views of reality as a continuum and a coordination of orientations within an organic system allowed the Chinese to accept and investigate magnetism. The Western theory of continua and fields that made possible the theory of electromagnetism stems from Leibniz in various respects. Leibniz's principle of continuity and his conception of a universe of monads each reflecting the universe from its own perspective integrated the Renaissance doctrine of microcosm/macrocosm with a relational theory of space and time and a conception of entities as centers of dynamic force that contributed in various ways to later field theory. Despite the limits of his knowledge of Chinese thought, Leibniz was correct in his claim that Chinese philosophy, particularly in its worked out, Neo-Confucian form, was parallel to his own. This harmony between Leibniz's metaphysics and Chinese thought shows how the organic-coordination view of the cosmos played a role in various stages of the prehistory of field theory.

PART III

Romantic Philosophy of Nature and Electromagnetism

CHAPTER 12

Philosophical Background of Romantic Science

The panpsychism of Leibniz, the monism of Spinoza, and the revival of such Renaissance occultists as Bruno and Paracelsus, rejected during the previous century, fostered Romantic theory.

From Spinoza to Romanticism

Monism is the most extreme form of holism, and Spinoza is certainly the greatest modern monist. His views are often noted to be congenial to a deterministic field theory. In fact, Spinoza's views were indirectly inherited and absorbed by the founders of field theory, such as Ørsted and Faraday. He was not read by either, but his ideas were transferred in a delayed, indirect manner through the mediation of Friedrich Schelling and German nature philosophy. The delay was partially due to the fact that Spinoza's views were closely associated with radical religious and political currents both in his day and in the succeeding century.

Spinoza in Social Context: Politics, Religion, and Monism

Lewis Feuer has summarized the case for Spinoza's philosophy of pantheist mysticism as the view of a repressed radical who forged his political connections once he was expelled from his synagogue.[1] Spinoza found readers and a discussion circle among the Collegiant Mennonites,[2] peacefully inclined descendants of the radical communist Anabaptists whose violent revolt (supported by Paracelsus) had been ruthlessly crushed. Spinoza moved to a Mennonite center, and his publisher in Amsterdam was likewise a Mennonite.[3] Spinoza was also in contact with English Quakers who preached a pacifism and doctrine of community at odds with his political world. An English Quaker, William Ames, arrested by the Dutch authorities, came into contact with Spinoza soon after the latter's expulsion and during the height of Quaker enthusiasm for communism.[4] The English radicals were favorable to Judaism, as they based their ideal society on the Old Testament, before, as Marx said, "Locke supplanted Habbakuk."[5] In seventeenth-century England many radical sects (such as the Levelers) subscribed to pantheist mysticism (the identification of God with the world, to be known through mystical intuition or insight). Robert Hooke associated Spinozism with the Quakers, and Oliver Cromwell's chaplain wrote a work including attacks on Spinoza's pantheism.[6]

Yirmiyahu Yovel attributes Spinoza's ability to criticize so radically current political and religious beliefs to his Marrano background.[7] The Marranos were Jews of Iberia who were forced to convert to Christianity, the alternative being either death or expulsion. Many Marranos secretly maintained their Judaism, but publicly followed the outward forms of Christian ceremony and profession of belief. Spinoza was descended from these Jews who, on emigrating to Holland, reverted to open profession of Judaism. This led to a dual consciousness and a tendency to express oneself in Aesopian language or metaphorical terms, whose double meaning could be grasped by those in the know without risking punishment by the Inquisition for reverting to Judaism. The situation of ex-Jews converted to Christianity, whether voluntarily or forcibly, led some to grasp Christianity in a more extreme form, whether institutionally or doctrinally, than those who were simply born into the faith. The ruthless inquisitor Torquemada and the intense mystic Teresa of Avila were both converts of Jewish origin.[8] Other Marranos (philosophers Juan Vives and Juan de Valdés) supported humanistic Catholicism, following Erasmus of Rotterdam, and opposed the Inquisition.[9] Some came to reject the differences between the faiths, and accepted a single worship independent of all dogmas.[10] A few became outright pagans or atheists, rejecting both faiths.[11]

Spinoza shows all the features of Marrano thought, although his struggle is one between rationalism and religion rather than between Judaism and Christianity. Spinoza was a "Marrano of reason," in Yovel's designation. The bifurcation in his life was between his early life as a Jew and, after his expulsion from the synagogue, his life as an outcast from Judaism living primarily among freethinking Christians and rationalists. Spinoza maintained his Jewish heritage, and he retained the impact of the study of medieval Jewish thought, while developing a secular and naturalistic philosophy. This philosophy contains aspects of earlier Jewish mysticism in its goal and fundamental attitude, but transformed into the rationalistic monism of a logical, systematic philosophy.

Yovel's claim is persuasive and illuminating; however, the Marrano background neither fully accounts for Spinoza's positions nor excludes a political explanation of the sort given by Feuer. And Yovel admits that Marrano heritage hardly explains the original and systematic rigor of Spinoza's thought.

Similarly, as an explanatory factor, the dual mind-set of the Marrano is not an exclusive alternative to allegiance with the republican and radical democratic movements. Interestingly, Yovel never so much as refers to Feuer's political account of Spinoza's beliefs, despite obvious familiarity with the English-language Spinoza literature. The only political account considered by Yovel is that Spinoza's excommunication was part of a strategy to eliminate any appearance of heresy among the Amsterdam Jews, so that they would be acceptable to the English for resettlement.[12]

This messianic link of Jewish thought with that of the radical English Puritans is much broader than the issue of the return of the Jews to England. It involves a number of currents of radical Mennonite and antimilitarist Quaker thought with which Spinoza associated himself. (Anne Conway, whose system resembles Leibniz's and integrates notions from the Cabala, converted to Quakerism.) Yovel argues that Spinoza's association with the Mennonites, radical Quakers, and democratic freethinkers was subsequent to his expulsion from the synagogue. Also, Spinoza's being tutored in non-Jewish European thought by the radical democrat and freethinker Francis Van den Ende was subsequent to his formulation of his heretical religious beliefs, according to recent historical research.[13]

Although this may show that the Marrano heritage of two-sided thought and expression may have stimulated Spinoza's earliest doubts about religion and conventional authorities, it does not show that his systematic philosophy, especially the *Ethics*, was not affected by his radical republican and democratic affinities. Although Spinoza's encounter with the modern scientific interpretation of nature in the works of Hobbes and Descartes was subsequent to his early critique of Judaism, it is obvious that the Latin works of early modern philosophy and science structured his categories and understanding in the mature cosmology and metaphysics found in the *Ethics*. Similarly, Spinoza's communal and radical republican contacts (which were, according to Yovel, subsequent to his early doubts about Judaism) could have influenced not only his later politics, but also his monistic cosmology and metaphysics.

The parallel between the political views and cosmologies of each of the major seventeenth-century rationalists is quite striking. Descartes allied himself with the absolute monarch, the founder of absolutism, Louis XIII, whose word was law. Similarly, Descartes's God wills the universe into existence from moment to moment, and the will of God has priority over the laws of logic. Leibniz's political allegiance was with the local royalty of the petty states within the loosely but ideally unified Holy Roman Empire in a Germany that had no real political unity. Similarly, Leibniz's God was the coordinator of the preestablished harmony of the isolated monads, just as the Holy Roman emperor was, at least ideally, the coordinator of the petty dukedoms of Germany. Finally Spinoza's political ideal was a republic in which the people, the community, constituted the governing body. Similarly Spinoza's God, like the populace as ruler of his ideal community, did not stand outside or above the universe itself, but was the universe.

From Spinoza to Schelling: Spinoza's Rehabilitation by German Romanticism

Both the religious and political doctrines of Spinoza were almost always presented in a negative light from the time of his expulsion (1656) to his death (1677) to the posthumous publication of the *Ethics*, almost to the beginning of the nineteenth century. Pierre Bayle, in a long article on Spinoza, in his *Philosophical Dictionary* (1697), portrayed Spinoza's philosophy as "the most absurd and monstrous hypothesis that can be envisaged." For nearly a century Bayle's portrayal of Spinoza as atheist was the major source of information about him. Even those skeptical of the religion of their day often defensively contrasted their own position favorably with the horrible atheism of Spinoza. Even so strong a critic of the abuses of organized religion as Voltaire portrayed Spinoza as a foolish metaphysician ignorant of the truths of physics. David Hume, who subtly undermined the proof of God from design in his *Dialogues on Religion*, famously referred to Spinoza as "universally infamous." One Enlightenment thinker, Denis Diderot, developed a naturalistic and monistic theory of the universe as one material substance with strong affinities to the views of Bruno and Spinoza, but in the *Encyclopedia* Diderot portrayed Spinoza in the standard, negative fashion based on Bayle's dictionary article. (Twentieth-century Marxist courses in the history of philosophy in the Soviet Union traced a line from Bruno through Spinoza and the eighteenth-century French materialists, especially Diderot with his active matter, to Marx.)

In contrast to this century-long denunciation, starting in 1785 Spinoza began to play a role in German transcendental philosophy through the Pantheism Dispute (*Panthesismusstreit*), which initially concerned Friedrich Heinrich Jacobi's claim that Gotthold Ephraim Lessing, an admired playwright and philosopher of art and of history, had secretly been a Spinozist.[14]

The Spinoza Controversy, or Pantheism Dispute, was a major event at the end of the German Enlightenment and the turn to Romanticism. By showing that Lessing, the universally respected paradigm of Enlightenment attitudes, was secretly a follower of Spinoza, who had consistently and logically drawn the conclusion from reason and the new science that there was no transcendent God, Jacobi undermined Enlightenment confidence that reason and traditional Judeo-Christian religion could be reconciled. The Jewish philosopher Moses Mendelssohn claimed to have reconciled Christian Wolff's Leibnizian rationalism (the standard academic philosophy in Germany) with religious belief in the God of the Bible and the immortality of the soul. Understandably, Mendelssohn reacted strongly to the intimation that Lessing, his friend, was a Spinozist. Lessing, widely believed to have portrayed Mendelssohn as the admirable Jew in the drama *Nathan the Wise*, had helped underwrite the cultural status and respectability of Mendelssohn.

Mendelssohn's responses to Jacobi's claims were passionate and desperate. There was an exchange of books, in which Jacobi was generally thought to have gotten the better of Mendelssohn. Mendelssohn contracted an illness in rushing the manuscript of his last reply to the publisher on a cold winter day and died from the illness. Literary reports blamed Jacobi for the death of the revered defender of reason.[15] Mendelssohn's death became a symbol of the collapse of Enlightenment rationalism's attempt at reconciliation with religion, indeed the collapse of the Enlightenment itself.

Jacobi's implication that the premises of Enlightenment rationalism, if their consequences were logically followed out, led straight to the determinism and naturalism of Spinoza, was a shock to Enlightenment complacency. Jacobi, in quoting Lessing's rejection of Jacobi's own *salto mortale*, suggested that the only way to avoid atheism and "nihilism" (a term first popularized in its modern sense by Jacobi) was to avoid rationalism as such. Lessing became the subject of lengthy reflections by the founder of existentialism during the next century. Sören

Kierkegaard, in his *Concluding Unscientific Postscript*, built his irrational fideism, his "leap of faith," in large part on the claims of Jacobi and of the later Schelling.

It is surprising that Spinozism could precipitate a minor scandal more than a century after Spinoza's death. Immediately thereafter, however, the major thinkers influencing the Romantic movement began to declare openly their allegiance to dynamized and temporalized forms of Spinozism. In these versions Spinoza's monism was accepted, but his denial of the fundamental reality of change was rejected. Goethe had a real interest in and affection for the work of Spinoza, although he read it in a manner congenial to his own organic and typological view of the world. During one phase of his life, Goethe carried a copy of Spinoza's *Ethics* in his pocket at all times. Philosopher, historian, and literary critic Johann Gottfried von Herder openly accepted the philosophy of Spinoza, which he transformed into a monistic metaphysics of forces as the fundamental realities. Herder borrowed Leibniz's notion of the ultimate reality as active, but combined it with Spinoza's monism, and divested it of the identification by Leibniz of the active centers of force with nonmaterial spirits. Herder explained nature, life, and human history in terms of a naturalistic metaphysics of force.[16] For Herder, human biology, in the form of vital forces, is the basis of the human mind and of the whole history of human culture.[17] Herder's metaphysics of force was developed with reference to the ideas of Leibniz and Roger Joseph Boscovich[18] and was justified in part by the discoveries of electrical and magnetic forces in the science of the day. "Before the magnetic, and the electrical forces were discovered, who would have suspected their existence in bodies, and what countless others may still lie dormant and undetected in them."[19] Human anatomy, in particular the upright stance, is the source of human mental capacity, related to the freeing of the hands for practical activity (a view opposed by Kant and developed by Ernst Haeckel and then further in Friedrich Engels's labor theory of culture). Rejection of the full-blown naturalism of Herder may have been a major stimulus for the second half of Kant's Third Critique, *The Critique of Teleological Judgment*.[20]

While Fichte could be said to have reversed Spinoza's objective monism of the universe = God into a subjective monism in which the Absolute Ego became the monistic unifying reality, Schelling moved from allegiance with a position close to Fichte's to a monism in which objective Nature held equal status with the Transcendental Ego. For Schelling, one could start either from Absolute Ego or from Nature to reach the Absolute. Schelling's own philosophy was much closer to Spinoza's than that of any of his predecessors. Schelling himself once described the goal of his system as "a system of freedom—as great in scope and at the same time as simple as Spinoza's, its complete antithesis—that would really be the highest."[21]

Schelling thought that *he* was being accused of pantheism because of his use of Spinoza's philosophy of monism (which would not have been an unfounded accusation, and pantheism was a position from which the later Schelling struggled to distance himself).[22] Jacobi's last hurrah, however, was to accuse Schelling of Spinozism in a derogatory sense, as he had successfully accused Lessing and

Mendelssohn and later Fichte (whom Jacobi accused of an "inverted Spinozism").²³ Schelling defended himself so well against Jacobi's attack that Jacobi's weapon of hurling the charge of Spinozism at his opponents was no longer effective, and Jacobi was discredited.

Herder's metaphysics of natural and vital forces had turned Spinoza's one substance into a unity of force. Kant criticized Herder's naive positing of invisible and unconfirmable vital forces to derive the human mind and culture from naturalistic and biological bases. Kant claimed that the mind constitutes experience in terms of Newtonian mechanics, and that purposes in nature are not constitutive, but merely regulative. Schelling, in turn, transcendentalized Herder's monistic metaphysics of force, by criticizing Kant's denial of real, constitutive status to purpose in the universe. Schelling reinstated both force and purpose from a transcendental perspective, as constituted by the Transcendental Ego. Thus Spinoza's objective but static universe = God was changed by Herder into a unity of force involving temporal process. Schelling took over this monistic unity and transformed it into the Absolute of transcendental philosophy.

Spinoza had gone from being Hume's "universally infamous" atheist to Novalis's "God intoxicated man." And Spinoza's monistic metaphysics of nature had gone through Goethe's general sympathy for it and through Schelling's dynamized and transcendentalized version of it to become nature philosophy. A Romantic, process-oriented, modernized version of Spinozism was indirectly transferred to the founders of electromagnetic theory, Ørsted and Faraday. When Einstein expressed his sympathies for Spinoza directly in relation to his monistic, deterministic universe of the unified field and of space-time, he was in fact expressing alliance with a figure who had, by a circuitous route, through Schelling, influenced the origins of the field theory that Einstein purified.

Revival of Renaissance Occultism and Natural Philosophy among the Romantics

During the Romantic era a number of the philosophers of the Renaissance who had been rejected by and even ridiculed during the Age of Reason were revived. These doctrines of the Renaissance hermeticists, alchemists, and philosophers of nature were incorporated into the ideas and writings of such figures as Goethe and Schelling who in turn supplied concepts and attitudes to the Romantic nature philosophers.

A general rise in appreciation for the Middle Ages was a feature of Romanticism. British gentleman Horace Walpole built a mansion, Strawberry Hill, in 1770 in imitation of a medieval castle,²⁴ decades ahead of the taste of the times and initially considered whimsical and eccentric; by the next century it was the norm. Goethe, in his description of a visit to the medieval cathedral in Strasbourg, gave expression to the new evaluation of medieval architecture. In the later part of the Romantic era, the image of the Middle Ages as an ideal of communal existence and historical continuity became a political ideal. Novalis, for instance, even during his period of support for the French Revolution, praised the medieval period, and in his *Christianity or Europe* of 1799 expressed some progressive ideas along with an extreme, almost caricatured veneration of the Middle Ages. Novalis claimed that members of the community ought to wear clothing with colors of their rank, and that the subject ought to pay taxes to the state as a lover gives gifts to the beloved. Novalis was not truly a reactionary, as he has often been portrayed. He did not reject philosophy out of hand, believing that its criticism of the Old Regime was salutary, and he did not pretend that the Old Regime was virtuous.[25] Other writers, such as Chateaubriand and de Maistre, took this veneration for medieval authority of king and church to reactionary extremes.

In the field of philosophy, the writings of Renaissance thinkers of nature were revived. Goethe's masterpiece *Faust* was, of course, based on a medieval legend and on medieval or Renaissance figures, such as Dee (who was likely a model for Marlowe's *Faust*), and Agrippa among others. Goethe admired the alchemical writings of Paracelsus and during his convalescence in Frankfurt, he studied alchemy as an avocation.[26] Schelling, in his later period, through the influence of Franz von Baader, took up the writings of Paracelsus in consequence of becoming interested in the doctrines of Louis Claude de Saint-Martin, an eighteenth-century occultist favored by Baader. Schelling had earlier used Giordano Bruno as his mouthpiece in the dialogue *Bruno*. In that dialogue, Schelling used doctrines of Neoplatonists and hermeticists extensively, as well as praising Kepler's physics and astronomy, influenced by the young Hegel's dissertation on the planets, which likewise praises Kepler over Newton.[27]

The Masonic movement influenced Forster, Soemmerring, Herder, and Fichte. Masonry was a significant force in the Enlightenment, insofar as it formed a seed of civil society, and was an arena in which the bourgeoisie and artisans mingled in a fraternity that suppressed their class differences.[28] At the same time it conveyed Renaissance hermetic and alchemical doctrines to its members. Elias Ashmole, devotee of alchemy, collector of alchemical tracts, and editor of the *Theatrum Chemicum Britannicum,* was one of the first nonartisan members.[29] Hermetic language appears in Masonic rituals.[30] Georg Forster, the "German Jacobin," the only major German intellectual who threw in his lot with the French Revolution, joined a Rosicrucian circle in Kassel in the Masonic lodge with his friend Soemmerring, an anatomist who was to write on the location of the soul.[31] Forster delivered a talk at the lodge that reveals his early liberal political ideas.[32] The power of the Masons is suggested by the fact that Forster and Soemmerring apparently fled Kassel in part because of the grip that the local Masonic lodge had over them (including financial debts and fears of reprisal), despite the fact that they wished to leave the Rosicrucians.[33] Likewise Goethe was evidently involved in Rosicrucian activities during his youth, which may have been absorbed into his alchemical poetry and contributed to his interest in Paracelsus.[34]

Both Herder and Fichte wrote on Masonry, in which both projected their conception of the good society.[35] For Fichte, Masonry was a way of advancing the vocation of the self.[36]

The Cabala was studied by Herder in his investigations of Hebrew literature. Schelling absorbed part of cabalistic doctrine through his study and appropriation of ideas of Jakob Böhme, who in turn had studied Cabala. Schelling in his later phase in Munich appropriated ideas from the Cabala, such as God's contraction, in his theosophy.[37]

The Romantic Reaction to Newtonian Science: Blake and Goethe

Two major poets of the Romantic era, Goethe and Blake, expressed in their writings strong antipathy to the mechanistic vision of the universe that they found in the physics of Newton. These two men, although neither was a professional scientist, expressed and epitomized most eloquently the Romantic rejection of Newtonian science.

William Blake

William Blake (1757–1827) was best known in his own day as an artist, although it is his poetry that is most valued today. Born of Dissenting parents, he made a meager living as an engraver and illustrator. He had visions that were the source of much of his mature poetry and art. His works combined poetry and illustration in an integral fashion. Apparently, he also composed and sang songs, but none of these has survived. Blake's own cosmology was much influenced by theosophist Jakob Böhme and by Paracelsus and Neoplatonism.[38] Blake read the hermetic documents that had been translated into English. His poetry contained powerful indictments of the London of the Industrial Revolution and of the stifling dogmas of traditional religion:

> In every cry of every Man
> In every Infants cry of fear.
> In every voice; in every ban.
> The mind-forg'd manacles I hear.
>
> How the Chimney-sweepers cry
> Every blackening Church appalls.
> And the hapless Soldiers sigh
> Runs in blood down Palace walls.
>
> But most thro' midnight streets I hear
> How the youthful Harlots curse
> Blasts the new-born infants tear
> And blights and plagues the marriage hearse.[39]

His poems contain as well an alternative religious and political vision of a higher innocence, including "The Marriage of Heaven and Hell," integrating opposites. He died singing songs and hymns.

Blake's complex but primarily negative attitude toward Newton was expressed in his famous picture of Newton drawing the geometrical patterns of the universe with a compass.[40] Donald Ault points out the contrast between the strength of Newton and the bending of his body, contorted to resemble the figure being drawn, while the muscles on his back become like parallelograms or scales of a serpent. Newton's body is seated on a rock that resembles the foot of a beast standing outside the picture frame, and Newton's own foot seems to have become part of the rock itself. The ambiguous shapes (faces?) in the lichens on the rock contrast with the sharpness of the geometrical figure being drawn by Newton. For Blake, according to Ault, Newton represented the tremendous powers of the imagination turned into a "demonic parody or disguise."[41]

Blake's strong rejection of atomism, and the irreligious materialism and Enlightenment attitudes thought to follow from it, is found in this poem:

> Mock on, mock on, Voltaire, Rousseau
> Mock on. 'Tis all in vain
> You throw the sand against the wind
> And the wind blows it back again. . . .
> The atoms of Democritus
> And Newtons Particles of light
> Are sands upon the Red Sea shore
> Where Israels tents do shine so bright[42]

Blake wrote that to propagate atomistic doctrines was "to educate a Fool how to build a Universe with Farthing Balls."[43] Blake in his murky way associated (correctly, I believe) British inductivism and atomism with the establishmentarian doctrines he opposed. "Bacon and Newton would prescribe ways of making the world heavier to me & Pitt would prescribe distress for a medicinal potion."[44] "I read Burkes Treatise when very Young at the same time I read Locke on Human Understanding & Bacons Advancement of Learning. I felt the Same Contempt & Abhorrence then; what I do now."[45] And concerning Newton's infinitesimals: "I know too well that a great majority of Englishmen are fond of the Indefinite which they Measure by Newton's Doctrine of the Fluxions of an Atom, A Thing that does not Exist. These are Politicians & think that Republican Art is Inimical to their Atom. For a line or Lineament is not formed by Chance: a Line is a Line in its Minutest Subdivisions: Strait or Crooked It is Itself & Not Intermeasurable with any Thing else. Such is Job, but since the French Revolution Englishmen are all Intermeasurable One by Another. Certainly a happy state of Agreement to which I for One do not Agree."[46]

Blake famously charged, "May God us keep / From Single vision and from Newton's sleep!"[47]

Goethe

In contrast to Blake, Johann Wolfgang von Goethe (1749–1832) not only rejected what he thought to be the Newtonian picture of the world but he attempted to construct an alternative science to it in Newton's greatest experimental stronghold, optics.

Goethe was a universal genius who did major work as an amateur scientist in botany, geology, anatomy, and the theory of color vision. Blake was far less educated in the literature of science and philosophy than was Goethe; nonetheless, he captured with profound intuition and expressed with an eloquent voice the connection between Newtonian science and the social and technological conditions of his time, which he wished to reject and transform. Goethe, on the other hand, admired Napoleon, not as legacy of the French Revolution, but rather as a world historical figure like himself, and was contemptuous of the French Revolution and its defenders. Especially in his later years as the "Olympian," Goethe had no interest in political events or in the sufferings of commoners.

Goethe was an extraordinary polymath who eventually became a kind of German national institution. Even in the later part of his own life, he had become a monument and spoke of himself as such with considerable detachment. During the Nazi era, members of the German resistance referred to Goethe as the symbol of the "good Germany." Thomas Mann often alluded to Goethe, as did historian Friedrich Meinecke, who naively proposed a series of local Goethe Communities to redeem Germany from World War II.[48]

Goethe, like Beethoven and Hegel, stood on the watershed between the Classical and the Romantic. Goethe and Hegel died within two years of each other, although Goethe had already written his first novel the year Beethoven and Hegel were born. They had something of a mutual admiration society, as when they exchanged a beer stein inscribed, "Beer turns flesh into spirit," and in gift giving Goethe recommended his own "primordial phenomenon" to Hegel's "absolute."[49] The young Goethe electrified Germany with *The Sorrows of Young Werther*, which caused a wave of suicides among German youth. Partly in critical reaction to this early work, Goethe shifted from this proto-Romanticism to an Olympian Classicism in his later life.

Goethe was in some ways too wise to be a philosopher,[50] but he incorporated elements of the philosophies of Spinoza, Kant, and Schelling in his own worldview. Goethe shares with many of the figures of Romanticism the combined interest in Kant and Spinoza. In book 14 of *Poetry and Truth* he says: "After I had looked around the whole world in vain for a means of developing my strange nature, I finally hit upon the ethics of this man. . . . Here I found the serenity to calm my passions. . . . Above all, I was fascinated by the boundless disinterestedness that emanated from him."[51] Goethe said that the time he spend studying Kant's *Critique of Teleological Judgment* was one of the "happiest periods of his life," claiming that Kant's work on teleology in biology and in art expressed his "most diverse thoughts" and attitudes.[52]

Goethe contributed to half a dozen scientific topics as a talented amateur. He wrote extensively on botany, basing his theory on the primordial plant, or *Urpflanze*. For Goethe, this was the archetypal form of the plant from which could be derived the various particular anatomies of the numerous species of plants.[53] This archetypal thinking was central to Goethe's approach to science.

Lisbet Koerner implicitly exposes the ambiguity in Goethe's science as "feminine" science.[54] On the one hand, Goethe pursued botany as an amateur, and amateur botanizing was a popular female preoccupation throughout the nineteenth century. Goethe's own antimathematical, qualitative style of science, which emphasizes empathetic participation in the object, and in his color theory did not make a sharp break between what we would today call subjective and objective aspects of the color phenomena, which would fit the characterization of some today of feminine science.[55] Also, Goethe is famous for praising the spirit of the feminine. On the other hand, Goethe was hardly a feminist; his portrayal of girls, women, and female nature, as in *Elective Affinities* (to be discussed for other reasons below) emphasized suffering as natural.[56]

In geology, he made a number of contributions. Indeed, in *Faust* the controversy between the Wernerian advocates of water as a shaping element of the landscape and the plutonist advocates of fire is dramatically played out in a debate between Faust and the Devil. Goethe considered granite to be the archetypal stone from which other stones could be built or derived. He defended gradual activity on the Earth in keeping with the principle of continuity and correctly argued for the glacial origins of erratic boulders. He identified the "ring of fire" of volcanic activity around the Pacific Ocean in Asia and the Americas.

In comparative anatomy, Goethe made one of his most important contributions: the discovery of the intermaxillary bone in humans. Goethe recognized this bone in the skulls of mammals, and considered on the basis of a notion of the unity of type that human beings should also have this bone in order to be unified with the rest of the animal kingdom. He noted that certain sutures (gaps between the bones of the skull) were more evident in children than in adults, and even more evident in embryos than in children. Goethe could find the palatal suture in this manner; the nasal suture could more easily be observed. The facial suture is never observed in humans, but Goethe inferred that since ruminants have the bone even when it is not needed to support the sharp incisor teeth of meat eaters, it must be present in humans as well who do have incisors.[57] In some respects, Goethe's claim concerning the intermaxillary bone and the resistance to it, insofar as it eliminated a supposed anatomical differentiation between humans and apes, is a forerunner of the better-known dispute a half century later between Thomas Henry Huxley and Richard Owen concerning hippocampus minor, according to Owen supposedly lacking in the brains of apes and thereby distinguishing humans from gorillas.[58]

Goethe's hypothesis of the intermaxillary bone in humans was first presented in 1784. It was rejected by the leading anatomists of the day, who claimed that it was not clearly visible. Goethe argued for it by inference and eventually won the day. In 1820 he published the original paper along with an account of the debate, although he mistakenly claimed that his initial opponents had come to agree with him.

Goethe's work in biology based on the notion of primordial or archetypal shapes is in many ways pre-Darwinian and harks back to a typological conception of species. Goethe, however, believed in a continuity of forms in nature and also in at least limited modifiability of species, although probably not in a full evolution. Nevertheless, Goethe's idea of the building plan (*Bauplan*) of organisms is a concept that did become central to comparative anatomy. Recent German anatomists and evolutionists, such as Rupert Reidl, appeal to the notion of building plan.[59] Evolutionists Stephen Jay Gould and Richard Lewontin make use of this idea to discuss nonadaptational features of evolution.[60]

In the modern evolutionary view, there are a number of basic building plans that are produced by major transitions between phyla (the largest groups of organisms below kingdoms), such as the four-leggedness of vertebrates. These plans may be relatively arbitrary but once present are built on by evolution. William James, an American philosopher and psychologist, recognized the role of features such as four-leggedness in constraining the paths of evolution. Goethe thought of these building plans as universal for the whole kingdom of plants or of animals, although their applicability was much more limited to flowering plants and vertebrates.

Goethe's major contribution to Romantic science was in his rejection of Newton's theory of colors. He supposedly started to read Newton's *Optics* and attempted to use a prism to follow Newton's experiments. However, he was clumsy and unsuccessful with the use of the prism and did not see the phenomena that Newton, one of the most precise experimenters of all time, had been able to exhibit. Furthermore, he was outraged by the Newtonian dissection of white light into a mixture of various colors and proceeded to develop his own alternative theory of colors, published in *The Theory of Colors*[61] years later.

There is a sociopolitical motive involved in Goethe's polemic against Newton.[62] Goethe identified Newtonianism with arbitrary tyranny of a closed circle of devotees and compared his opposition to Newton to the religious reformation of Martin Luther, portraying Newtonianism as a theology.[63] Goethe was a conservative supporter of enlightened despotism of nobility, and he associated Newton with Napoleonic dictatorship that was the outcome of the French Revolution. He saw himself in this period as battling the artificiality of Newtonian abstraction and tyranny on one side and the anarchistic arbitrariness of the early Romantics, who supported the early radical phase of the French Revolution, on the other. He saw his own methods as more open to the dilettante as well as to the artisan than Newton's mathematical abstractions and procrustean experiments.

Goethe claimed that light and dark are the fundamental colors and that the turbidity of a liquid medium produces colors out of light and dark. He based this on observing light shined into liquids. Goethe developed his own theory of the relationship of the various colors, claiming that orange was a compound of red and yellow because the human eye can see red and yellow within orange. Similarly, he claimed that green was a compound of blue and yellow even though the eye cannot see the blue in green as well; in the mixture of paints one makes green from blue and yellow. Goethe claimed that purple is an intensified red. Psychologically, the sort of purple that Goethe concerns himself with is indeed an intensified red in comparison with which ordinary red appears to be yellow.

Goethe thought that he was developing a physical theory of colors to oppose Newton's *Optics*. In fact, he was primarily developing a psychological theory of color that had not been scientifically developed previously. Goethe wrote on such phenomena as the compounding of colors from warm and cool hues and the role of colors in afterimages produced by staring at a color and then staring at a white sheet. Goethe also discussed a variety of psychological phenomena, such as the emotional reaction to colors and the role of colors in fashion.

Goethe is often claimed to have totally lost his dispute with Newton. Certainly within physics, if not psychology, this is the case. (Although some of Germany's leading physicists of later periods gave talks attempting to evaluate the psychological and humanist focus of Goethe's account of vision in relation to the physicists' account.)[64] Even Goethe's acquaintance, physicist and fellow aphorist Georg Lichtenberg, would not take his objections to Newton (for instance concerning colored shadows) fully seriously.[65] The only scientist to commit himself totally to Goethe's program in optics was Thomas Seebeck. Seebeck is remembered for his discovery of thermoelectricity, which shows us that he was not ignorant of mainstream physical science. Nevertheless, Seebeck's commitment to Goethe's color theory led to "crisis" for him.[66] The philosopher of science Joseph Agassi, says, "There is something cockeyed about Goethe. . . . There is no need for any historian of science to mention Goethe."[67] The discovery of the intermaxillary bone shows that it is not true even in the narrowest sense of science, and the psychological descriptive accuracy and suggestiveness (if not physical explanatory power) of his color theory shows that it is not so even of this disputed project.

Goethe's theory of colors had some resemblance to Edwin Land's theory of color vision, developed in the 1950s.[68] Edwin Land, the inventor of Polaroid lenses, was experimenting with color filters on two slide projectors when he discovered that full-color images could be produced by certain pairs of colors. This went against the accepted theories of color vision, which claimed that at least three colors were needed to produce a full-color photograph.[69] Land explained these phenomena on the basis of a theory of warm and cool pairs of colors similar to that of Goethe. His discovery, like that of Goethe, aroused a strong reaction from the scientific community of his time. In part, this was because Land, already a multimillionaire inventor, had stumbled by accident upon a discovery that had eluded the professionals in the Optical Society.

The initial reaction to Land was to say that the phenomena did not exist. It turned out that Land had a very precise apparatus not available to those who originally attempted to replicate his work. The next reaction was to claim that Land's results were well-known to optical scientists. It was claimed that simultaneous contrast phenomena (in which a color against a background of another color would look different from the original color) could explain Land's results. A computer reconstruction of simultaneous contrast phenomena does map onto the pairs of colors that Land claimed were sufficient for full-color images.

Another peculiarity of Land's color-vision demonstrations was that they seemed to depend on pictures of objects whose colors we generally know, as such as bananas and apples. Thus a factor may be our expectations concerning the "correct" colors of objects. This does not seem altogether to be the case. In experiments I performed, the addition successively of three-dimensionality, shadows, and textures to a set of similar cubes (which give no clue as to color by their shape or size) enhances the visual effect of Land's two-color projections.

Another area of science that was of great interest to Goethe was chemistry; however, Goethe's interest in this area was expressed in a nonscientific manner in the form of a novel, *Elective Affinities*.[70] The novel describes the relationships of two couples and the mutual attractions of the members of each couple both to their partner and to the member of the opposite sex of the other couple. One of the couples manages to stay together and the other breaks up. The various attractions and pairings are driven by largely unconscious forces, or "affinities," which can either be accepted or disciplined. Goethe implicitly criticizes the rationalistic approach to emotional attractions and repulsions by presenting the character Mittler (middleman, or mediator), who resembles Jeremy Bentham. Mittler acts as a go-between for the couples, but only makes their relationships more difficult by his well-meaning meddling. Goethe is here implicitly criticizing the rationalistic treatment of interest, pleasure, and pain in Bentham's philosophy.

The title of the novel is taken from a concept in chemistry. Goethe had read works of the early alchemists, as is shown by his work, *The Parable*. He was also familiar with the chemistry of his own day and had read the chemist Torbern Bergman on elective affinities.[71]

Bergman's elective affinities are attractions of various chemicals for each other. The notion is a precursor of the later concept of chemical bond, or valence; however, Bergman's chemistry is earlier than the atomic chemistry of John Dalton and others. Different chemical elements and compounds have different attractions for one another. Goethe considered these attractions to be analogous to the mysteriously overpowering emotional attractions between the couples. They are consciously recognized by the lovers to be irrational, immoral, yet at the same time irresistible.

We shall see the role of the notion of chemical affinities in the work of Humphry Davy and Michael Faraday. In Davy and Faraday the affinities are treated in terms of electricity and are related to Romantic concepts of electrical polarity. Bergman's affinities precede this electrical theory of affinity.

Goethe's novel had an interesting, indirect impact on sociology. Max Weber read Goethe's *Elective Affinities* in the classroom as a young student, hiding it behind his textbook. Weber borrowed the concept of elective affinity as an unexplicated but crucial term in his sociology. Apparently, he first applied it to sociology in discussing the relationship of the ideas of socialist intellectuals to the German labor movement.[72] Weber applied the concept in a number of other areas. Most importantly, he used it to describe the relationship between religious movements and economic systems in a famous passage in his most famous book, *The Protestant Ethic and the Spirit of Capitalism*.[73]

Weber's English translators variously render the term as "correlation," "relationship," and "bond." Thus, the concept largely disappears from the English-language versions of Weber's works. This is unfortunate, since the term plays a crucial role in replacing a Marxist notion of the relation between material base and intellectual superstructure. In place of the concept of "reflection" or "expression" of the base in ideas, Weber uses this notion of affinity, which is less deterministic, essentialistic, or causal. Affinity, however, is a stronger tie than a mere correlation.

Weber's concept of elective affinity in sociology has structural similarity with concepts we have been examining in correlative cosmology and philosophies that aided the development of electromagnetic field theory. Weber has often been mistakenly accused by empiricist critiques of being an extreme holist. John Watkins, a follower of Popper, does so citing quotations that do not prove his point. Weber's notion more resembles Leibniz's preestablished harmony or Chinese correlative cosmology than it does the extreme holism of German historicists.

Weber wishes to treat society as in a state intermediate between a rationalistic, individualistic atomism and an all-encompassing holism. He also wishes to treat the sociological relationships of human beings at a status intermediate between a purely mechanistic or behavioral account and a purely conscious rationalistic account. Chemistry, in the period of Bergman, was in an intermediate state between earlier alchemy and later atomism. Weber's sociology is perhaps in a similar position.

The range of Goethe's work, from successful discoveries in comparative anatomy and the psychology of color to suggestive speculations in geology and botany, shows that Romanticism, despite its critiques of Newtonian physics, far from attempting to limit science, expanded its role.

CHAPTER 13

Social Background to Romanticism: Revolution and Women

———••••———

*T*he French Revolution (1789–1815) was the major political and social context for the ferment of radical new ideas about human freedom, art, sexual relations, and nature. Just as religious Taoism surfaced during a period of widespread peasant revolts in China, and just as English hermeticism and the alchemical philosophy were associated with currents in the English Revolution and Civil War of the 1640s, so too did the French Revolution set the tone for the early utterances of German Romanticism. Just as Taoism, soon after its explosive appearance on the political scene, in large part accommodated itself to the establishment in Chinese society, so too Romanticism within a few years (or at most, a few decades) became for the most part extremely conservative. This shift in the political associations of either movement does not, however, distract from the politically radical origins of both.

The French Revolution charted a course from moderate reform to radical terror to moderate dictatorship and finally to imperialistic nationalism. The Revolution began in 1789 with the king's calling of the Estates General, a parliamentary body that had been moribund for a century and a half. The demands of this body were at first very moderate compared to the later course of the Revolution. The three estates were the nobility, the clergy, and everyone else. The so-called Third Estate demanded voting by numbers rather than by estate, and called for a National Assembly. After the king disbanded the body, and after finding themselves locked out of the assembly hall (apparently by accident), the members of the Third Estate refused to leave and reconvened in a tennis court, taking the famous Tennis Court Oath. The king acknowledged the new situation, but the earlier dismissal of the finance minister Jacques Necker, who had made possible the doubling of the number of representatives of the Third Estate and the moving of troops into the Paris suburbs, led to the storming of the Bastille, a former political prison.

The Revolution in its early stages was led by reformist nobles who generally had no desire to overthrow the monarchy but simply wished to institute a less arbitrary rule along the lines of the British constitutional monarchy. Political conflicts with the king, however, combined with economic problems and food shortages led the Revolution to take a much more radical turn, culminating in the rule of the Jacobins. One of the major events leading to the full-scale break with the monarchy was a march for bread led by market women to the palace at Versailles. The court went to Paris. (Interestingly, the 1905 Russian revolution was similarly touched off by the brutal suppression of a peaceful march of mothers—in that case, more ironically, actually led by a priest who was a police agent.) The ill-conceived flight of the royal family eventually led to their execution.

A number of events fed the growing fears of subversion and overthrow of the revolutionary assembly. These included the flight of the nobility to England and Germany, rumors of plots of the emigrés to invade France with the help of foreign powers, and a foolish declaration by German princes (who were too weak to conquer France), threatening an invasion to overthrow the new democracy.

The fear of foreign agents and foreign invasions led to the Terror, in which opponents of the Revolution as well as previously moderate supporters and political opponents of the present ruling faction were executed. The Reign of Terror reached its culmination in 1793/1794, when Robespierre was beheaded and a less radical regime, the Directory was installed. During the Terror, foreign military threat became a reality, and during the Directory France conquered neighboring regions of Germany and Italy, supposedly in self-defense. The brilliant military leader of the highly motivated popular army of France, Napoleon, rapidly rose from general to military dictator (or first consul) in 1799. During the succeeding years, Napoleon's *Grande Armée* conquered most of western Europe.

In 1804 Napoleon crowned himself emperor. He expanded French rule through much of eastern Europe but, after an ill-fated invasion and winter retreat from Russia, was finally defeated by various European alliances at Waterloo. At the Congress of Vienna in 1815, Europe returned, at least outwardly, to monarchical rule. Prince Metternich of Austria oversaw the forging of the new order in Europe. The Holy Alliance of the Catholic king of Austria, the czar of Russia, and the Protestant ruler of Prussia, soon joined by all the kings of Europe, pledged itself to the suppression of any democratic or liberal religious activity. Despite the reinstating of the traditional nobility, Napoleon's freeing of the serfs in the German states and the institution of a modern, rational legal code in the countries he had conquered remained after his defeat. Even the most reactionary rulers feared a return to the anachronistic and universally despised features of feudalism that had been abolished by the Napoleonic armies.

The course of the Revolution to an increasingly radical position, in part driven by the threat or reality of foreign invasion in support of emigrés, has its parallels in the English Puritan revolution of the previous century and the Russian revolutions of the twentieth century. Likewise, the execution of the king, the rise of the Terror, the eventual turn to military dictatorship, and the conquest of neighboring countries has its parallels in the rise of Cromwell in England and the English occupation of Ireland, as well as in the rise of Stalin in Russia and the Soviet Terror and the conquest of Eastern Europe after the Nazi invasion (although the latter was decades later than the Russian Revolution and a decade after the Great Purge).

The French Revolution and the Romantics

The reaction of the German Romantics to the French Revolution likewise shifted with the nature of the revolution itself. The early years of the French Revolution were greeted almost unanimously by German thinkers as a positive development in the realization of human freedom. It was only the growth of the Terror and the slaughter of many of the Revolution's supporters by the revolutionaries themselves that led many previously fervent German supporters to turn against it.

Among those who remained loyal to the ideals of the French Revolution, the French invasion of Germany in the name of "liberty" turned many into passionate nationalists opposed to the French. Later, others were disillusioned by Napoleon, who initially claimed to be a military dictator protecting the gains of the Revolution against its enemies, betrayed democracy by becoming emperor, and made peace with the Catholic Church. When Beethoven heard that Napoleon had had crowned himself emperor, he removed the dedication of his Third Symphony to Napoleon.

One common misinterpretation of the politics of the German Romantics is the identification of Romanticism with conservative political ideology. It is true that by the end of Napoleon's rule most Romantics had turned to political conservatism, though virtually all the Romantics were enthusiastic supporters of the Revolution in its early years as a fulfillment of the development of human freedom. Many continued to support it to some degree even after the Terror and after Napoleon's imperial designs became clear.

For instance, it is often mistakenly assumed that the early opponents of rationalism in eighteenth-century Germany were political conservatives. This is a mistaken reading-back of the conservatism in middle or old age of figures such as Schlegel and Schelling from many decades later. The German proto-Romantics, even the "Magus of the North," Johann G. Hamann, were all Enlightenment liberals. Herder continued to defend the French Revolution into the early 1800s, after many of the Romantics had abandoned it. Jacobi, despite his attacks on reason, was Germany's greatest expert on the newer economics and an advocate of the doctrines of the French Physiocrats. He is often mistakenly credited with having introduced the ideas of Adam Smith to Germany.[1]

A further factor in the mistaken claim that the Romantics were all politically conservative is that fact that many of the Romantics died young. The Romantics who were alive after 1815, such as Schelling and the Schlegel brothers, were indeed conservative; however, figures such as Shelley and Forster died decades earlier than Wordsworth and Coleridge, who became apolitical and ultraconservative in their later years. Perhaps the distinction between radical Romantics such as Shelley and Fichte and conservative Romantics such as Coleridge, Schelling, and the Schlegels is simply a matter of longevity.

The most radical figures in Germany among the major intellectuals in the period of the French Revolution were Forster, Fichte, and Kant, in order of their radicalism. Forster actually joined the radical Jacobins during the French occupation of Mainz and retreated with the French troops when the Germans retook the city. Forster became progressively more negative about the course of the Revolution while living in France but continued to support it until his death.[2] Forster himself radicalized Caroline Böhmer, successively the spouse of August Schlegel and Friedrich Schelling, who in turn increased the sympathies for the Revolution of both these figures, but especially of August's brother, Friedrich.

Fichte defended not only a right of revolution but a duty of revolution to overthrow unjust governments. He was not a full-fledged Jacobin but did defend the radical developments of the Revolution. Among major German intellectuals, Fichte was certainly the most favorable to all aspects of the course of the Revolution up until the French occupation of large portions of Germany. At that point, Fichte shifted from his Jacobin phase to his nationalistic phase. A century later, Fichte's *Addresses to the German Nation* was printed in an edition to be carried in the knapsacks of German soldiers during World War I. Fichte's later nationalism and its popularization in the twentieth-century led many writers in the English-speaking world to consider him a forerunner of the most extreme German chauvinism.[3] This neglects the extent to which Fichte was the strongest defender of the French Revolution's slogans of liberty and equality prior to his nationalistic phase, and also the extent to which Fichte's nationalism was not yet the fully political nationalism of the national state, as Germany would not be unified for more than another half century.[4] Fichte's nationalism was one of language and culture. The alleged superiority of the German language was based simply on its supposedly closer relationship to its prehistoric roots than were believed to be possessed by the Romance or the Scandinavian languages.

Perhaps most surprisingly, given the image of his highly routinized and morally rigorist old age, Kant defended the French Revolution until his death. Two of the rare occasions when Kant allegedly missed his daily walk were when he received a copy of a book by Rousseau and when news of the French Revolution reached Königsberg. He took offense at the execution of the king but did not reject the Revolution even after learning of massacres and the Terror. Kant did not live to see Napoleon defeat many of the German states or threaten the far eastern part of Prussia where he lived.

Herder has much more unjustly than Fichte been considered a source of later German nationalism. In fact, Herder, in spite of his emphasis on the importance of national language and customs, was not a defender of the superiority of one nation over another. He introduced notions of historicism and of a plurality of equally valid national cultures, but in many respects his attitudes were those of the Enlightenment. He considered humanity to be unified and rejected the major distinctions among the intellectual capacities of the races claimed by such other philosophers as Hume and Kant. Herder, like Kant, defended the French Revolution to the very end of his life in 1804, though he was upset by the December Massacres and the Terror.[5]

It is a mistake to associate the Enlightenment with democracy and Romanticism with the conservative support of the Church and the monarchy, despite the fact that criticism of religion, defense of materialism, and utilitarian objections to monarchical government (especially in France) contributed to the climate of opinion that supported the French Revolution. Most of the defenders of the Enlightenment, especially in Germany, defended so-called enlightened despotisms of Prussia and Russia. The ideal state was rationally organized and ruled by a despot who had been educated in the ideals of the Enlightenment. Thus, in Germany many of the Enlightenment liberals, although initially briefly supporting the French Revolution, rapidly turned against it. The early Romantics, such as Friedrich Schlegel and Novalis, were writing their essays aware of and opposed to the political philosophy of enlightened despotism defended by the German Enlightenment political philosophers (including Christian Wolff, Johann Eberhard, and Christian Garve).[6]

Likewise, most of the early Romantics were not political reactionaries who from the start opposed the French Revolution. Later developments of conservatism and the growth of reaction after the defeat of Napoleon led writers even in the early nineteenth century to associate Romanticism with political reaction. Such radicals as Heinrich Heine, Arnold Ruge, and Karl Marx tended to identify Romanticism with obscurantism and defense of religious reaction. Similarly, in the twentieth century, many of the Nazis embraced the irrationalism and intuitionism of the Romantics as congenial to their cause and celebrated the nationalism of Fichte and others as support for German expansionism and racism.

During the the French Revolution, however, critics of rationalism were not necessarily defenders of political reaction. Throughout his life the proto-Romantic Jacobi criticized the rationalism of the Enlightenment as leading to religious nihilism in polemics against Moses Mendelssohn. Despite this fact, Jacobi was a liberal and was actually the most knowledgeable person in Germany about the modern economic theories of the French Physiocrats. Johann Herder and Johann Hamann were both liberals *and* critics of Enlightenment rationalism. Even Novalis, who wrote *Christianity or Europe*, extolling the virtues of the Middle Ages and the Catholic church, was in fact an earlier enthusiast of the French Revolution. He had praise for the cult of Robespierre as a model for civic religion and made an early defense of the Terror. Even when Novalis praised the major attack on the French Revolution by Edmund Burke, he called it a revolutionary work against the Revolution. Novalis's defense of the Middle Ages was not based on irrational dogmatism but rather on the view that the Middle Ages manifested an admirable sense of community and that the Catholic Church successfully integrated aesthetics and religion in the manner Novalis desired. The "Romantic School" (in the stricter sense of Novalis and the Schlegels) supported these features of the Middle Ages without rejecting other positive aspects of radical Protestantism.[7]

Friedrich Schlegel, the central figure of the literary Romantic movement and leader of circles in Jena and Berlin successively, was in his youth a very strong supporter of the Revolution. His association with Caroline Schlegel-Schelling lead to greater sympathy for the Revolution. He wrote one of the earliest and strongest defenses of Georg Forster as a major literary figure at a time when Forster was almost universally condemned as a dangerous political radical and a traitor to Germany in his support of the French occupation.[8] Schlegel, in his earlier phase, defended the French Revolution as making possible the "aesthetic state" and the sort of community that he and many other Romantics worshiped in the ancient Greek city-state. The early Schlegel not only took radical positions in defense of the French Revolution but defended equality for women as well as sexual freedom, again inspired by Caroline Schlegel-Schelling. Schlegel defended equal educational opportunities for women and emphasized the common humanity of men and women. In this early period and even later Schlegel defended freedom of sexual union and in one controversial fragment defended marriage of four people.[9] Perhaps his brother's tolerance of his wife's involvement with Schelling would be an example of this, although by then Friedrich and his wife had become more critical of such sexual freedom. It is certainly true that by the first decade of the nineteenth century Friedrich and Dorothea Schlegel had converted to Roman Catholicism and were defenders of reactionary religious despotism. Indeed, Friedrich Schlegel was working for the prince of European reaction, Metternich; however, this later reactionary trajectory of the Schlegels has misled many into thinking that the early Friedrich Schlegel, and indeed the Romantics in general, were reactionary. Schelling was associated with the circle of the Schlegels, Novalis, and others at a time when they were strong supporters of the French Revolution.

This early, more radical period of the Romantics was also the era of Schelling's nature philosophy, which was important for Romantic natural science. In the college at Tübingen, Schelling, along with his friends the poet Friedrich Hölderlin and the philosopher G.W.F. Hegel, erected a liberty tree in honor of the Revolution, and Schelling was accused of having translated the verses of the French revolutionary anthem, the "Marseillaise," for the students to sing. Indeed, Schelling himself was so critical of Novalis's sympathy for Catholicism that he wrote a poem in favor of pantheism satirizing Novalis's work.[10] The works of the later, more conservative (or reactionary) Schelling were not published in his lifetime and did not have an impact on nature philosophy. Positivists and later materialists may have considered nature philosophy reactionary in the sense that it was antiempirical and speculative, but the political atmosphere in which Schelling's nature philosophy was forged was one of enthusiasm for the Revolution.

Women and the Romantic Movement

In each of the major social-intellectual movements we have examined in tracing the development of holistic alternative approaches to science—the Chinese Taoist, the Renaissance hermetic, and the German Romantic—we find a more positive role for women socially and a more positive evaluation of the feminine than is to be found in most of the other, opposing movements of each respective period and civilization. We have seen the higher status of women in early religious Taoism, in which women as well as men could hold all priestly offices except the highest. There is the positive evaluation of the "female" in the *Tao Te Ching*, inverting and deconstructing the traditional Chinese value system based on domination of the female by the male.[11] We have seen the role of witchcraft and of women healers, alchemists, and philosophers (such as Maria the Jewess and Anne Conway) in the hermetic movement. Similarly, in the Romantic movement, we find an important social role for and influence of women, as well as, a (not wholly consistent) positive evaluation of the "feminine."

Women played an important role in the circles in which the Romantic movement originated. The early Romantic movement that centered around Friedrich Schlegel involved much collaborative work and writing. Schlegel and Novalis called their joint work "symphilosophy." Although the major works of Romantic philosophy and literature were written by men, the social circles that contained the discussions and personal contacts that built the Romantic movement as a movement were largely organized and hosted by women. These women also contributed importantly to the literary production of the Romantics through collaborative writing and translations, although they more often than not failed to receive credit for their contributions. The most notable example of this is in the works of Dorothea Schlegel, which were published by her second husband, Friedrich Schlegel, and republished in his collected works. The literary contributions of Caroline Schlegel-Schelling are even less recoverable, but her role in the early radical views on politics and sexual equality of her admirer Friedrich Schlegel, the brother of her second husband, as well as her importance as an inspiration for her third husband, Schelling, are evident. After the divorce of Caroline from August Schlegel, both August and Friedrich became progressively more conservative, in the latter's case encouraged by his wife Dorothea. After Caroline's death, Schelling virtually ceased to publish.

The impact of the Romantic women on the ideas and social connections of the Romantic movement has sometimes been credited by historians and commentators in a backhanded way. The emotionalism and social disagreements of the Romantics have sometimes been attributed to the women. For instance, a biographer of Fichte claims that the emotionalism and sentimentality of the Romantics as well as their frequent disagreements and schisms were the fault of the wives and female salon leaders of the Romantics. "In Berlin, as in Weimar, the leaders and directors of the new Romantic school were in truth the women who stood in such close and ambiguous relation to the better known men of letters. Henriette Herz, Dorothea Veit, and Karoline Schelling were the most potent factors in the disturbed chaotic movements of the literature of the time; and the dismal quarrellings and bickerings of men like Schlegel, Schleiermacher and Schelling, can only be understood when their relations to these leaders are taken into account."[12] Rather than discrediting negative aspects of the Romantic movement, as does Adamson, one can rather attribute the creativity and originality of the Romantic movement (its "disturbed and chaotic movements") to the positive presence of the women associated with the male Romantic writers, who themselves did not write the major political and literary works of the period, but conceptually contributed to their collective work as well as inspired them.

Some of the important leaders and hosts of the Romantic literary salons were Rahel Levin, Dorothea Schlegel, Henriette Herz, and Johanna Schopenhauer. One feature of note of the Romantic salon organizers was that a surprising number in Berlin were Jewish.[13] This was a period when Jews were assimilating into German culture and society, and women were beginning to attain intellectual status. Levin, Schlegel, and Herz were all highly creative, intellectual women who participated in the philosophical and literary discussions of the day despite the prohibition of higher education for women and the only partial assimilation of the Jews. It is significant that in the Jewish salons of Berlin, those led by men emphasized the Classical, Enlightenment viewpoint, while those led by women emphasized the ideas of the Romantic movement.[14]

One of the functions of the literary salons besides the discussion of philosophy and the arts was the social networking of artists and intellectuals. The salons also fulfilled the function of informal matchmaking between Jewish intellectual women and non-Jewish philosophers and artists. Several of the Jewish Berlin salon participants met their future Gentile husbands or lovers in the salon.

Two of the Jewish Berlin salon hosts were well connected with important Jewish philosophical figures. Dorothea Veit was the daughter of Moses Mendelssohn, the leading Jewish philosopher of the Enlightenment in Germany and a universally respected figure. Henriette de Lemos was engaged at the age of thirteen and married at fifteen to Marcus Herz, a physician who became a professor of philosophy to whom Kant confided some of his most important discoveries in a correspondence that also describes hypochondriac worries about his ailments.

Dorothea Mendelssohn was likewise married at a very early age to a wealthy banker, Simon Veit. She eventually left this relatively unhappy marriage to marry Friedrich Schlegel. She converted to Protestantism upon marrying Schlegel. Later, both of them converted to Catholicism as they became more conservative in the wake of the defeat of the French Revolution.

Rahel Levin did not have the parental or marital connections with important philosophical figures of the other two; however, she was more widely and profoundly read in the works of the major philosophers of the period. In contrast to the more conservative and religious Dorothea Schlegel, Rahel was a strong proponent of democratic self-expression and women's equality. She had several platonic attachments to noblemen and diplomats (who would not marry her because of her Jewish background) and to a soldier who died before their relationship could develop. She eventually married a diplomat, Karl Varnhagen von Ense, who supported her intellectual and political activities and salons.[15]

A number of the major salons were in Berlin. Rahel Levin's salon included educational, political, and linguistic theorist Wilhelm von Humboldt, conservative political thinker Friedrich von Gentz, the Schlegel brothers, Protestant religious philosopher and educational organizer Friedrich Schleiermacher, and writer Jean Paul.[16]

Henriette Herz was involved early in a salon that included Dorothea Veit and Moses Mendelssohn as well as many Gentiles, with visitors such as Madame de Staël (who described German intellectual life to the French), Friedrich Schlegel, Fichte, and Jean Paul. Henriette Herz was also a friend of philosopher Christoph Nicolai, and Friedrich Schleiermacher; late in life she was close to nature philosopher

Henrik Steffens and was supported by a pension recommended by naturalist and explorer Alexander von Humboldt.

Johanna Schopenhauer ran a salon in Weimar that was graced by visits from Germany's foremost literary figure, Goethe, as well as by other luminaries of the so-called Athens of Germany.[17] She was mother of philosopher Arthur Schopenhauer, who as a child met some of the most famous intellects of the era.

The list of participants in these salons is a roll call of almost all the major intellectual figures of the early Romantic period of Germany. The role salons played would later be filled by academic conferences and regular professional contact within universities and research institutes.[18] Indeed, it was one of the later Romantic nature philosophers, Lorenz Oken, who initiated associations for the advancement of science.[19]

Two individuals can serve to exemplify the importance of literary collaboration as well as political and philosophical inspiration by the women of the Romantic movement. Dorothea Schlegel worked tirelessly to help her somewhat disorganized and procrastinating husband, Friedrich, in his literary activities. Friedrich emphasized the importance of the style of the literary fragment, and his own life and intellectual output was fragmentary to a large extent. He was for a long time unable to find a permanent academic position. During these decades of financial need, Dorothea produced numerous translations under her husband's editorship and auspices to help support him, including a Romantic novel of Madame de Staël, translations of Dante, and various French medieval works. She also wrote a novel, *Florian*. Most of her own work was not recognized as hers until considerably later.

Caroline Böhmer was born Caroline Michaelis, the daughter of a professor of Eastern languages and Indic studies. While she was a child visitors to her home included Benjamin Franklin, Comte de Mirabeau, as well as Lessing, Goethe, Schiller, and Novalis.

Caroline married a doctor, Johann F. W. Böhmer, and bore him several children, but he died at an early age. After his death, she moved to the city of Mainz and was present when the French revolutionary army took the city. She became a close friend Georg Forster, who supported the French occupation partly out of economic considerations but also due to his own genuine admiration for the Jacobins. Forster had studied with Caroline's father, the orientalist Johann Michaelis, and had translated William Jones's version of the Indian *Sakuntala* into German, a work that fascinated Goethe and led to adaptations by Herder.[20]

Caroline Böhmer was jailed for collaborating with the French revolutionaries once the Germans had reconquered the city. She was also pregnant by a French soldier, and appealed to August Schlegel, with whom she had earlier been acquainted. He was able to secure her release from jail and eventually married her. Early in her marriage to August Schlegel, she had contact with his brother, Friedrich, who once had a crush on her and was now strongly inspired by her support and defense of the French Revolution. The young Friedrich had written an article in defense of Georg Forster at a time when the latter was universally condemned as a dangerous radical and traitor.[21] His essay "Diotima" on women in ancient Greece[22] and his "Athenaeum Fragments" contain unusually advanced advocacy of education for women: "Women will probably have to remain prudish as long as men remain sentimental, stupid and bad enough to demand for them eternal innocence and lack of an education."[23] He also criticizes Rousseau's theory of the nature of women as made to be pleasing to and ruled by men.[24] Friedrich Schlegel later became much more conservative politically and religiously, in part due to the opinions of and encouragement by his wife, Dorothea.

Caroline Schlegel meanwhile had met Schelling. At first, the young Schelling fell in love with Caroline's daughter. The daughter, however, soon became ill, and it was rumored that her death was hastened by Schelling's alternative medical treatments for her. After the death of the daughter, Schelling became involved with Caroline Schlegel. August Schlegel, true to the Romantic ideal of free love and friendship, held no hostility toward Schelling and remained friends with Caroline and her new lover. Friedrich and Dorothea Schlegel were highly hostile to Schelling and Caroline, which led to estrangement from brother August.

Caroline's adventurous life as a revolutionary and her numerous love affairs led the more traditional, moralistic Schiller to brand her "Madame Lucifer." Caroline Schlegel-Schelling, well known enough simply to be called "Caroline" by later literary historians, has been the subject of at least six novels and is now recognized as an important figure in the political developments of Germany as well as in the history of the emancipation of women.[25] All that remains explicitly of her own writings, however, are several volumes of lively and insightful letters.[26]

CHAPTER 14

Schelling's Romantic Philosophy of Nature

*F*riedrich Wilhelm Joseph von Schelling's philosophy has connections and affinities to a number of the philosophies previously discussed in relation to the rise of electromagnetic theory. Schelling is part of the tradition of German idealism and is usually treated as such; however, he made use of a number of figures in the monistic and holistic traditions we have examined, such as the early Greek pre-Socratics, the Neoplatonists, the Gnostics, Bruno, Spinoza, and Leibniz.

Schelling's Philosophy

In Schelling's idealism, the world is structured and/or constituted by the activity of the mind. Schelling worked his way back from the extreme idealism of his older contemporary and initial philosophical source, Fichte, toward a philosophy of objective nature.

Schelling published his first essay while still in his teens, his first book at the age of twenty, and held a university chair by the age of twenty-three. He was close friends with Hölderlin and while at Tübingen Stift. Hegel and Hölderlin were five years older than Schelling, who entered college at fifteen. Schelling produced system after system of philosophy in rapid succession in such a way as to lead Hegel to say that Schelling "conducted his philosophical education in public."

In a very early work, Schelling contrasted the spirit of criticism with the spirit of dogmatism, a contrast that had earlier been developed by Kant. Schelling's early criticism was directed at Fichte and Spinoza's dogmatism. He defended the critical point of view in which the world is construed as constituted by the ego or mind. He opposed the dogmatic view of Spinoza in which a metaphysical system of objective nature is deduced from certain axioms.[1]

Even this early work foreshadowed the synthesis that would appear a bit later in Schelling's philosophy. Despite the fact that Schelling defended Fichte's subjective point of view against Spinoza's dogmatism, Schelling was clearly interested in making room for something like Spinoza's philosophy of the universe as an objective totality with a transcendental idealism.

Schelling developed a twofold approach to the account of philosophy. In his *Philosophy of Nature*, he sketched the development of a series of levels of objective nature toward the rise of subjectivity or spirit. On the other hand, in his *System of Transcendental Idealism* he worked from the assumption of a transcendental mind that constitutes the world toward an account of objective nature.[2]

Schelling differs from the other transcendental idealists, such as his predecessors Kant and Fichte, as he does wish to present a metaphysical philosophy of nature. Kant's last writings were moving in that direction to some extent. Kant's *Metaphysical Foundations of Natural Science* was a successor to his critical treatment of knowledge. Additionally, in Kant's posthumous works there is an attempt to make the transition from the metaphysical foundations to a deduction of the principles of physics. In particular, there are the passages attempting the so-called aether deduction.[3] Similarly, in Kant's Third Critique there is the development of a purposive, or teleological, view of nature; however, Kant denies that this teleological account can ever be truly constitutive of nature itself (that is, is really the source of the structuring of the phenomenal world in the same sense as are the categories of substance or causality). Schelling, on the other hand, develops a metaphysics of nature, a speculative physics, in which principles of physics are deduced from metaphysical principles, and an account of nature itself as manifesting genuine purposiveness or teleology. Schelling claimed that sections 76–78 of the *Critique of Teleological Judgment*, which he called one of the richest series of paragraphs in all of philosophy, on the *intellectus archetypus* and *intellectus echtypus* held the key to returning to the view that there is real purpose in nature that can be grasped through intellectual intuition.

Although Schelling is part of the tradition of transcendental idealism, his philosophy of nature has more affinities with early Greek conceptions of nature, Renaissance nature philosophy such as that of Bruno and the hermeticists, and the general spirit of the Romantic conception of nature of Schelling's own time. He wrote a book in which his protagonist in the dialogue is the Renaissance magician and hermeticist Giordano Bruno.[4]

Schelling's conception of nature gave rise to the development of purposes or teleology, and was governed by fundamental polarities. In this respect, Schelling's conception of physical nature was close to the early Greek conception of fundamental polarities and analogies.[5] Indeed, this conception of nature, based on a system of polarities, is similar to that of even earlier so-called primitive thought. It involves primary binary oppositions of the sort found among precivilized people as well as in Chinese correlative cosmology and Renaissance hermeticism.

Indeed, the whole philosophical tradition we have been examining involves a return to patterns of thought similar to those of the so-called primitive variety. This also helps to explain the affinities and connection of many parts of this tradition to the underground tradition of peasant and preliterate thought contemporary with the various intellectual figures.

The use of polarities was the basis for inferences drawn by Schelling and his followers within so-called speculative physics. Speculative physics involved the inference of unobserved forces of nature as opposing poles to forces of nature already observed. Ritter's discovery of ultraviolet radiation as well as Schelling's denial of the existence of isolated magnetic poles, or monopoles, stems from this approach. Indeed, the discovery of the interconnection of electricity and magnetism by Ørsted and Faraday arises from this general approach.

In his earlier, more idealistic, or Fichtean stage, in his *System of Transcendental Idealism*, Schelling developed the categories of nature parallel to the categories of mind. He used the Fichtean scheme of the generation of the nonego from the ego as a model for the description of the forces of nature. In Fichte's scheme, the Absolute Ego posits an external reality in order to have something to resist its own activity. Schelling, in a rather naive and straightforward way, takes these activities and presents them in a geometrical form. He speaks of both activities as infinite but as infinities of the number sequence in opposite directions— that is, as a positive infinite and as a negative infinite. He claims to give a "deduction of matter" in parallel to these psychic activities.

He presents the two basic forces of nature as corresponding to the psychic activities. On the one hand is an expansive or repulsive force that, if unhindered, would expand to infinity. On the other hand is a negative restraining force that, if unhindered, would contract to the infinitely small. He claims that these two forces correspond to the active power of the self and the limitation of the self. When these two forces coexist, they will result in the limited structure of matter. Schelling bases himself on the two opposing forces that Kant posited in the *Metaphysical Foundations of Natural Science*; however, he does not identify gravitation with the first general attractive force, but considers it a specialization. Through the interaction of the two forces, the construction of matter is possible and takes place.

Schelling also derives the dimensions of matter in space from the forces. He claims that the polarity of magnetism accounts for a one-dimensional structure. This makes magnetism the most fundamental and general force of matter. Electricity, on the other hand, he claims, produces a boundary and from it arises two-dimensionality. Furthermore, chemical forces in which opposites interpenetrate and every point is simultaneously attractive and repulsive produces the third dimension. According to Schelling (in the *System* phase), galvanism is a process in which all three forces operate.[6]

The Romantic poets' conception of nature as organic unity was similar Schelling's. An entire work has been written on the parallels between Schelling's philosophy and the vision found in Wordsworth's poetry.[7] Indeed, the poet Hölderlin was a friend of Schelling at the university at Tübingen. Schelling is the systematic transcendental philosopher most closely associated with the Romantic movement. (Friedrich Schlegel is the philosopher most involved with Romantic literature, but he was not a systematic transcendental philosopher.) Kant's philosophy was very much a product of Enlightenment rationalism despite its foreshadowing of Romantic philosophy. Hegel, although borrowing ideas from and early collaborating with Schelling to the extent that he was thought of as a disciple of

Schelling during his twenties, was a strong critic of the whole attitude of Romanticism. Schelling placed art in a central place in philosophical cognition. Kant's Third Critique, in particular, Section 77, was central to Schelling's thought, but he did not move judgment of taste of genius (which Kant emphasized) to center stage, but rather presented aesthetics and teleology as an afterthought to complete the system. Furthermore, Kant had an important place for the transcendental imagination (A102) and its role in accounting for the affinity of appearances (A123). Similarly, he alluded to the unknown but fundamental "common root" "hidden in the depths" behind both sensory intuition and understanding,[8] but did not draw out the full systemic implications of this claim. Schelling put aesthetics at center stage, unlike Hegel, who claimed that art presents the content, in inferior sensuous form, that philosophy presents in superior conceptual form. Schelling claims that art, by freeing the mind from bad abstraction, allows the highest understandings to be expressed.[9]

Schelling's philosophy, despite its setting within the framework of transcendental idealism, has many affinities with Spinoza's. Like Spinoza, Schelling treats the world as a unified totality, a single organic entity. Just as Spinoza treats the world as a monadic, single substance behind both mind and matter (in contrast to Leibniz's treatment of the world as a plurality of individual, spiritual substances), so Schelling's universe, which is objective nature as well as spirit, contrasts with the universe of Kant or Fichte, which is constituted by minds.

Indeed, Schelling refers to the ultimate system of philosophy as one that has the scope and simplicity of Spinoza's system, but is a system of freedom rather than one of determinism, "a perfect counter-image [*Gegenbild*] of the Spinozist system."[10] Schelling also claims that "probably no one can hope to press to the true and complete in philosophy without having at least once in his whole life sunk to the depths of Spinozism," and (in his "Darstellung meines Systems der Philosophie") "I have in this work taken Spinoza as my pattern."[11]

A central concept of Schelling's philosophy is the Absolute. The Absolute is an entity in which all differentiation and distinction ceases. This notion has similarities to Parmenides's One or to the One of the Neoplatonists. It is the highest being and functions as the ground of all further reasoning. It also has similarities to the God of mystical theology. The late medieval German mystic Meister Eckhart and the early modern German mystic Jakob Böhme emphasized the extent to which God is beyond all traditional, humanly understandable attributes. The late Schelling was introduced to Böhme by Franz von Baader, a Catholic mystic, and appears to have borrowed much from Böhme.

It is this Absolute that Hegel is generally claimed to have ridiculed in the preface to the *Phenomenology of Mind* as "the night in which all cows are black." The criticism of Schelling in this preface by Hegel, who had earlier been Schelling's ally and coworker in the publication of a philosophical journal, led to a break between the two men. The famous phrase itself, or something very close to it, had already occurred in Schelling's own writings, describing something he wished to avoid.[12]

Schelling and Asian Philosophy

There is a definite similarity between Schelling's approach and that of Eastern philosophy. Schelling's own interests were in Indian philosophy, but his monism and emphasis on polarity have parallels to the Chinese Neo-Confucian systematization of Taoist nature philosophy. Schelling's Absolute and his general philosophy have a great deal of similarity to the philosophy of Hinduism in India. His notion of an Absolute reality in which all distinction and differentiation ceases resembles the India notion of Brahma. Indeed, the identification of the Absolute with the self in Schelling's idealism parallels the claim that Brahma equals Atman. The similarities are so striking that one is tempted to believe that Schelling's views may have been reinforced by acquaintance with Indian philosophy. Schelling was aware of the major literature on Indian philosophy and translations of Indian works available in his time. Schelling knew the *Bhagavad Gita* in translation and knew Humboldt's commentary on it as well as some of the work of William Jones and the writings in *Asiatic Researches*.[13]

There had been an influx of studies and translations of Indian work in Germany in the late eighteenth and early nineteenth centuries. Herder referred to Indian work in his studies of folk songs and myths. In his universal history of humankind, he emphasized the importance of Indian religion and myth, although he fitted it into his own scheme of historical progress in a way that was not true to the Indian materials. The Schlegel brothers learned Sanskrit and translated Indian writings. In his conversations with Johann Eckermann late in life, Goethe sneered that the Schlegels could not compete in European literature and so turned to Indian literature as a diversion, despite the fact that in *East-West Divan* he himself had used the conceit of a Middle Eastern setting and characters. Goethe had planned an enormous encyclopedic expansion of the *Divan*, which he never finished, and spoke highly of the *Sakuntala* translaion by Georg Forster. Friedrich Schlegel was the most dedicated among his peers in the group in Paris that attended lectures on Sanskrit. August Schlegel pursued his interest in Indian literature far beyond that of his brother and became a professor of Sanskrit.

Others engaged themselves in Sanskrit literature. Humboldt studied Sanskrit and translated the sacred writings as well as working on comparative linguistics. Goethe showed interest in Indian literature, but he withdrew in horror from what he considered the formlessness of Indian religion. Various other Romantic poets, such as Ludwig Tieck and Novalis, worked Indian scenes and myths into their poems and stories. Schelling was familiar with and sympathetic to all this. For instance, he referred to August Schlegel as "our Brahmin." He also praised Indian religion to his student Max Müller and showed himself familiar with both the English and German translations of the *Baghavad Gita*.

It may be that he used Indian conceptions in his earlier work without crediting them or even being aware of the extent of their presence. Interestingly, Schelling's first wife, Caroline, his major inspiration (Schelling may have collaborated with her on *The Nightwatch of Bonaventura*,[14] and all but ceased to publish after her death), was the daughter of an Indologist and had worked collaboratively with the Schlegels during their period of interest in Indian philosophy. Schelling's father, who was a pastor, after retirement became a professor of Sanskrit and Indic studies. Schelling knew Arabic by age fifteen, and his father, a Hebraicist, had already moved in the comparative direction.[15] These family ties suggest that interest in Indian philosophy may have been stimulated by conversations with either wife, father-in-law, or father. Although Schelling's well-known works on mythology, which contain his discussions of Hinduism, date from the last period of his life, it is well to recall that his very first work was "On Myth," written before even his earliest philosophical articles on the Romantic transcendental philosophy of Fichte. He also gave lectures on mythology that touched on Hinduism in 1799, in the midst of his work on the philosophy of nature.

Schelling's Philosophy and the Feminine

Schelling's work, like Taoism and hermeticism, shows a more positive evaluation of the "female" principle in his metaphors and metaphysics of nature and reality. In Schelling's case, the contrast with his immediate predecessors and contemporaries Kant, Fichte, and Hegel is particularly strong. Schelling made the female principle a positive and fundamental element of his metaphysics, while Kant and Fichte, and to a lesser extent Hegel, portray the female as inferior to the male.[16] For instance, while Hegel claimed that males corresponded to the animal and females to the plant, Schelling used this analogy but emphasized the equality of both partners.[17] Schelling is said to have used a kind of procreative causality in his dialectic,[18] and the metaphors of pregnancy and birth occur in *The Ages of the World* and *Of Human Freedom*.[19] Schelling even goes so far as to say that early modern philosophy from Descartes on, by rejecting the realism of nature, tends to "a dreary and fanatic enthusiasm which breaks forth in self-mutilation or . . . self-emasculation."[20] (Schelling claims, just prior to this, that idealism is the mind of philosophy, but realism is the body. Thus rejection of realism is rejection of one's own body.) Here, as with the hermeticists, the gender metaphors in Schelling's philosophy for the description of the relation of the mind (male) to the world or matter (female) are of more significance to gender issues in science than are Schelling's personal behavior toward women, although in Schelling's case, his relation to Caroline was an egalitarian one with a highly intelligent and opinionated partner. Contemporaries were unsure in the case of several reviews whether Schelling or his wife had written them.[21]

There is some parallel between the progression in rationalistic philosophy from Descartes to Leibniz and Spinoza with the progression in German idealism from Kant to Fichte and Schelling. Just as one can say that Kant made sense of those features of Descartes which Hume made nonsense of, so one can see Fichte's ego-centered world constitution as analogous to Leibniz's spiritual universe replacing the dualistic universe of mind and matter in Descartes and Schelling's monistic universe of the *Philosophy of Nature* as analogous to Spinoza's pantheistic monism. In a sense, one can say that Hegel synthesized the views of Kant, Fichte, and Schelling in the manner that Kant synthesizes the views of the previous rationalists.

Engels, Schelling, and Magnetic Monopoles in Modern Physics

Friedrich Engels, in contrast to the religious philosophers, took away an interest in Schelling's philosophy of nature of a very systematic and metaphysical sort. In his writings, Engels oscillates between two views of metaphysics: a positivistic denunciation of metaphysics that claims that philosophy should simply summarize the results of the sciences and a speculative system of philosophy of nature that criticizes the standard results of the sciences of his day. This latter enterprise is strongly based on Schelling's philosophy of nature, despite its explicit filiation from Hegel's philosophy of nature. For instance, Engels emphasizes the polarity at the basis of magnetism. On these grounds, he rejects the existence of isolated poles of a magnet. That is, a north pole of a magnet must always be united with a south pole. This was simply an empirical fact in Engels's day and thus was supported by the evidence and theories at hand.

One of the consequences of Schelling's and nature philosophy's doctrine of polarities is the notion that the north and south poles of a magnet in some sense demand each other or are complementary. This idea was taken up by Engels in his *Dialectics of Nature* (around 1878, published 1925): "Dialectics has proved from the results of our experience of nature so far that all polar opposites in general are determined by the mutual action of the two opposite poles on one another, that the separation and opposition of these poles exists only within their unity and interconnection. . . . There can be no question of a final canceling out of repulsion and attraction, . . . consequently there can be no question of mutual penetration or of absolute separation of the two poles. It would be equivalent to demanding in the first case that the north and south poles of a magnet should mutually cancel themselves out or, in the second case, that dividing a magnet in the middle between the two poles should produce on one side a north half without a south pole. . . . Two poles whose activities did not completely compensate each other would indeed not be poles, and also have so far not been discovered in nature."[22]

Later, in the twentieth century, after the development of James Clerk Maxwell's equations of electromagnetism, it was noticed that there is an asymmetry in the equations, which allow isolated electric charges—that is, a negative charge can exist without a positive one. Oliver Heaviside developed a version of electromagnetic theory that was completely symmetrical with respect to electricity and magnetism. He called this his "duplex" method. Heaviside did not take this theory literally or realistically, but thought that it was mathematically elegant and convenient. One could always neglect the magnetic terms or set them equal to zero. Several other British physicists in the early twentieth century, such as S. B. McLaren, developed theories that involved magnetic monopoles by means of a magnetic current. The well-known physicist and historian of electromagnetism E. T. Whittaker developed a monopole theory within the context of quantum mechanics, although it was rejected by such physicists as Hendrik Lorentz. P.A.M. Dirac may have developed his own account from these early but unsuccessful theories. Dirac had studied electrical engineering and thus probably was familiar with the work of Heaviside. His mathematical style, involving the formulation of informal mathematics appropriate to physical concepts, even when the mathematics is unrigorous is very similar to that of Heaviside. Dirac's "delta" function is mathematically very similar to some of Heaviside's constructions and indeed is made rigorous within the same general mathematical theory of distribution.[23] Dirac noticed this asymmetry and proposed a symmetrical version of Maxwell's equations in which isolated magnetic monopoles can exist.[24]

An isolated magnetic monopole is a north pole of the magnet existing on its own as a particle without a south pole, or vice versa. Such isolated monopoles predicted by the symmetrized theory did not turn up. With Dirac's prediction of the positron, or positive electron, on the basis of his relativistic quantum theory, the notion that particles should exist both in a negatively charged version and a positively charged version became plausible within quantum mechanics. Dirac's original prediction was based on the so-called hole theory, in which positrons existed as holes in a sea of negative energy. This fairly strange model did in fact correctly predict charged particles resembling electrons but with a positive charge. The hole theory was later rejected as at best of purely heuristic value after quantum field theory was developed.

Dirac's 1931 derivation of magnetic monopoles involved a symmetrization of Maxwell's equations that solved the problem of the fact that only integral electrical charges exist in nature. Electrons, positrons, and protons all come with charges of -1 or $+1$ but never with fractional charges. (Even quarks, postulated later, which have fractional charges, do not undermine this, as free quarks are not observed, only particles resulting from combinations of several quarks, such as protons or neutrons, which have a charge of $+1$ or 0, respectively.)

This integral, or whole number, value of charge is, in terms of quantum mechanics, the quantization of charge. Quantum mechanics did not in itself predict this quantization of charge in its original form. Dirac found that one could deduce an inverse proportion between the strength of the electric charge and the hypothetical magnetic charge. The constant of this proportionality involved basis units of the quantum theory, such as the velocity of light and Planck's constant. Once magnetic monopoles were introduced, the quantization of both magnetism and electricity was produced by this relationship. The number of actually existing monopoles was not important. In order for this relationship to exist, only a single monopole in the universe was necessary.[25]

Quantum field theory starts with a quantitized field rather than with the wave-particle complementary of ordinary quantum mechanics; however, quantum field theory can also incorporate monopoles. In fact, in the so-called gauge theory, in which each point in space has associated with it a rotation, it turns out that monopoles are a consequence of the theory. In the gauge theory, as formulated in terms of topology, one basically associates with each point in space another little space. In so-called fiber-bundle theory, a fibered space is locally a cross-product of the two spaces. That is, at each point in one space there is a "fiber" of the other space; however, the space as a whole is not a cross-product of the two spaces. For instance, an ordinary two dimensional graph could be looked at as the cross-product of the horizontal and vertical lines.

A Möbius strip is a paper strip bent around into a cylinder, but instead of being directly bent it is twisted first. This surface has only one side rather than two like an ordinary flat strip of paper circled around into a cylinder. The Möbius strip is the simplest example of a space that is locally a cross-product but is not globally so. The base space is a circle, the fibers are the vertical lines attached to the circle; however, the vertical lines are arranged with a twist.

The so-called electro-weak theory, which unifies electromagnetic and weak nuclear interactions, is formulated as a gauge-field theory. Dutch physicist G. t' Hooft and Russian physicist A. M. Polyakov in 1974 independently derived monopoles from the electro-weak gauge theory. This theory is most elegantly formulated in the topology of fiber bundles. It turns out that the particles that correspond to magnetic monopoles come out as Möbius strip–like formations in this field.[26]

Because of this, and the unification of this form of theory with the theory of the early universe, the prediction of so-called grand unified theories, which incorporate strong nuclear interactions with weak ones and electromagnetic ones, is that numbers of north and south monopoles were produced at the origin of the universe. In the breaking down of the symmetry thereby separating the electromagnetic from the weak interactions, magnetic monopoles are theorized to be produced. None of these monopoles has been found as of this writing, however. Either there are no monopoles or they are exceedingly rare. Insofar as there is an asymmetry in the original early state of the universe, it follows that there should be left over monopoles possessing one of the polarities. Theories that use this symmetry breaking predict the existence of monopoles. Because there is great interest in finding such monopoles (because they are predicted by symmetry-breaking theories), reports of their discoveries are often ballyhooed in the scientific press. An example of this was the "discovery" of a monopole by Blas Cabrero on Valentine's Day 1982. The event occurred while no one else was in the lab; and no one has been able to find another one.[27]

In 1975 announcement of the discovery of monopoles by a group of experimenters generated great excitement.[28] This discovery was soon rejected by the community of physicists.[29] This alleged discovery was made in observations of cosmic rays and was recorded in a stack of plastic Lexan sheets. Various theoretical calculations showed discrepancies of the event recorded with theory. Experimental criticism suggested that there are several alternative heavy nuclei that could have made the track recorded. Luis Alvarez, who had earlier unsuccessfully sought monopoles in moon rocks, criticized the experimenters for not having eliminated alternative explanations before making their announcement.

The search for magnetic monopoles is also motivated by the so-called missing-mass problem, which is that there does not appear to be enough matter in the universe to account for its curvature. The old Dirac monopoles were only of the mass of an ordinary proton but the new gauge-field monopoles are quite large, equivalent to 10^{16} ordinary particles. Thus their presence, even if relatively rare compared to electrons and protons, would help a great deal with the missing-mass problem. (The suggested alternatives to WIMPs, or Weakly Interacting Massive Particles, are MACHOs, or Massive Astronomical Compact Halo Objects.)

Some dialectical materialists, basing themselves on Engels, and ultimately if unknowingly on Schelling, have criticized and rejected the claim of the existence on monopoles.[30] Soviet physicists were not sympathetic to such philosophical claims, so far as I know. This may seem like a dogmatic and metaphysical or ideological (or both) intrusion of theory upon hard science, but if the failure to discover monopoles persists, and the announced discoveries continue to turn out to be spurious products of wishful thinking, this might suggest the existence of an anomaly in the standard theory, and make it worthwhile to formulate and explore a modern field theory that does not predict monopoles. Although some physicists suggested this in the wake of the discrediting of the 1975 announcement, most did not take it seriously. One reason is that one cannot prove that monopoles *do not* exist; one can only confirm the fact if they *do* exist. The hypothesis of the existence of monopoles, like other existence statements, is not falsifiable.[31] Thus the empiricist cannot be satisfied that monopoles do not exist, and the theorist, given that monopoles' existence is tied to the one great, classical physical theory (Maxwell's equations) that has not been modified or overturned by the revolutionary developments of twentieth-century physics the way Newton's theory was, is loath to reject a major structure of theory, or even to invest a great deal of time and effort investigating an alternative theory, purely on the hypothesis that monopoles *do not* exist, without there being major problems in the powerful and successful theories that have the consequence that they *do* exist.

CHAPTER 15

Coleridge: Poet
of Nature

———❦———

Samuel Taylor Coleridge (1772–1834)
was a leading poet of English Romanticism and a major figure in the introduction
of German idealism into England. Coleridge, with his eminence and persuasive
powers, as well as his friendship with Humphrey Davy, mentor of Michael Faraday, and his acquaintance with William Rowan Hamilton, was a major conduit to
England of German philosophy of nature. Coleridge built on the tradition of biblical criticism, which incorporated German scholarship into English work[1] and
imbibed German philosophy from Thomas Beddoes (1760–1808)[2] and possibly
indirectly from Joseph Priestley (1733–1804).[3] Coleridge claimed that he was able
to grasp German thought because he had already been led to similar ideas by means
of English Platonism. Some might doubt this route to preparation for appreciation of the Germans, but Coleridge did indeed have a strong background in the
writings of the Cambridge Platonists of the seventeenth century.[4] If the loose formulations of the earlier Cambridge Platonists seem hardly preparation for the rig-
ors of Kant, it may be noted that one major intellectual historian thought the
similarities great enough in content to look (in vain) for sources of Kant in the
British writers of over a century earlier.[5]

Literary critics and historians of science who have discussed his philosophy
in the past have tended to be condescending or dismissive of Coleridge's knowledge of science.[6] Coleridge studied major scientific textbooks, attended lectures
of one major German scientist, Johann Blumenbach, and had philosophical impact upon Hamilton, who took the ideas of Coleridge seriously. L. Pearce Williams, in his *Michael Faraday*, instigated controversy by emphasizing the role of
Coleridge in the ideas of Faraday concerning field theory. Some have since argued against this impact, and others have attempted to downplay Coleridge's *philosophical* impact on Davy and his chemistry, but none can deny the closeness of
Coleridge and Davy, or the actual use of ideas of Coleridge in the metaphysics of
Hamilton, which guided the latter's researches in physics and mathematics. Recently, scholarship on Romanticism and science, in particular a book by Trevor

Levere, has made clear the extent of Coleridge's serious interest in and knowledge of the sciences.[7] I believe that Coleridge's contribution to the direction of the development of British philosophy is far greater than what even his partisans claim.

The young Coleridge was a follower of David Hartley, the founder of associationist psychology. Hartley worked in the tradition of Locke and Hume, building up experience from elements of sense impressions, very much in the tradition of psychological atomism. For associationism, all "higher" or complex psychological states are built up out of elementary, primitive sense impressions by means of the constancy of their copresence or occurrence in sequence. Coleridge retained his interest in and use of Hartley even after he had become sympathetic to the ideas of such German philosophers as Kant and Schelling, who rejected the atomistic and associationist approach to mind.

The young Coleridge was an adherent of typical British empiricist approaches to knowledge and also a fervent believer in the Enlightenment and the ideals of the French Revolution. He was a follower of William Godwin, an extreme proponent of human perfectibility. Coleridge also was an admirer of Priestley, a Socinian, as was Isaac Newton. Coleridge was himself a Unitarian at this period, and rejected the notion of the Trinity. Thus, he was sympathetic to Priestley's even more radical Unitarianism, and to his support of the French Revolution. Thomas Beddoes, a physician who left Oxford for Bristol because of his radical political views in support for the French Revolution, was the center of a circle that included Robert Southey as well as Coleridge and Davy. Beddoes was concerned with the cure of respiratory ailments and thought that the new gases, or "airs," that were being discovered by such chemists as Priestly would be useful in the treatment of lung diseases associated with the Industrial Revolution. Beddoes's experiments led to those of Davy and to the psychological experiments of Davy, Southey, and Coleridge with laughing gas.

As Jan Golinski has powerfully described, Beddoes's public exhibitions with laughing gas, in which large numbers of people felt and acted upon the euphoric effects of the gas, led to the discrediting of such chemistry and such public science. For one thing, the mass gatherings and mass participation in science were considered threateningly associated with the mob and its revolutionary threat. Furthermore, the crowds imbibing the gas were associated with the irrationality of radicalism and revolution. "Gases" had come to be associated with radicalism through the work of Priestley, whose laboratory was broken into by an angry mob, raging against the Jacobin threat. Beddoes's own radical sympathies were associated with and attacked in terms of the mass activity and hallucinatory euphoric enthusiasm of the laughing gas demonstrations. After Beddoes's debacle, science retreated from mass participation to limited experiments by an elite in cloistered laboratories.[8]

Beddoes also followed in part the medical theory of John Brown, the so-called Brunonian theory, which was based on the notion of the excitability of the nervous system as the basis for life. Beddoes criticized Brown's theory, but did attempt to treat people with opium. Coleridge's later opium habit was the result of this treatment.

Coleridge and his circle were enthusiastic advocates of the French Revolution. Coleridge, Shelley, and a number of others planned a pantisocratic organization that was to be a community of a dozen people on the Susquehanna River in the United States. It never came to fruition, but through it Coleridge met Southey's fiancée's sister, who became his wife. Coleridge later fell in love with Sara Hutchinson, whom he met through the Wordsworth family. Coleridge's work *Dejection: An Ode* was an expression in metaphysical form of his frustrated love.

Coleridge soon became disillusioned with the French Revolution, and turned away from it toward a religious conservatism. He also renounced Unitarianism for the Trinity. This may in part account for his later interest in Germanic triads in the organization of philosophical principles. With Coleridge's political rejection of the French Revolution came his philosophical rejection of the Enlightenment and of atomistic philosophy. He was very much a part of the conservatism of the religious Romantic movement, which had many French representatives in reaction to the Terror, and to which some German figures belonged, such as the once very radical Friedrich Schlegel in his last years.

John Stuart Mill wrote two essays, "Bentham" and "Coleridge," in which he portrayed the atomistic and mechanistic conception of the mind, society, and ethics in the philosophy of the utilitarian Jeremy Bentham who had been Mill's father's associate and leader. Mill contrasts the rationalistic and reformist philosophy of Bentham with the organic traditionalist and historical conservative philosophy of the mature Coleridge. Mill attempts to extract what is of value from each of these opposite thinkers and to develop his own position independently of the rigid Benthamism his father, James Mill.

John Stuart Mill had been indoctrinated by his father from a very early age in the principles of Bentham's utilitarianism. According to this doctrine, the goodness of an action is to be judged in terms of its consequences for pleasure and pain for sentient beings (both human and animal) in general. Pleasures were measured on a scale of intensity and duration. Mill had a nervous breakdown in his twenties when he realized that his frenzy of utilitarian activity held no meaning for him. He was brought out of his depression by reading a poem by Wordsworth, interestingly involving the death of the latter's father. Mill then developed a less rigid utilitarianism in which there are higher and lower pleasures. "It is better to be Socrates dissatisfied than a pig satisfied."

Mill's portrayal of Coleridge tends to play down Coleridge's conservatism and the dangerous tendencies within a highly state-oriented authoritarian political Romanticism. Mill emphasizes the value of considering history, tradition, and social context in making proposals for social betterment. Bentham, with his extreme Enlightenment rationalism, totally neglected these considerations in planning an ideal legal system.

Coleridge developed a philosophy at odds with the main tendencies of British philosophy of his time and of the previous century. He claimed that everyone was born either a little Platonist or a little Aristotelian. For Coleridge, Plato represented the emphasis on the mind and ideas, while Aristotle represented empirical observation. Coleridge allied himself with Plato and with the conception of reality that he attributed to Plato but that really comes from the later Neoplatonists of ideas in the mind of the God. Coleridge was largely responsible for introducing the term Neoplatonism into British philosophical discourse, but his understanding of Plato was dated insofar as it tended to conflate Plato with the later Neoplatonic interpretations.[9] Coleridge allies himself with the Neoplatonists, the Renaissance hermeticists (such as Bruno) and with the mystics (such as Böhme). He opposes himself to the tradition of the atomists and to Newton. Coleridge denounces atomism as a product of Abdera, which he claims, apparently to deny it Western status, is in Crimea in Russia. He now denounces Newton, as do Blake and Goethe, although Coleridge in his early phase had praised Newton.

He does praise Francis Bacon, whom he at times sees as a sort of British Plato. He supports Bacon's inductive method but claims that his followers allowed this to degenerate into a directionless fact gathering. His defense of the sophistication of the original Bacon as opposed to eighteenth- and nineteenth-century British "Baconianism" as a buzz word is reminiscent of various recent writers, such as Yehuda Elkana and Peter Urbach.[10] Most innovatively, Coleridge introduces the ideas of Kant and the German Romantics. Thus, the tradition of philosophy that he supports is very much the one described in this book.

We must be careful, however, not to identify the tradition Coleridge personally supports with the totality of the tradition described in this book. Because of Coleridge's conservatism, those aspects of the tradition which were associated with lower-class revolutionary movements would be distrusted by the older, more aristocratic Coleridge. For instance, the *Pimander* of Hermes Trismegistos was translated into English by a member of the Ranters, one of the radical sects in the Puritan Revolution. As we had seen, much of the appropriation of the hermetic and Paracelsian tradition in seventeenth-century England was by revolutionary movements.

Similarly, in the period before the French Revolution, there was a movement of organic materialism associated with the revolutionary currents. Such figures as Denis Diderot were supporters of an organic materialism that was conceptually opposed to the mechanistic materialism of Descartes and Hobbes. Although the organic materialist tradition in the eighteenth century was conceptually similar to the vitalistic and organic traditions that Coleridge advocated, the association of many of its partisans with the French Revolution and Enlightenment would make it suspect to Coleridge. Also, Coleridge rejected the atheistic construal of the active-matter tradition in alchemy, hermeticism, and organic materialism insofar as the activity of matter was appealed to in order to deny the need for a deity. Coleridge was centrally motivated in his rejection of atomism and the Enlightenment by his religious convictions, and opposed Newtonianism because of its supposedly atheistic consequences.

The depth of Coleridge's philosophic knowledge, the seriousness of his scientific interests, and the extent of his scientific erudition have often been underestimated in part because of Coleridge's own self-deprecation and honesty. He often presented himself in the worst possible light, exhibiting his faults in print for all to see. He emphasized his own laziness, procrastination, and disillusion. He also described his addiction to opium and his hypochondria. Coleridge's self-portrayal as a lazy and undirected character fitted in part with the Romantic image of genius; however, this sort of presentation of self was offensive to Victorian critics and historians later in the century.

Further compounding the tendency to downplay Coleridge's philosophical acumen and the depth of his scholarship was the very real issue of plagiarism. In his *Biographia Literaria* Coleridge summarized, epitomized, or recounted passages from Schelling in his own language. He often wove them together with ideas of his own and of other philosophers, often in an original manner. In other passages, however, he did genuinely lift close paraphrases or near quotations from Schelling, often lines or paragraphs in length. Because of this there has been a century-long dispute about the accusation of plagiarism against Coleridge, in recent decades revived and presented as a fresh discovery by Norman Fruman in *Coleridge: The Damaged Archangel.*[11] As Elinor Shaffer says, Fruman's work is based on century-old secondary sources, "'plagiarized' from secondary sources down to the last charge of 'plagiarism,'" and "'plagiarism' is another name for the history of ideas, but not the most illuminating one."[12]

Even if the worst charges of plagiarism are true, they do not diminish the stature of Coleridge as someone who against the current of almost two centuries of British mainstream philosophy brought the ideas of Kant and the German Romantics for the first time to Britain. Indeed, Coleridge was slightly less isolated in reviving the Neoplatonic ideas that had been translated into English by Thomas Taylor, himself a somewhat isolated and eccentric figure in the previous century.

Coleridge's emphasis was not on the physical sciences, despite his involvement with Hamilton and Davy. Besides chemistry, Coleridge's major focus was on the medical sciences. His interest in both chemistry and biology was an outcome of his own fascination with the medical theories and disputes of the day, in part stimulated by his own illness and hypochondria. Coleridge's collaborator in life and editor for decades after his death was a medical man, J. H. Green. Coleridge wrote lectures for Green's medical courses concerning the value of logic and general methodology. He wrote a systematic medical article on the history and nature of the disease scrofula. Coleridge believed himself to have this disease, which was at the time very broadly defined as swellings in various parts of the body.

Scrofula was also of interest to Coleridge because of the political and religious implications of its history. Scrofula was the so-called king's evil,[13] the disease that supposedly could be cured by the royal touch. The decline of belief in the efficacy of the king's touch in curing scrofula was connected with the decline of belief in miracles as well as with the decline of the divine right of kings. Coleridge, as a born-again political and religious conservative, was concerned with the defense of royal authority and the existence of miracles. He certainly did not defend the reality of the royal touch as a cure but engaged in a detailed and insightful conceptual and social history of the disease, pointing out that it was a disease of the lower classes before the rise of capitalism, at which point it gained medical notice because it became a disease of the middle and upper classes as well. Perhaps surprisingly, this essay on apparently obscure medical and historical points was the jumping off place for what is probably Coleridge's deepest and most comprehensive philosophical work, originally titled *Theory of Life*.[14]

Coleridge's general theory of life from a biological point of view was based in part on his respect and admiration for John Hunter. Coleridge utilized the Hunterian Collection of medical literature and objects at the Royal College of Surgeons. He referred to it a number of times and used it as an example of the arrangement of concepts as well as of scientific method. Hunter was a central figure in the British orthodox medical tradition who claimed that vital principles played a role in the nature of life. Hunter's own vitalism involved an appeal to active principles of the sort to which Newton appealed. Recall that Newton appealed to active principles not only for phenomena of gravitation, electricity, magnetism, and chemistry, but also for processes, such as fermentation and vegetable growth. Newton's active principles, as we have seen, could be interpreted (and had been by Newton himself at various times) either as refined, "imponderable," or weightless fluids, such as the ether, or as pure forces.

One commentator on Hunter was John Abernethy, who interpreted his vitalism in a direction amenable to Coleridge. Abernethy associated Hunter's vital principles with such phenomena as electricity and magnetism. For Abernethy, in the tradition of mesmerism, electricity and magnetism were principles of life. Abernethy's "Hunterian" oration presented this controversial, somewhat Romantic, interpretation of Hunter. It was immediately attacked by William Lawrence, who claimed that Abernethy's exposition was a distortion and misrepresentation of the fundamental ideal of Hunter. Coleridge defended Abernethy's interpretation against that of Lawrence.

Coleridge was familiar with theories of magnetic medicine in the school of German nature philosophy, which extended and developed the theories of Franz Mesmer on the role of animal magnetism. Adam Eschenmayer turned to magnetic medicine and the study of hypnotism, somnambulism, and other such phenomena after having been a follower of Schelling. Eschenmayer was a disciple of Schelling and remained his good friend even after breaking with him over the a priori theory of nature and advocating a "nonphilosophy" as the only possible philosophy. This is a term and issue later revived by French phenomenologist Maurice Merleau-Ponty in the mid–twentieth century. Merleau-Ponty had earlier, in his *The Structure of Behavior*, made a Schelling-like attempt to move toward consciousness from "below" to supplement the phenomenological approach from "above." Coleridge was a follower of the magnetic biology of Eschenmayer.

Despite the title of the work, *Theory of Life*, and the coauthorship with James Gillman,[15] a central theme is that of electricity and magnetism. "Life," in Coleridge's terms, is much broader than the usual definition or characterizations in terms of plants and animals alone (something to which the editor of the earliest printed edition objected).[16] For Coleridge life is unity in multiplicity. The greater the number of elements in the multiplicity unified, the higher the degree of life. "Life *absolutely* as the principle of individuation . . . the living power will be most intense when that individual, which, as a whole, has the greatest number of integral parts presupposed in it."[17] Life is to be found throughout the physical world. Even if a lump of gold is not an organism, the essence of gold manifests life. Coleridge praises William Gilbert as "a man of genuinely philosophical genius"[18] for his discoveries concerning magnetism. Coleridge uses magnetism as a prime example of his conception of life, of reality as power, body as product.[19]

Coleridge follows Schelling and Steffens in his exposition of life as unity and multiplicity and of the nature of magnetism, electricity, and gravitation. Critics of Coleridge as plagiarist point to *Theory of Life* as a prime example of his borrowing ideas without attribution (including very close paraphrases and reports of experiments without reference in such a way as to give the reader the impression that Coleridge had performed the experiments himself).[20] Certainly Coleridge, who cites numerous authorities, is careful not to refer to Schelling or Steffens, from whom his ideas and even some turns of phrase are taken. But for our purposes, in tracing the filiation of these ideas from Germany to England, the issue of Coleridge's honesty is irrelevant. Coleridge did communicate accurate and detailed accounts of the writings of Schelling and Steffens to an English audience, even if he led the reader to attribute the ideas wholly to himself. Certainly a New Critic or deconstructionist, allegedly unconcerned with biographical and auctorial details, should be totally oblivious to the issue of Coleridge's plagiarisms. The historian of reception of ideas need not be concerned with plagiarism as a means for ideas to be propagated across cultures.

Coleridge uses the Schelling-Steffens account of the arrangement of magnetism, electricity and gravitation. According to this, magnetism is a one-dimensional force,[21] electricity two dimensional, and gravitation three dimensional. Coleridge totally rejects all accounts of magnetism, electricity, or gravity as ethers.[22] He criticizes mechanistic theories for "the exclusion of all modes of existence which the theorist cannot in imagination, at least, *finger* and *peep* at."[23] As his alternative he uses the theory of matter as a result of opposing powers, "the product, or *tertium quid* of the antagonistic powers of repulsion and attraction."[24]

The *Theory of Life* was a late product, posthumously published. Nevertheless, it shows the knowledge that Coleridge possessed of Schelling's and Steffens's philosophy of nature. It also shows Coleridge putting the consideration of magnetism (followed by electricity and gravity) at the conceptual center even of his treatment of biology. It, along with the borrowings from Schelling in the *Biographia Literaria*, gives some idea of the sort of ideas over which Coleridge enthused with his friend Humphry Davy on his return from Germany and during his subsequent studies of German literature on natural philosophy.

Coleridge's conversation was notable for its eloquence. Pearce Williams quotes an obituary reminiscence that characterizes Coleridge's conversation as "unlike *anything* that could be heard elsewhere; the kind was different, the degree was different, the manner was different. The boundless range of scientific knowledge, the brilliancy and exquisite nicety of illustration, the deep and ready reasoning."[25] This oral fluency more than his writing is what introduced the leading chemist and the leading physicist of his day, Davy and Hamilton, to the ideas of German Romanticism.

CHAPTER 16

Ørsted: Romanticism, Nature Philosophy, and the Discovery of Electromagnetism

$\mathcal{H}$ans Christian Ørsted (1777–1851), Danish chemist and physicist, is by far most famous for his discovery of electromagnetic induction in 1820. Ørsted noticed the effect at the end of a demonstration experiment. He happened to align a compass needle with an electric wire and saw it move slightly. The motion was too small to be significant, but that summer he built a much larger battery than what was available to him before and performed more rigorous experiments.

He showed that the current in an electric wire would align a compass needle perpendicularly to the wire. One of the most peculiar features of this at the time was that the magnetic force induced by an electrical current was at right angles to the current itself. He collected five scientist-witnesses (presumably because of the controversial nature of the finding) and dispatched a four-page Latin report of it to numerous academies in Europe. Soon translations appeared in European languages.[1] Some of the leading scientists of the day at first rejected the result as ridiculous or impossible. Laplacian physicist Pierre L. Dulong[2] initially proclaimed that the result was nonsense, calling it "just another German dream" in a letter to chemist Jöns Berzelius. Mathematical physicist Christian Heinrich Pfaff agreed.[3] Another notable French physicist, François Arago, at first claimed the same as Dulong but then replicated the experiment in a more detailed fashion proving Ørsted correct.

Ørsted's Romantic Physics

Ørsted's discovery was originally viewed by historians of science as a paradigm of accidental discovery. The fact that he made it at the end of a laboratory demonstration was supposed to highlight its accidental nature. The basis of this account was a letter written by Christopher Hansteen, a colleague of Ørsted, to

251

Michael Faraday;[4] however, this letter was written thirty-seven years after the fact, and the details of the account may be in doubt. Hansteen claimed that Ørsted was a clumsy experimenter, although this is certainly not the case.[5] Ørsted had in fact been looking for some effect of electricity on magnetism for at least seven years, although he had previously conjectured the wrong experimental conditions for it. He proposed that such an effect existed in his book on chemistry, *Researches on the Identity of the Chemical and Electrical Forces*.[6] Although Ørsted made the initial discovery in the class demonstration, the effect was weak and made little impression on the students.[7] Ørsted then constructed a more powerful battery during the early summer and performed the experiment which he then published.

In recent decades, investigations of Ørsted have supported the maxim that fortune favors the prepared mind. Ørsted not only had been working on this and related problems for a long time, he also was working within an unusual philosophical framework that motivated him to look for connections between apparently different forces of nature. Ørsted was a *naturphilosoph*. Nature philosophy was the doctrine of the followers and associates of Schelling, discussed in Chapter 14. Ørsted himself had studied with Fichte as well as the Schlegel brothers, met the nature philosophers Henrik Steffens and Franz von Baader, and wrote his dissertation at Copenhagen on Kant's *Metaphysical Foundations of Natural Science*. Ørsted also published a survey of recent physical studies in the journal of arch-Romantic literary critic Friedrich Schlegel.

The Fortunes and Misfortunes of German Nature Philosophy

For a long time, nature philosophy was cited solely as a prime example of the bad influence of philosophy on science. Starting in the mid–nineteenth century with the rise of the "vulgar materialism" of figures such as Karl Vogt, Christian Ludwig Büchner, and Jacob Moleschott, the philosophy of the Romantics and the dialectical historical philosophy of Hegel was strongly rejected, especially within the scientific community and its philosophical counterpart. The watershed seemed to have been the failure of the revolutions of 1848. The pessimistic philosophy of the previously neglected Schopenhauer was suddenly popular. Also, the optimism and sweeping paeans to freedom of the German Romantic movement and the liberal side of Hegelianism suddenly seemed unrealistic. By 1870 Marx could write that Hegel was "a dead dog" in Germany.[8]

In the latter half of the nineteenth century, as industrial capitalism deepened its roots in the semifeudal Germany, empiricist philosophies finally began to become popular. In England, empiricism had arisen with the Puritan Revolution of 1640 and the Glorious Revolution of 1688. During this period, the rationalistic philosophy of Descartes and his followers was still dominant in France. With the development of capitalism in France in the eighteenth century, rationalistic, Cartesian philosophy was replaced with an empirical and sensationalistic philosophy as with Voltaire as popularizer of Locke and Newton. Further, in *Treatise on Sensations*, based on a British empiricist approach to philosophy, Etienne de Condillac developed and popularized the notion that learning was based solely on sensations.

During the period of the rise of the bourgeoisie culminating in the French

Revolution in 1789, the dominant philosophy in Germany was still the rationalistic philosophy of the followers of Leibniz led by Christian Wolff, who gave a lecture in praise of Chinese civilization. Even when the rationalism of the followers of Wolff was superseded by the philosophy of Kant, the main forms of interpretation of Kant's philosophy by such Romantics as Schelling and Hegel took the form of a partial return to rationalism and a rejection of Kant's high respect for mathematics. It was not until the late nineteenth century, with the rise of the chemical industry and other capitalistic and technological enterprises in Germany, that the empirical philosophy that had dominated Britain since the seventeenth century and had predominated in France since the mid–eighteenth century came to dominate relatively backward Germany. The revolutions of 1848 played the role in the shift of German thought that the Puritan and Glorious Revolutions had played in Britain in the 1600s and the French Revolution had played in France in the late 1700s. Yet in Germany the revolution of 1848 was a failure, and the development of a truly democratic bourgeois state could not be said to have started until the end of World War I. This development was partial and hesitant, and was soon overwhelmed by a return to more primitive thought in Nazism. Perhaps the end of feudalism occurred in Germany finally only with the end of World War II in 1945. Because of this tortured political history, philosophies of intuition and organic unity that resembled earlier feudal, even primitive, forms of thought continued in Germany long after economic and political individualism and mathematical formalism had submerged such views in England and France. Of course, in the United States, which lacked a feudal past, such views barely or rarely arose.

Within the scientific community in Germany the post-1848 rejection of nature philosophy was extremely strong. Liebig, a German agricultural chemist, called nature philosophy a "black death."[9] Hermann von Helmholtz, who surreptitiously incorporated and utilized aspects of German nature philosophy, likewise spoke strongly against its excesses, although he later paid tribute to Goethe's science in standard German fashion.

In the twentieth century, the logical positivists, in their defense of empirical scientific knowledge as the only genuine knowledge, made reference to German nature philosophy as a prime example of precisely the sort of bad metaphysics they wished to expunge from science and philosophy. The work of the German nature philosophers varied considerably in quality. Much of it was indeed the sort of speculative and loose theorizing that the later German scientists and the logical positivists despised and ridiculed. Schelling and the mainstream of German nature philosophy did concern themselves with surveying the empirical results of science and did not profess to deduce specific scientific results from pure philosophical principles. Rather, their goal was to order and systematize the results of science in terms of their philosophical principles.

Ritter: The Romantic Physicist

There was one nature philosopher who was truly a "Romantic physicist" in the full sense: Johann Wilhelm Ritter. Ritter, like a good Romantic, lived only to his mid-thirties (1776–1810). He was both a genuine Romantic, in life and in philosophy, and a genuine physicist. Ritter's discovery of ultraviolet radiation is the one truly solid major discovery unquestionably motivated and suggested by Schelling's nature philosophy.

It has recently been argued that Ritter's investigations in animal electricity, or galvanism, were not Romantic but were grounded in the previous, eighteenth-century work in the field.[10] Certainly Ritter's experimental results in this field were very much an extension of the tradition of experimental technique available; however, no one can deny that the later extension of his electrical work into speculation concerning animal electricity as a feature of the universe as a whole and of an organism, and his work on dowsing, do not suggest the Romantic associations of his work. Ritter certainly did closely associate with such figures as Novalis (whom he considered a soul-mate) and Caroline Schlegel-Schelling. Even if Ritter's relationship with Schelling was not on the most friendly terms, his very close personal association with the Romantic circles in Jena, and later in Munich, can hardly be denied.

Schelling and the nature philosophers organized the universe in terms of polarities in the manner of primitive thinking. With the discovery of infrared radiation by William Herschel, the nature philosophers inferred that there must be an opposite pole to an extension of the spectrum beyond red—that is, there must be a light beyond violet. Ritter experimentally pursued this notion using silver chloride, which turned darker as light was shifted beyond the blue end of the spectrum. Ritter was able to show that this compound was turned even a darker shade by an invisible radiation beyond violet.[11] (Oddly enough, the first book in English in the twentieth century that discusses the work of the nature philosophers in a positive manner fails to mention Ritter's most famous discovery, the most important genuine physical discovery by the nature philosophers.)[12]

Ritter did other very important work in galvanic electricity and bankrupted himself in paying for the expensive silver batteries he needed for his very accurate electrical experiments. Luigi Galvani had discovered animal electricity and had become involved in a dispute with Alessandro Volta over the role of the animal itself in this electrical phenomenon. Galvani believed that the animal nervous system itself generated the electricity. For Volta, on the contrary, the legs of the frog were merely a makeshift electroscope that transferred the electricity from the battery. Ritter, with the vitalistic Romantic nature philosophy of Schelling, was fascinated by the concept that all nature is alive and electrical. This was apparently the motivation for his very careful electrical experiments.

Ritter had a lifestyle that made him virtually a caricature of the Romantic. He supported himself by borrowing money and then tended either to squander it on generous entertainment for his friends or to expend it on the costly equipment for his researches (such as silver and platinum for his electrical piles). He had to give up his family and died at an early age from poverty, drink, and disease. Ritter made a (coincidentally correct) forecast concerning Ørsted's 1820 discovery of electromagnetism in a sort of horoscope based upon a temporal pattern of previous electrical discoveries.[13] Unfortunately Ritter did not live to see Ørsted's discovery.

Late in his short life, Ritter became involved with the extreme mystic Franz von Baader. A number of figures from the Jena circle of Romanticism, such as Schelling and his wife, Caroline, had moved to Munich. Bavaria's ruler had made a treaty with the French, and the city was peaceful and unharmed compared to many northwestern German cities occupied by the French. Since the ruler of Jena had become distrustful of intellectuals and academics after the French occupation, Caroline Schlegel-Schelling and Ritter hoped that Munich would become a new Jena, and that the youthful hopes and exuberance of the early Romantic movement would be rekindled. Ritter mistakenly believed that Ernst Chladni was coming to Munich and was energized by the prospect.[14] Chladni was developer of "Chladni figures," two-dimensional wave patterns of tones visible on metal plates (the sort of union of art and science that Goethe and the Romantics loved). Munich had its own mystical and occultist tradition of an even more intense hothouse atmosphere than that of the Romantics in Berlin or Jena. Paracelsus and Böhme were major sources for Munich Romantics such as Baader. Saint-Martin was the conduit for the occultist ideas of Paracelsus to Baader and his circle.[15] The Romanticism of Munich was highly conservative, indeed reactionary, in contrast to that of Jena, which had been inspired by the ideas of the French Revolution.[16] Schelling had already turned in a much more conservative direction (and had been the least political of the Romantics to begin with). He adapted his philosophy to the failure of utopian hopes concerning the French Revolution, the collapse of Prussia (center of the German Enlightenment), and he heightened his attack on all forms of Enlightenment thought.[17]

Ritter engaged in experiments concerning phenomena similar to what we would call dowsing. He investigated the sensitivity of people to water and minerals using wands and pendulums that supposedly turned in the direction of the body being sought. To Ritter, this was just as scientific as the investigation of animal electricity in the legs of a frog, but it led to the total discrediting of Ritter as a serious scientist. He discovered a peasant dowser, Campetti, in Italy and immediately applied to the Munich Academy for funds and brought him back to Munich. Ritter performed the function of the bohemian intellectual in attempting to bring peasant lore within science in too crude and naive a way. Ritter was initially financed (somewhat clandestinely) by Baader and some other members of the Munich Academy, keeping more hardheaded members of the academy, such as Soemmerring, from learning of the project until the last minute,[18] but soon the academy demanded verification. Ritter drew up plans for testing, but the academy balked at them. The academy decided that Ritter should write a report, and the report should be circulated to foreign academies for evaluation. Campetti became ill and went back to Italy, and Ritter never finished the report.[19] At one point he proposed replacing the human body with an experimental device, just as he and Volta had removed the frog from galvanism. This would have made the experiment less dependent on the peculiarities and subjectivity of the dowser. He suggested that the Lichtenberg figures (alluded to below) would be a means to this.[20] Two years later Ritter died in poverty.

The dowsing experiments, despite their disastrous effects on Ritter's reputation, caught the interest of Romantic intellectuals. Hegel corresponded with Schelling (just before the break between them) concerning the experiments, and Hegel attributed his lack of success with them to his own lack of skill, mentioning Goethe's curiosity combined with jokes about the experiments.[21] Goethe portrayed such an experiment in his novel *Elective Affinities*.[22] Despite the interest of philosopher and poet, Ritter's reputation was destroyed by this episode.

Part of the motivation for the discrediting of Ritter was the development of physics as a separate discipline. *Annalen der Physik* had originally been edited by Friedrich A. Gren, a pharmacist (the same profession in which Ritter had originally been trained and which had been followed by Ørsted's father and by the young Ørsted for a time), but Ludwig W. Gilbert, the new editor, was more of what we should consider a pure physicist. He waged a campaign of vilification against Ritter. It is interesting that electrical investigations continued in Germany, not within the realm of physics, which one might think after Volta's defeat of Galvani on animal electricity, but within the realm of chemistry, often by physicians and pharmacists.[23] Physics chairs were sometimes held by pharmacists, and more orthodox physicists sometimes received their original education in pharmacy (Johann Poggendorff) or edited pharmacy journals (Gustav Fechner).[24] It is also worthy of note that the Romantics (such as Schelling and Ørsted) often proposed chemistry as a higher scientific synthesis than physics, repeating the position held by the Paracelsians, whose strength had been among the pharmacists in earlier centuries.

Some respectable scientists—such as Seebeck, discoverer of the thermoelectric effect, founder of thermoelectricity, and friend of Goethe—were positive toward Ritter's proposals for the testing of "siderism," or dowsing. Seebeck had a certain skepticism about Ritter's claims but thought that the plan presented to the Munich Academy was good, while realizing that he was going too far in his claims.[25] In noting the fact of Gilbert's campaign against the later occultist dowsing turn of Ritter, Maria Trumpler suggests that even the notion of "Romantic Science" itself is a retrospective construct of later-nineteenth-century German physicists who denigrated earlier cross-disciplinary speculation to characterize their own position as professionals.[26]

Even after the dowsing episode Ritter continued to do valuable experiments and make worthwhile speculations. He performed a series of experiments on electrical reactions in sensitive plants (types of mimosas), making comparisons with the work of Galvani and himself on frogs. Until the 1920s Ritter's bioelectric experiments with plants had not been surpassed. He can be considered the founder of research in general bioelectricity. (It is interesting that the sensitive plant had for decades played a role in literary and psychological speculation.)[27] Ritter also wrote up his speculations, which included the intertransformability of matter and energy and the possibility of construction of electrical and magnetic telescopes.[28]

Ritter and Ørsted

Ritter collaborated with Ørsted during two extended periods in Jena, and the two scientists, both sons of pharmacists, remained friends for the rest of Ritter's short life.[29] Ørsted, unlike Ritter who was self-taught, possessed a university background with a dissertation on Kant's *Metaphysical Foundations of Natural Science*. Ørsted, hearing of Ritter's electrochemical discovery, traveled to Germany, and then to Paris to demonstrate Ritter's galvanic battery (which differed from Volta's in being constructed solely of copper plates, not of alternating zinc and copper) to the savants of Paris, with great success. He attempted to gain a Napoleon Prize in electricity for Ritter. This attempt, which was welcomed by Ritter because of his usual accumulation of debts, at first looked promising; however, another speculative claim of Ritter concerning detection of an electric polarity of the earth by means of a new instrument met with disaster when Ørsted attempted to demonstrate this new, overreaching claim of Ritter's. This led to partial discrediting of Ritter and Ørsted. Nonetheless Ørsted had gained contacts with leading French scientists and published several papers in the major French journal of experimental physics (as opposed to rational mechanics).[30] Ritter's interest in the Chladni figures of sound patterns on plates mentioned above was shared by Ørsted. These patterns, as well as the electrical discharge patterns of Georg Lichtenberg, were treated by Ritter and Ørsted as "hieroglyphics" according to Christensen, who mentions the obvious similarities to the sort of thinking we encountered in hermetic science.[31] Ritter wrote concerning the Lichtenberg figures, "My aim was to rediscover, or else to find the original or natural script by means of electricity."[32] Ørsted, like Ritter, pursued the relationship of electricity to biology, although Ørsted mediated this through chemistry more than Ritter.[33] Ørsted, although bound both by friendship and allegiance to the Romantic program of nature philosophy, had orthodox academic training and credentials and was respectable in his lifestyle and formal interaction with academic and governmental figures. He did speculate in the manner of nature philosophy, but was more restrained in speculative empirical or premature experimental claims than was Ritter. Unlike Ritter, who despite several important discoveries (ultraviolet radiation, electrochemistry, bioelectricity) died in disgrace, Ørsted became a respected figure in his university and country.

Ørsted and Nature Philosophy

Ørsted was thought to be a nature philosopher after his European tour because of his early teachers and associates. His attendance of lectures by Fichte and August Schlegel, his friendship and publication with Friedrich Schlegel, his acquaintance with Friedrich Schleiermacher, Henrik Steffens, and Franz von Baader, and his collaboration with Ritter certainly link Ørsted to the circles of nature philosophy. Ørsted's dissertation was on Kant, not on Kant's more Romantic successors. Ørsted read Schelling, but not uncritically. He praised the "beautiful and great ideas" of Schelling, but criticized Schelling's erroneous empirical claims.[34] Ørsted learned to temper his speculative ideas with experimental tests after his initial enthusiasms and disillusionment with the chemistry of Jakob Joseph Winterl, who postulated two substances—a principle of acidity and a principle of alkalinity—from which all matter is constructed.[35] Nevertheless Ørsted retained his interest in speculative theory and developed one based on conflict and

258   Romantic Philosophy of Nature

polarities. In discussing light he envisages that his conflict theory stands between the wave and particle theories the way that the dynamical theory of heat stands between the kinetic and caloric theories. "Schelling in his *Weltseele* has recognized the possiblity of such a theory."[36] When Ørsted announced his discovery of the creation of magnetism by electricity, he presented it in the framework of his conflict theory: "To the effect which takes place in the conductor and the surrounding space we give the name of the *conflict of electricity*," and "the electric conflict is not confined to the conductor but is dispersed pretty widely in the circumadjacent space."[37] Although Ørsted's theory of conflict of polar forces is not fleshed out in detail, it supplied a framework for his grasping one of the great experimental discoveries of all time. Previous forces, such as Newton's gravitation or Coulomb's electrostatic attraction or repulsion, acted along a straight line between the attracting points. Ørsted's force acted transversally or perpendicularly. This was a force different from the Newtonian model.

CHAPTER 17

Davy, Faraday, and Field Theory

———※———

$\mathcal{M}$ichael Faraday was first to develop a full-fledged concept of a physical field. Although there were precursors, these earlier attempts were fragmentary and incomplete.[1] Faraday, the Cinderella of science, was an uneducated son of a blacksmith who was apprenticed to a bookbinder. He had the good luck to be hired by Humphry Davy at Davy's institute where public lectures were given. Faraday started as assistant in 1813 and by 1821 was overseeing the Royal Institution laboratories. Faraday learned chemistry and physics from Davy, but never mastered advanced mathematics. Faraday's science was a combination of careful, ingenious experiments and qualitative metaphysical speculation.

Humphry Davy: Chemist and Romantic Poet

Davy and Faraday could hardly have been more different. Davy was a wealthy man and flamboyant lecturer and demonstrator who lived a life of luxury. Faraday was an impoverished, highly puritanical member of an austere religious sect, the Sandemanians.[2] Davy was very much a typical Romantic, a friend and intimate of Coleridge who appeared in the poet's dreams.[3] At times, strains in the relationship between Faraday and Davy arose from their contrasting circumstances. While on a tour of Europe, Faraday ended up acting as a sort of valet to Davy and his wife, and felt "debased" by her treatment.[4] He thought that Davy did not give him sufficient credit and praise for his work. At one point Davy accused Faraday of plagiarism of the work of William Wollaston and of not sufficiently crediting Davy himself in his work on magnetic rotation. This was a particularly hurtful accusation, as the work in question was Faraday's first major electromagnetic investigation and his first major publication. Faraday's discovery was far more original than Davy recognized, as Faraday was conceptualizing magnetic rotation, not in terms of Newtonian forces, but in terms of circular relationships.[5] Faraday also believed that he did not receive adequate credit from Davy for work on the liquefaction of gases, which was work in Davy's own area of research. The last straw was Davy's attempt to block Faraday from membership in the Royal Society, leading to cooling in their relationship.[6]

Humphry Davy was a major figure in the history of chemistry, but his contributions were irregular. He was not educated by a technical chemist of the first rate. At the end of his life, Jöns Berzelius claimed that Davy had produced "brilliant fragments" because he never worked industriously in all areas of chemistry. In Davy's early days, he made major contributions to both experimental and theoretical chemistry. One of his important achievements was to show that Antoine Lavoisier, the founder of modern chemistry (who isolated and identified oxygen as an element), had erred in claiming that oxygen was the basis of acidity. Davy also did fundamental work in electrochemistry before Faraday made his major discoveries.

Davy became overextended once he was involved in supporting himself through the Royal Institution, which had been founded by Benjamin Thompson, discoverer of the mechanical equivalent of heat who left America for Britain because of the American Revolution and his own royalist sympathies. Thompson, also known as Count Rumford, wished the Royal Institution to be a public educational foundation as well as a place of research, but the dour, combative, and crabby Thompson was hardly a crowd pleaser.[7] Davy was able to help the finances of the Royal Institution by giving spectacularly successful, dramatic public lectures.[8] This led to Davy's renown and position in high society but also distracted him from full-time chemical research.[9] In his position, Davy also became involved in the practical applications of chemistry, wrote on agricultural chemistry, and worked in geology.[10] These fields were of great social interest and public demand, but since chemistry was in a rudimentary state, Davy could not make the sort of theoretical advances in conceptual clarity that he had made in classification of the elements and discovery of new elements in fundamental chemistry.

Davy's chemical researches were tied to some of the psychological experiments of the Romantics. Thomas Beddoes prescribed opium to Coleridge, Davy, and other Romantics. Beddoes followed John Brown, who made excitability the center of his system, and claimed that the goal of medicine was adjusting the degree of excitement of cells.[11] Davy and Coleridge were involved in experiments with nitrous oxide, or laughing gas. This became a fad in the nineteenth century and an object of satirical cartoons, though Davy and Coleridge thought that they could achieve mystical experiences through its use.[12] Robert Southey claimed that nitrous oxide was a fantastically powerful energizer. "Davy has invented a new pleasure for which language has no name. It makes me strong and happy! . . . oh excellent airbag."[13]

Dr. Beddoes's group experiments with laughing gas, with dozens of people in a state of hallucination and euphoria, were taken by "respectable society" as showing the dangers of mass participation in science. The irrational behavior of the subject-participants was associated with the irrationalism and enthusiasm of the supporters of the French Revolution, and gases were associated with the radical Priestley, who had discovered a number of "airs" including nitrogen and had made oxygen, without interpreting it to be such, calling it "dephlogisticated air." The association of nitrous oxide with riotous behavior, irrationalism, and the French

Revolution discouraged the medical profession from making use of it as an anesthetic for another half century.[14] Conservative political philosopher Edmund Burke used chemical imagery to attack the evils of the French Revolution and associated it with such chemists as Priestley who supported it and the French chemists who worked for its war machine. Popular cartoons took up Burke's imagery of the Revolution, melding it with Priestley and his experiments. Priestley's laboratory was burned, and he moved to the United States. Beddoes was generally discredited, and Davy left Bristol, where he was associated with Beddoes's demonstrations, to enter the Royal Institution in London.[15] Davy soon shed any radical associations and became a social lion.

In his later years, Davy more or less retired from active chemical research and wrote several books of Romantic meditations. One of these is *Salmonia*, a treatise on salmon fishing.[16] He also wrote *Consolations in Travel; or, The Last Days of a Philosopher*.[17] In this, he presented his general philosophy of life in the form of a travelogue of the ruins of Italy. The older Davy presents a general philosophy of nature as process.

In several places Davy advocated the atomism of Boscovich. For instance, in his journal in 1813, he writes: "By assuming certain molecules endowed with poles or points of attraction & repulsion as Boscovich has done & giving them gravitation & form, i.e. weight & measure . . . but we may suppose inherent powers (thus we suppose iron naturally polar with respect to the magnet."[18] At the end of his life, in a fragment added to the *Consolations*, Davy again states his allegiance to Boscovichian atomism: "You mistake me if you suppose I have adopted a system like the Homooia of Anaxagoras, and that I suppose the elements to be physical molecules endowed with the properties of the bodies we believed to be indecomposable. On the contrary, I neither suppose in them figure nor color . . . ; I consider them, with Boscovich, merely as points possessing weight and attractive and repulsive powers."[19]

Before Davy went to the Royal Institution he lived in Bristol, and had close contacts with Romantic poets—including Southey, Coleridge, and Wordsworth—and wrote poetry taken seriously by them.[20] Davy helped with the proofs of the second edition of the *Lyrical Ballads* of Coleridge and Wordsworth. Coleridge praised Davy as "the father and founder of philosophical alchemy. The man who *born* a poet first converted poetry into science and *realized* what few men possessed the genius enough to fancy."[21] William Godwin, an anarchist and theorist of progress, married Mary Wollstonecraft (whose daughter became Mary Shelley, author of *Frankenstein*, and whose optimism was the target of Thomas Malthus's *Essay on Population*), was quoted by Coleridge as saying of Davy, "What a pity that such a man should degrade his vast talents to chemistry."[22]

Nature Philosophy as Metaphysics and as Method

The German nature philosophers were reassessed several times. For a brief period in the early nineteenth century, they were part of the enthusiastic Romantic movement. By the time of political and scientific disillusionment with utopian speculation in the mid–nineteenth century, the nature philosophers were ridiculed and considered to be unscientific. This attitude continued within the history of physics up until the late 1950s through the 1960s, when countercultural views and Romantic conceptions of nature grew in popularity. The atmosphere of this period was conducive to yet another reassessment of the views of the nature philosophers.

In the English-speaking world from 1957 to 1969 a number of articles emphasized the positive contributions of Romantic nature philosophy to the science of Ørsted and others. Robert Stauffer emphasized that Ørsted's experiment was no accident, but was the result of a quest focused by nature philosophy.[23] It was suggested by Thomas Kuhn and others that Romantic views on the unity of forces contributed to the multiple simultaneous discovery of the conservation of energy.[24] Pearce Williams presented the thesis that the views of Kant and Schelling had influenced Faraday via Coleridge and Humphry Davy.

Since the early 1970s, just as there was a neoconservative political reaction to the counterculture, there was a reaction in the history of science to the favorable evaluation of the nature philosophers presented by Stauffer, Kuhn, Williams, and others.[25] This reaction has largely been an attempt to show that successful scientists, such as Davy, Faraday, and Ørsted, were not really affected by the nature philosophers in their scientific activity. In the case of Davy, for instance, Yehuda Elkana in *The Discovery of the Conservation of Energy*[26] argued that Davy was not really led by ideas of the nature philosophers in his valuable scientific theories and discoveries. Elkana barely conceals his contempt for Davy's lifestyle and manner of presentation. Elkana grants that Davy made a number of major chemical discoveries, but claims that Davy's supposed contributions to the theory of the conservation of energy are bogus, and sneers at attempts to find suggestions of the conservation of energy in Davy's poetry. He also hints that Davy's experiments concerning heat are both ineffectual and plagiarized. Elkana claims that one cannot find any philosophy of science in Davy's works, citing those who have found a scientific instrumentalism in Davy as well as those who have found him to be a nature philosopher. He notes that the later Davy became more cautious and empirical in comparison to his earlier, more rash, and speculative work, and he quotes Davy himself saying this.

Similarly, Timothy Shanahan argues that Ørsted was not affected by nature philosophy in his discovery of electromagnetic induction.[27] Against Stauffer and others, Shanahan argues that Ørsted was primarily influenced by Kant. This Kantian influence makes Ørsted seem at least comparatively more respectable than if he were influenced by the Romantic Schelling. No one can claim that Ørsted was an empiricist, as some of the revisionists claimed for Faraday and Davy; however, the association of Ørsted with Kant will at least partially cleanse him from the taint of being influenced by nature philosophy.

What both these revisionists must do in order to make their case with respect to Davy and Ørsted is, first, to separate the undoubted impact of personal and literary style upon these figures from what are claimed to be different sources of their science and, second, to separate methodology from metaphysics and emphasize that the methodology was primarily influenced by more experimental, empiricist, or at least Kantian currents. For instance, no one can deny that Davy was in close contact with Coleridge, with whom he exchanged an intense personal correspondence for several decades and wrote openly of the intimacy of their friendship. Elkana argues, however, that Davy was not at all following Coleridge in the area of philosophy and emphasizes Davy's experimental methods and his increasing skepticism with respect to theories as he grew older.

Similarly, one cannot deny that Ørsted was in certain respects a follower of Romantic nature philosophy. Ørsted attended lectures by Fichte and the Schlegels and was personally acquainted with Schelling, though he was nowhere as close to Schelling as Davy was to Coleridge. One can hardly deny that Ørsted was philosophical in a Romantic, nature-philosophical manner. Ørsted wrote works, among them *The Soul in Nature,* that clearly present nature philosophical views. What Shanahan must claim is that Ørsted's literary and philosophical outpourings concerning nature philosophy had little relationship to his science. To do this, Shanahan emphasizes the empirical and experimental aspect of Ørsted's scientific technique, and also quotes those statements of Ørsted in which he criticizes the nature philosophers for being too speculative and insufficiently empirical. Maria Trumpler, in her study of Ritter, similarly emphasizes the extent to which Ritter's experimental technique was very much a product of his non-Romantic, eighteenth-century predecessors.[28]

One central fallacy shared by these attempts to purify the great Romantic scientists of German nature philosophy is the way in which methodology and metaphysics are distinguished. Elkana emphasizes Davy's pronouncement against speculation and his "instrumentalist" view that theories are merely formal tools to make predictions and do not necessarily represent ultimate reality. Similarly, Shanahan emphasizes Ørsted's empiricism and experimentalism in order to ally Ørsted with Kant and against Schelling. Actually, it is harder with Ørsted since Kant is certainly much less empirically oriented that the British empiricists. At least Kant denies that one can deduce particular empirical facts a priori.

Showing that Ørsted carefully checks his hypotheses against experiments, or that Davy is both experimentally oriented and considers his theories to be mere instruments for prediction, does not show that either man was uninvolved with nature philosophy. One can hold a certain metaphysical view of the world and yet subject the scientific theories based on that metaphysics to tests. This is precisely what Karl Popper and the Popperians had emphasized. Agassi studies precisely this aspect of Faraday's work as a kind of experimental metaphysician.[29]

In the case of Ørsted, one can admit that he criticizes the other nature philosophers for being too vague and ambiguous in their theories or for not checking their theories sufficiently against the experimental data. This does not mean that Ørsted did not share the worldview of the nature philosophers. Ørsted was concerned with a view of nature that involved dynamic polarities of forces such as Schelling depicted. This is different from the somewhat static balance of forces that Kant described. Similarly, Ørsted was searching for the unity and interconnection of the various forces of nature in the manner of Schelling and was not committed to a Newtonian account of these forces. Rather, Ørsted described the relationship of the electric and the magnetic forces in terms of "conflict." Shanahan and the other revisionists do nothing to show that Ørsted's metaphysics of nature did not incorporate ideas of Schelling and other nature philosophers.

The case is slightly different with Elkana's revision of Davy. Davy was working within the context of British empirical science and the praise of Baconian induction. Even if Francis Bacon is methodologically a more sophisticated figure than both his more simpleminded epigones or his violently antiempirical critics allow, the Bacon whom late-eighteenth- and nineteenth-century British methodology praised was for the most part the simpleminded Bacon. Again, Davy's appeal to inductivism does not prove that he lacked a Romantic metaphysics. His lack of acquaintance with or interest in the German philosophers whom Coleridge had propagated in England does not show that Davy did not speculate in the manner of Romantic philosophy of nature. Davy presents a Romantic process philosophy in his last work, *Consolations of Travel; or, The Last Days of a Philosopher.*[30] Anti-Romantic revisionists claim that this is a work of senility, as many have mistakenly claimed of the late philosophical works of such physicists as Schrödinger and Heisenberg. This late work actually reflects views that Davy had held earlier in his life in his personal associations with the Romantic poets.

Davy's claim that scientific theories are mere instruments and not representations of ultimate reality shows nothing about the nature of the content of these theories. That is, Davy may treat the theories as purely heuristic or predictive, but the theories themselves present models of nature that have strong affinities with the views of the Romantic nature philosophers. It is true that the nature philosophers generally took their theories as direct revelations of reality and claimed at times to be able to deduce the structure of nature a priori. This is a separate issue from the model or picture of nature that the theories contained. Here Boscovichian as opposed to Daltonian atoms reigned, and the precursors of Faraday's fields of force replaced the Democritean view.

Faraday and the Discovery of Field Theory

Writers in the nineteenth century portrayed Faraday as the model of a pure empiricist and experimenter. Faraday's humble upbringing, hard work, and extreme moral purity made him an ideal subject for inspirational biographies. Faraday himself, who did not have children although he loved them very much, presented Christmas lectures for children at the Royal Institution. Like many other great popularizers and educators in science, he was himself childlike, able to communicate the enthusiasm that he had as a child for scientific investigation. Faraday gave a famous lecture on the candle that was used to illustrate all the forces of nature.[31]

The portrayal of Faraday in the public image of science and the history of science was very much like that of Newton. Just as Newton was portrayed as a pure empiricist and follower of the inductive method, so Faraday was portrayed as a Baconian inductivist. Just as Newton's alchemical and theological speculations were hidden and suppressed by the historians and popularizers, so Faraday's far-reaching and brilliant metaphysical speculations were played down in favor of his equally ingenious but more socially acceptable experimental successes.

Faraday was extremely unusual in that he engaged in experiments of the most exacting and ingenious sort as well as theorizing of the most daring and far-seeing kind, and yet was ignorant of higher mathematics and of the subtleties of Newtonian mechanics. In spite of this, he developed a physical concept—the field—that was the greatest conceptual breakthrough in physics since the theories of Newton. His ignorance of analytical mechanics may have aided his originality; his speculations and models were not dilettantish and ignorant alternatives to Newton, but were conceptually and experimentally on a par with those of Newton himself. Since Faraday's physics was not deductive mathematically but rather was in terms of models and qualitative conceptualizations, he has been treated as a metaphysician by Popperian philosophers.[32] One must recognize here that Faraday's metaphysics was an experimental metaphysics, his theories were metaphysical in that they were not hypothetico-deductive mathematical ones, but they were scientific metaphysics in the sense of being controlled and rigorous speculations concerning models, at least in intent, tied to descriptions and predictions that could be experimentally tested.

The nature of Faraday's thought, yielding conceptual innovations in physics of the most powerful and fruitful sort, while being all but devoid of mathematics, makes his work difficult to pigeonhole. For instance, Clifford Truesdell, who is an advocate of "rational mechanics" in the eighteenth-century style, and who sneers at philosophers and historians of science unequal to himself in mathematics (that includes almost everyone), nevertheless recognizes Faraday as one of the world's greatest physicists. Truesdell saves his own purely mathematical characterization of physics, as well as the nonmathematical Faraday's status as one of history's great physicists, by claiming, as did Maxwell, that Faraday thought mathematically even if he did not use equations.[33]

Faraday certainly thought in a geometrical and pictorial style, and ingeniously linked his diagrams to physical experiments. The issue is then whether this highly worked out pictorial thinking should count as geometry, as genuine mathematics. Certainly a pictorial thinker, such as an artist, works out rigorous patterns of shapes, but is not considered a mathematician. On the other hand, geometrically oriented mathematicians often communicate informally by means of diagrams and words in a manner quite different and much easier to understand than what appears in finished mathematical papers.

I believe that Faraday's pictorial description of electric fields and currents is more adequately represented by the mathematics of differential forms developed by Elie Cartan in the 1920s than it is by the mathematics that Maxwell and Heaviside developed in the 1800s. It was Maxwell who laboriously mathematicized Faraday's conception of lines of force in one of the major developments of mathematical physics since Newton. Maxwell's mathematical notation was extremely complex, as he often represented the three components of each quantity by three separate equations. It was only in the work of Oliver Heaviside that the vector notation and operator notation that is used today was developed. This is the precursor of the standard vector calculus of the electromagnetic field.

The struggle to propagate the vector calculus took decades. In this case, it was William Hamilton's quaternions that were defended by the mathematical establishment. Such figures as algebraicist James Sylvester defended quaternions against vector analysis. Meanwhile, such figures as Heaviside and American physicist Josiah Willard Gibbs defended vector analysis.

The triumph of vector analysis over quaternions did not occur until the early twentieth century;[34] however, I believe that Faraday's own pictorial and physical intuition were not fully captured even by the mathematics of the vector version of Maxwell's equations. Cartan, who was like Faraday the son of a blacksmith,[35] developed the differential-form notation that, geometrically interpreted, represents tubes of force and "egg carton" patterns of circulation. This mathematics became fashionable among practical physicists only in the 1970s. Many working physicists still do not use it. It is in this notation, which began to be popularized at the undergraduate level only in the 1980s, that Faraday's intuitions find their formal fulfillment.

Faraday utilized currents of thought of his day that have been sketched in previous chapters. He did not read these Romantic Continental thinkers but rather heard about them primarily through Humphry Davy. Davy had been informed by his friend Coleridge, who brought the ideas of Kant to England and who transcribed large portions of Schelling as if they were his own work. Faraday presumably heard about Kant's theory of forces and matter from Davy.

Faraday had not read Boscovich (whose theory was available only in Latin at that time), although talk about the ideas of Boscovich was common in the scientific circles with which he was associated. In one article, Faraday speaks favorably of Boscovich. It has often been said by historians of science that Faraday was a Boscovichian; however, in his main reference to Boscovich, Faraday presents both the solid atom of Dalton and the point atom of Boscovich in a critical manner.[36] A number of British physicists during the nineteenth century spoke of themselves as Boscovichians. Often they meant merely to emphasize their belief in point atoms or in fields of force around a point atom. Faraday was not really a Boscovichian, as his concept of field was not dependent upon Newtonian point sources.

Similarly, Faraday can be said to have indirectly received concepts of Kant. Faraday's ideas are actually closer to those of Kant than to those of Boscovich, for Kant, unlike Boscovich, did not really have point atoms at all, but only the fields of attractive and repulsive force. Faraday's fields were quite different from Kant's, however, which were modeled solely on Newtonian forces.

P. M. Heimann has criticized Pearce Williams's claims that Faraday's ideas were influenced by those of Boscovich and Kant.[37] Heimann first argues that the Faraday of the 1830s (contra Pearce Williams's account) held a particulate conception of matter, not a Boscovichian one.[38] This certainly was not borne out by William Rowan Hamilton's impression of Faraday in 1834 as having been "led to almost as *anti-material* a view as myself," and as having rejected the view that electricity consists of particles constituting a fluid or fluids.[39] Heimann also claims, correctly, that there were other views available in British natural philosophy (such as those of Robert Greene as well as James Hutton, Gowin Knight, and several others).[40] Thus Faraday could have gotten his later ideas of atoms as centers of fields of force from sources other than Davy's Boscovichianism or Coleridge's Kantianism (via Davy). This does not disprove that Davy and Coleridge were a source and it does not eliminate the fact that Faraday was close to Davy and Davy to Coleridge. Also, Faraday may have gained ideas from British writers about atoms as centers of powers, but those British Newtonians and Lockeans accepted atoms themselves as hard, indivisible entities of a Newtonian sort residing in empty space with central forces extending to infinity. These were views that Faraday denied.[41] Heimann then goes on to surmise that Faraday got his "Boscovichian" ideas from Joseph Priestley's *Disquisitions Relating to Matter and Spirit.* (Priestley used the Boscovichian atom as a way of breaking down the dichotomy of matter versus spirit.)[42] Priestley is said to have received his conceptions of atoms as centers of force from British sources. Heimann theorizes that Faraday cites Boscovich rather than Priestley as a source because of the ill fame of Priestley's political radicalism and materialism; however, one wonders if allying with a Jesuit would be a way of achieving religious respectability for an austere Protestant of the Sandemanian sect. He notes that Faraday's Sandemanian religion may have led him to avoid associating himself with Priestley.[43]

Trevor Levere is more willing to grant the effect of Davy on Faraday's shared rejection of Daltonian atomism, although Levere notes the ambiguities of Davy's Boscovichianism.[44] In order to discredit the Kantian influence on Davy through Coleridge, Levere notes that Davy associates Kantianism with Neoplatonism.[45] In the light of the similarities between Kant and the Cambridge Platonists, who Coleridge claimed made his mind already attuned to Kantian approaches, and whose similarity to certain ideas of Kant made intellectual historian Arthur Lovejoy search in vain for historical connections,[46] perhaps Davy's comparison is not so inapt as Levere thinks. Levere believes, more strongly than Heimann, that the Sandemanian influence is the source of Faraday's particular use of the force theory of matter.[47]

Geoffrey Cantor, in his extensive study of Faraday's Sandemanian religious beliefs, claims that Faraday's belief that the universe is one of powers created by God lay behind his scientific theorizing concerning atoms as centers of force and electricity and magnetism as fields of force. Sandemanians emphasized the attribute of God as all-powerful.[48] God was not an artisan or crafter.[49] Creation was by God of powers. According to Cantor, Faraday's theology of nature placed electricity as the highest power and gravitation as the lowest.[50] In contrast to the Enlightenment view that electrical science removed the divine mystery from the lightning (as in a French engraving representing Benjamin Franklin taming the lightning and conquering superstition), Faraday saw the power of the Creator in electrical storms.[51] Faraday rejected the void, or absolute vacuum, and believed that space was filled with the powers of the "sphere of influence" of particles.[52] By the principle of plenitude, the powers created by God fill all.

L. Pearce Williams shows the complex and fascinating road to the concept of the electromagnetic field in the work of Faraday.[53] Faraday moved from his interpretation of Ørsted's experiment through work in electrochemistry to the discovery of diamagnetism to the full-fledged conception of an electromagnetic field. This was in advance of even the great physicists and mathematicians who translated Faraday's work into sophisticated equations. Faraday's idea of space, rather than matter, as the location of the energy and tensions of the electromagnetic and other fields, was not realized in mathematical physics until the work of Einstein in the twentieth century. With an almost complete ignorance of formal mathematics, Faraday was able to imagine what the most educated and mathematically advanced physicists of his day, and even those of two generations thereafter, were unable to comprehend fully.

Faraday started with the interpretation of Ørsted's experiment of the induction of magnetic effects by an electrical current. He interpreted this result in terms of circular forces, which alone undermined or opposed the whole edifice of Newtonian theory based on straight-line forces acting at a distance. The transformation from circles to straight lines as the natural motion of the planetary bodies was a struggle that took several centuries. From the time of Plato through the time of Copernicus, the circle was revered as the perfect form of motion in the heavens. Even Galileo, who proposed that natural motion on Earth was that of a straight line and gave no particular priority to rest over motion, assumed that motion in the heavens was circular, and only Kepler with his ellipses broke the tyranny of the circle. It was not until Newton that the motions on Earth and in the heavens were fully unified and the straight line became the natural path of bodies in both.

Just as straight-line motion was the natural, unconstrained form of motion for Galileo on Earth and for Newton in both the Earth and the heavens, so the lines of gravitational and other attractions were for Newton and the Newtonians the basis of all physical theorizing. With one stroke, Faraday rejected this whole conception. Even before he interpreted Ørsted, Faraday had looked on the notion of action at a distance between bodies as a speculation and as a "philosophical" conception that might possibly be wrong.[54]

When André Ampère, a great mathematical physicist, reconstructed the phenomena of Ørsted's experiment as a complicated mathematical resultant of straight-line forces, Faraday was not impressed. For Faraday, the circular motions and circular forces were experimentally and intuitively evident. Ampère's mathematically complicated reconstruction in terms of straight-line attractions appeared to Faraday like the epicycles of the medieval astronomers. It was an ingenious mathematical construction, but it was not in itself suggested by the experimental evidence.

One of the peculiar sources of Faraday's way of conceptualizing was his work as a chemist under Davy. Although Davy told Faraday about Boscovich and (via Coleridge) Kant and Schelling, the fact that Davy was a chemist and Faraday's early researches were more in chemistry than in physics influenced Faraday's attitude toward Ampère and the other Newtonians. Ampère had to hypothesize that the magnetic currents in matter were produced by the circulation of a fluid. These atoms with whirling fluids only made chemistry more complicated. Faraday was closer to the notion of an atom of Boscovich and Kant than to these whirlpools.

Faraday was able to show that a bent wire produced the effect of a magnetic pole. Thus, one could use the circular-force notions to reconstruct those of the opposing theory. He conceived of lines of force of the magnet traveling in circles through the magnet itself, and of rows of point atoms undergoing strain the way a stretched chain or rope would. Faraday imagined the electromagnetic force as varying with the number of lines cut. He demonstrated this by putting a floating magnetic needle inside of a helix and showing that the needle travels down the length of the helix when electricity flows through it and comes to rest at one end of the helix, which functions as the north pole of the magnetic forces. Faraday envisioned the floating needle as traveling down a magnetic line of force.

Faraday's chemical researches led him to further conceptions in relation to lines of force. He wished to show that there was one electricity, whether static electricity, electricity generated by moving magnets, or electricity produced by contact of different metals. Faraday ingeniously showed that static electricity could decompose water just as electricity was produced by a voltaic pile. The decomposition of water into hydrogen and oxygen had excited the nature philosophers as a proof that electrical and chemical forces were the same. Ritter had spent much of his time and money on experiments measuring the amounts of resultant hydrogen and oxygen produced by the decomposition of water, but no one had been able to decompose water using static electricity. The electricity needed had to be produced by a gigantic spark discharge, whose explosion disturbed the water and apparatus to the extent that the small amounts of hydrogen and oxygen gas produced could not be measured.

Faraday ingeniously conducted the electricity along a wet string to slow down its travel considerably. In this less violent process, the hydrogen and oxygen produced could be observed on a glass plate. Faraday went further and showed that a set of plates connected by wires could also decompose water. In this situation, there were no positive and negative poles at which the hydrogen and oxygen could accumulate. This evidence persuaded Faraday that the positive and negative poles were not essential for the separation of the elements as other people had thought. He did a typically clever and simple experiment involving a sandwich made out of acid-detecting paper (like litmus paper) and disks of gelatin that contained a salt solution. He was able to show that the indicator paper turned color only at the top and bottom ends of the sandwich where the electrical wires were attached. An indicator paper in the middle did not turn color, showing that the elements separated by the electricity did not travel past one another to the opposite electrical poles. Faraday conceived of the situation in terms of chains of molecules, oriented heel to toe, moving past one another.

This conception of chains of polarized molecules in electrochemistry gave Faraday the notion of a general conception of continuous action in electromagnetism. In the chemical process of disassociation of compounds, the disassociation did not take place by action at a distance from the positive and negative electrical poles. Rather, the electrical force was transferred from molecule to molecule, neighbor to neighbor. Faraday later went on to generalize this chemical conception of the transfer of electrical force to a general conception of the transfer of the electrical and magnetic force. Finally, the chains of point atoms were eliminated along with the material objects, and the series of points in space that constituted the electromagnetic lines of force became the bearers of continuous action.

One important modification that Faraday made in 1838 of the equation for electrostatic force was to find that different substances had different inductive capacities. He developed this idea in analogy to the notion of the different chemical affinities that different substances have. This was an important empirical discovery, but even more than that, it undermined the apparent theoretical power of those who, like Ampère and Coulomb, had developed the equation for electrostatic attraction in perfect analogy to Newton's equation for universal gravitation. Part of the power of this equation, in which the force of attraction between two bodies varied as the inverse square of the distance between them (just like gravitation), was that this law exactly resembled Newton's law and appeared to be equally general. In it the amount of charge that bodies carried took the place of mass in Newton's equation.

What Faraday showed was that the constant of proportionality (called Coulomb's constant) was not a universal constant like Newton's gravitational constant. It was different for each kind of substance. This made Coulomb's law appear to be much more specific and context dependent than Newton's law of universal gravitation. Thus, it could not bask in the glory of Newton's law by virtue of its formal similarity to it.

In 1838 Faraday also introduced important new terminology. He spoke of the "electro-tonic state," a name that he settled on after consultation with other scientists. This state was thought of as analogous to the strain in metals. Just as a break in a band of metal sends shock waves up the band, so Faraday thought that overloaded bodies in the electro-tonic state shot an electric discharge. Faraday thought that this strain in the electro-tonic state was produced by the particles in the line of force being polarized like the molecules in the electrochemical processes on which he had experimented. Thus, he located the strain or tension in the physical medium rather than in the particular object that might have been the source of the strain. This strain was still thought of as located in some sort of physical substance. When Faraday tackled the more mysterious forces of magnetism, he was led to locate the strain in empty space itself.

The freeing of the conception of lines of force as lines of material points from the notion of matter and its replacement by geometrical points of empty space was stimulated by Faraday's discovery of diamagnetism. Magnetic materials, such as iron, are called *paramagnets* as they align themselves parallel to the magnet that acts on them. Other less well known magnetic materials were discovered that align themselves perpendicularly to the magnet that acts on them. These materials were called *diamagnets*. Faraday was convinced that magnetic force, like electricity, should be universal; however, the only clear and common exemplar of a magnet was iron. Other sorts of crystal magnets were known, but their behavior had not been closely observed. Some had observed that certain magnetic materials aligned themselves, not parallel to the north and south pole of the magnet, but perpendicular to it, but little attention was paid to these observations.

Faraday worked to produce a general theory of diamagnetism. Ørsted and others had thought that diamagnets simply result from the production of reversed poles while paramagnets reproduce the orientations of the poles of the original magnets. Followers of Ampère and the Newtonian approach (most notably Antoine César Becquerel) reconstructed paramagnetism in terms of straight-line-action-at-a-distance forces. In order to account for the peculiar behavior of paramagnets, Becquerel used a notion similar to that of the Archimedes's notion of buoyant force. Wood has weight, but since its weight is less than that of water, it floats on water.

Becquerel claimed that a magnet put in a medium even more magnetic will behave oppositely, or diamagnetically, to one that is put in a medium that is less magnetic, where it will behave paramagnetically. One difficulty with this is that a diamagnet in empty space will behave as a diamagnet. Thus, Becquerel must assume that the ether that fills supposedly empty space is more magnetic than are diamagnetic substances. Furthermore, diamagnets do not have poles.

Faraday searched for effects of the strain in crystals to show the strain involved in magnetic lines of force. One way to discover strain of an ordinary mechanical sort was to shine a ray of polarized light through the substance, and the strain would depolarize it. Polarized light has all the planes of waves oriented at one particular angle, while ordinary, depolarized light has the angles of the planes of the waves distributed at random. Faraday had tried for two decades to depolarize light by means of magnetic strain. He finally was able to do so with a piece of borate glass. The strange thing was that the rotation of polarization was absolute in space and was not affected by the rotation of the piece of glass. This was initially very puzzling, but eventually led Faraday to the idea that the strain produced by magnetic lines of force did not exist in the glass but existed in the lines in space itself.

Faraday went further and came up with an original conception of the behavior of diamagnets and paramagnets that totally rejected the notion of magnetic poles. He did an experiment that proved to his satisfaction that magnetic lines of force travel right through the physical magnet and out the other end. He built an apparatus consisting of two hollow, box-shaped magnets and ran a wire from a copper collar connecting them in such a way that he could connect the wire at various positions. He also set up the wire so that it could rotate when contacting his system at each of the varying positions. He found that rotation of the magnet itself had no effect on the direction of the polarity produced in the wire; however, the rotation of the wire, whether above the magnetic system or inside of it, did produce an electric current. He took this to show, at least with some degree of probability, that the magnetic lines of force exist inside the magnet as well as outside.

Faraday concluded that the so-called north and south poles of the magnet are not really centers of attraction the way that the Newtonian theory would have it. Rather, the north and south poles are arbitrary positions that correspond to the regions of greatest density of the magnetic lines of force. Faraday entirely eliminates magnetic poles from his theory.

Faraday's move from the notion of actual physical strain in crystals to the notion of the strain of electromagnetic lines of force in empty space interestingly replicates what independently occurred in the evolution of the tensor concept in mathematics. Faraday was totally ignorant of the higher mathematics that included vectors and calculus, let alone tensors; however, the mathematical development of physical tensors occurred in the description of the shapes of crystals and actual physical tension in solid bodies. These early physical tensors (in the literal sense) were Cartesian tensors insofar as they were based on a system of right angles in ordinary Euclidean space. These tensors are much easier to understand than the abstract tensors dealing with the curvature of space in non-Euclidean geometry such as was used by Einstein.

Using the conception of greater or lesser density or frequency of magnetic lines of force, Faraday is able to explain the peculiar behavior of paramagnets and diamagnets when placed between two ordinary magnets. A paramagnetic material, such as iron, will seek the point of greatest density of the magnetic lines of force. Thus, if it is displaced from this point or placed away from it, it will move back to that point. On the other hand, diamagnets seek the place of least density of magnetic lines of force. This point happens to be on the inside of the diamagnet itself. The lines of force from the magnets tend to go around the piece of diamagnetic material. Thus, the point of equilibrium for a diamagnet will be its own center of gravity, and the diamagnet will stay put wherever it is placed.

With his mathematically untrained conceptual brilliance Faraday was able to draw the inference from physical tensions in crystals to electromagnetic tensions in empty space. This notion was too much even for Faraday's mathematical defenders and reconstructers, such as William Thomson (Lord Kelvin) and James Clerk Maxwell. These men constructed the mathematics of Faraday's theories but did so by treating the lines of tension as existing in the ether that supposedly filled all space. The elaborate and complex development of electromagnetic theory in the late nineteenth century was based on the mechanics of the ether. Even Faraday's best friend and defender, John Tyndall, smiled at the naïveté and vagueness of Faraday's conception of lines of force in empty space as real things.

In the early twentieth century, Einstein got rid of the ether with his special relativity theory. Einstein was able to deduce more simply the results that were gotten by the late-nineteenth-century ether theorists by assuming the constancy of the velocity of light from the point of view of an observer traveling with any velocity whatsoever with respect to it. Now that the ether was dropped, the electromagnetic forces and tensions had to exist in empty space. When Einstein introduced curved space in generalizing relativity theory to account for gravitation, he resorted to tensor analysis as a way of discussing his curved spaces. These tensors are much more difficult to imagine or visualize. For this reason, Cartan's mathematics of differential forms developed after Einstein's work has recently become popular with some physicists and most mathematicians as a more intuitive way to treat Einsteinian phenomena.

Interestingly, the mathematical analogy between the simpler Cartesian tensors of the physics of stress and strain in solid bodies and the mathematics of the Einsteinian Riemann tensors describing fields of force as the curvature of space itself were in a certain sense reconceptualized by Faraday's analogy between the strain in a crystal and the strain of magnetic lines in empty space.

Faraday gave the world the concept of field. His extraordinary ability to imagine spatial structures of force in empty space was probably aided by his artisan background, similar to twentieth-century mathematizer of fields, Elie Cartan. It was also inspired by a number of ideas he encountered. Davy's relaying of the German ideas of Kant and Schelling, introduced to England by his close friend Davy, and Priestley's presentation of Boscovich's ideas of atoms as centers of force aided Faraday in his quest for an understanding of electromagnetic phenomena. Faraday's experimental arrangements and visual models were handed on and translated into mathematical form by another extraordinary thinker, James Clerk Maxwell.

CHAPTER 18

Maxwell, Field Theory, and Mechanical Models

In the history of the rise of the concept of the electromagnetic field, John Clerk Maxwell holds pride of place. Maxwell's equations of electromagnetism are a part of classical physics least modified by the introduction of twentieth century theories. Maxwell's equations trumped both Newtonian mechanics and classical thermodynamics in the formulation of relativity theory and quantum mechanics. Einstein modified Newton's laws of motion in order to preserve Maxwell's equations in special relativity. Max Planck modified the laws of classical thermodynamics (making energy discrete rather than continuous) to preserve Maxwell's theory of electromagnetic radiation in early quantum theory.

Maxwell was a physicist who could quote the Latin poet Horace in criticizing Riemann's non-Euclidean geometry. Maxwell wrote tolerable Victorian verse and, while studying physics and mathematics at Edinburgh and Cambridge, pursued literature, philosophy, and theology in considerable depth, writing, "My metaphysics are fast settling into the rigid high style, that is about ten times as far above Whewell as Mill is below him, or Comte or Macaulay *below* Mill."[1] Besides his work in electromagnetism, Maxwell was an important contributor to the kinetic theory of gases and basically invented the modern theory of color vision.[2] His mathematical formulation of Faraday's field concept and experiments, as well as his unification of this with the work of the mathematical physicists of the early nineteenth century, such as Joseph Lagrange and William Rowan Hamilton, is the central watershed in the development of electromagnetic field theory. As a mathematical physicist, Maxwell clearly stands above Ørsted and Faraday in that he transformed the results of Ørsted's vague Romantic ideas and Faraday's often speculative, qualitative ones into a rigorous and powerful formal framework.

The conceptual affinities of Romantic nature philosophy to Ørsted's and Faraday's concepts are quite obvious. Both were reacting against a Democritean atomic worldview and were presenting views based on polarities and fields. Maxwell is famous, however, both for his role in the mathematical formulation of field theory and for his construction of detailed mechanical models to explain phenomena later accounted for in a pure field theory, hence playing an anomalous and controversial role in the history of holistic conceptions of physics. Einstein, in a tribute to Maxwell on the centenary of the latter's birth, claimed that Faraday and Maxwell's introduction of the field concept and the accompanying major reorientation of the worldview of physics was "the greatest change in . . . the basis of physics since Newton" and that "the lion's share in this revolution fell to Maxwell."[3] Maxwell was not thoroughly a field theorist, however, and Einstein truly deserves credit for the development of a pure field theory. For Einstein the only true reality was the numbers of various quantities associated with points in space-time. These are the solutions of the partial differential equations of Maxwell's and Einstein's own theories. A field is a set of physical quantities associated with points in space or, since Einstein, in space-time. It does not involve intrinsic reference to physical objects. Rather, the field is simply the physical quantities at various locations, such as the electric and magnetic intensities. Although Maxwell used the partial differential equations, he did not attribute ultimate reality to the numerical solutions corresponding to the field quantities.

There is great controversy among students of Maxwell's physics regarding to what extent Maxwell was a mechanist. He clearly used mechanical models in describing the electromagnetic field, but the question is to what extent Maxwell took these models literally as real mechanisms. He clearly did not think that his mechanical models of the field were literal, precise, and accurate descriptions of the hidden mechanisms underlying the field.

As a child of three, James Clerk Maxwell was involved with tracing physical, mechanical connections. In his childhood home was a series of bells, used to summon various servants, connected by wires to different rooms of the house. Contrary to the Piagetian child developmental theory of conservation for children of this age, Maxwell became involved in figuring out how the various wires were connected to the various rooms, and in particular the mechanism (pulleys) by which the wires managed to turn corners.[4]

Near the end of his life, Maxwell used an amazingly similar metaphor of physical connections to explain the Lagrangian method in the example of a group of bell ringers pulling the ropes of a carillon. The ringers cannot see above the hole in the ceiling where the ropes connect to the bells, although they can, of course, hear the different bell tones. Furthermore, there are connections of a complicated sort in the belfry between the various bells. The bell ringers have no opportunity to observe these connections. They can, however, tug on various ropes and gauge the tension on each rope when various combinations of ropes are pulled, and also can cease tugging various ropes once the bells are ringing, and gauge the pull on their own arms as the bells above continue to swing, connected by unknown mechanisms.[5]

Maxwell's methodology has been portrayed as both mechanistic and antimechanistic, as both hypothetical and induced from experiments. His finely balanced position makes these opposing interpretations understandable. Maxwell correctly claimed to be translating Faraday's ideas of fields and lines of force into mathematical terms. As a disciple of Faraday, he held to the primacy of fields over electric charges, and treated charges as artifacts of interruptions of field lines. The notion of a particulate electron came after Maxwell's death and was in conflict with the pure field theory of the strict "Maxwellians." Maxwell also rejected the action-at-a-distance approach to electromagnetism, modeled on Newton's theory of gravitation and developed by Coulomb and Ampère, and most comprehensively by Maxwell's contemporary, Wilhelm Weber (1804–1891). While Weber took Faraday's experiments and explained them in terms of action at a distance, Maxwell took both Faraday's experiments and his theoretical concepts and mathematically described his field, lines and tubes of force. In his opposition to the quasi-Newtonian action-at-a-distance approaches, Maxwell rejected atoms of electricity.

Maxwell was quite explicit in holding that his mechanical models were mere analogies that help in understanding the physical processes under investigation. As an undergraduate at Cambridge, Maxwell spoke to a student club on the nature of analogy.[6] He read and utilized the ideas of the Scottish philosophers who emphasized the role of analogies in explanation, such as Dugald Stewart and Sir William Bart Hamilton.[7] Scottish commonsense philosopher Thomas Reid's criticism of analogies was rejected by Stewart and Hamilton, while Reid's notion that all thought begins in analogies was accepted by them and developed into a favorable assessment of analogies properly used. Sir William Hamilton combined a concern with models and a Kantian conception of the contribution of the mind to our theories, and considered models to be our only means to discovery, while remaining critical of the reality of the models. From recent physics Maxwell borrowed the method of "physical analogy" that William Thomson had applied in using "resemblance of form" between heat and electrostatics and between light and a vibrating solid.[8] While certainly far from the realm of protoscientific and pseudoscientific theories of analogies of micro- and macrocosm in medieval China and Renaissance Europe, it is significant that the man who finally fully mathematicized the field concept had extensive recourse to the theory of analogy, which dominated those earlier theories of polarity and analogy. Maxwell, the consummate mathematical physicist, not only developed the theory of analogy, which has a role in the theory of induction as well as in literature, but even on occasion referred to allegory.[9]

John Herschel, a leading British physical scientist and philosopher of science of the early nineteenth century, emphasized in his methodology that a good explanation needed a good analogy.[10] For Herschel analogy is the route from direct experience to a *vera causa*—a true, unobserved cause. Another Cambridge product who used Herschel's doctrine is Charles Darwin, who applied the analogy of artificial selection to explain natural selection in evolution.[11] Maxwell likewise attempted to illuminate the workings of electromagnetic processes through mechanical analogies.

Mary Hesse has rightly emphasized the importance of analogy in Maxwell's work, but described Maxwell's use of analogy as a form of induction from experiments.[12] Maxwell himself sometimes spoke in the Newtonian manner of deducing laws from phenomena,[13] but these claims—like Darwin's claims to follow the strict Baconian method, or the depiction of Faraday, who made some of the boldest physical speculations ever, as always most cautiously generalizing from the facts—are part of the misleading ideology of nineteenth-century science, not its real nature. Maxwell's masterful use of analogy as a means of inferring properties of electrical systems is better seen within the framework of the formulation of hypotheses on the basis of previous physical theory, not direct generalization from experiment.

One might think that Maxwell's rejection of electric particles and his doctrine of priority of fields to charges would unequivocally put him in the antiatomistic and antimechanistic camp; however, Maxwell is also famous for his use of mechanical models of the electromagnetic field. His first great electrical paper, "On Faraday's Lines of Force," follows Faraday in treating lines of force as the fundamental reality.

In Maxwell's second great paper, "On Physical Lines of Force," Maxwell introduces his "idle wheel" hypothesis as part of an elaborate mechanical model. Maxwell treats "molecular vortices" as rotating quasi-circular entities illustrated as hexagons (generalized in three dimensions to rotating blobs). Since neighboring vortices rotate in opposite directions and would block each others' motion when treated as material objects with friction, little disks or particles, like idle wheels in an engineer's system of rotating wheels, fill the spaces between the vortices. Thus, rotation can be transferred between neighboring vortices rotating in opposite directions. These idle wheels, far from being an ad hoc device to save the model, turn out to be useful in accounting for Maxwell's most important contribution to electromagnetic theory—the displacement current—which enables Maxwell to make consistent his versions of the equations of Coulomb, Ampére, and Faraday, and to account for static electrical phenomena as well as dynamical ones. Ultimately Maxwell's elaborate mechanical model ran into trouble, and was rejected by him in his last works, where he returned to Faraday's lines again. Noting the ultimate rejection of a mechanical substratum for electromagnetism after Maxwell, some commentators have claimed that his model-building methodology was unhelpful to him.[14] Daniel Siegel, however, has convincingly shown how fruitful this mechanical model was both for Maxwell's introduction of the displacement current and for his unification of light and electricity, optics and electromagnetism.[15]

If we grant that the now somewhat bizarre-looking idle-wheel model was tremendously fruitful in Maxwell's discoveries, and that its details were useful in predicting important new phenomena, ought we conclude that Maxwell was a mechanist? Siegel thinks so.[16] Having shown how useful the idle-wheel hypothesis was for Maxwell, Siegel denies that Maxwell could have thought of it only as an analogy or model, and must have been committed to its reality. Ironically, A. F. Chalmers, who erroneously rejects the fruitfulness of the idle-wheel model, claims that Maxwell was a realist with respect to it, but this assumes that if one takes into account the fine details of a model or analogy then one must be committed to its reality. Apparently a detached or skeptical use of models should treat them only in broad and partial terms. An intellect like Maxwell's could draw out the detailed ramifications of a model without being committed to its ultimate reality. Maxwell held that none of our representations or models are an exact and perfect fit with nature.[17] He presented his incompressible fluid of the lines-of-force model as "merely a collection of imaginary properties" and his idle-wheel hypotheses in the physical theory of the lines-of-force model as "provisional," and "awkward," "not a mode of connection existing in nature," despite his very detailed and fruitful development of its consequences.[18] His later vortex treatment of magneto-optical rotation is presented as "of a provisional kind."[19]

William Thomson, who had first developed a mathematical account of Faraday's notion of electric strain, believed in the necessity of mechanical models for understanding and demanded that a true physical explanation be reducible to mechanics. Thomson claimed that he could not really understand Maxwell's electrical theory insofar as it was not presented mechanically. The usual story of the revolutions in early-twentieth-century physics was that the mechanical worldview of nature, as exemplified by physicists such as Thomson in the 1890s, was replaced by the nonmechanical views implicit in Einstein's field equations and in quantum mechanics with its indeterminacy. This is a considerable oversimplification of the situation.

By the late nineteenth century, Thomson, despite his eminence, was somewhat of a holdout. Other views besides the mechanical worldview were competing for dominance in physics. There was, most famously, the interesting and short-lived "electromagnetic worldview" fostered by Hendrik Lorentz and Max Abraham.[20] In this view, the electromagnetic ether rather than atomic particles was the ultimate physical reality. Similarly, the doctrine of "energeticism" of Pierre Duhem and chemist Wilhelm Ostwald (1863–1932), claimed that energy rather than matter was the ultimate reality.

To what extent Maxwell was a mechanist hinges on what is meant by mechanism or mechanistic physics. Major scholars of Maxwell claim, in effect, that Maxwell was a "vacillating and inconsistent mechanist."[21] A case can be made for this, depending on what one includes in mechanism. Early in the nineteenth century there was the so-called *dynamical* view of nature, which emphasized energy and potential energy in the Lagrangian and Hamiltonian equations in contrast to the emphasis on three-dimensional physical atoms in the *mechanical* view.[22] (These nineteenth-century terms need to be distinguished from the general sense of *mechanics* and *dynamics* in physics texts today. The mechanical view emphasizes the reality of atoms and their interactions. The dynamical view denies the reality of solid atoms and emphasizes the reality of force instead of matter.) The dynamical view stresses the central role of energy, and potential energy, in dynamical physics as well as its tendency to abstract into multidimensional phase spaces rather than directly treating physical atoms in three-dimensional space. This dynamical approach to physics, stemming philosophically in part from the Romantics and mathematically in part from Lagrange's abstract analytical mechanics, avoided hypothesis concerning the internal structure of mechanical systems, but formulated equations that could deal with the observable outputs of such systems. It combined an epistemological rationalism (deductive approach) with antimechanist theory of reality.

Maxwell absorbed dynamical conceptions both at Edinburgh and also at Cambridge, where he read and absorbed the doctrines of William Whewell, the polymath who introduced the terms *scientist* and *physicist* in the English-speaking world. Whewell was even more Kantian than Hamilton, and emphasized the contribution of theory to our conception of the physical world. Whewell and the other Cambridge methodologists who influenced Maxwell accentuated the deduction from hypotheses as opposed to the Scottish model-theory approach. Maxwell absorbed both traditions and held them in a fine balance.

Whewell was influenced by the dynamical conceptions of William Rowan Hamilton, who rejected atomism and materialism and treated physics at the utmost level of mathematical generality. The Hamiltonian approach to mechanics made no assumptions about the physical constitution of bodies, whether of particles or waves. Indeed, Hamilton developed a formalism that applied both to optics and mechanics—the optical-mechanical analogy—which a century later, like the Hamiltonian function, found an important role in quantum theory.

The dynamical approach to physics, in which highly abstract, general equations for the system were formulated dealing with an abstract multidimensional space and without making particular hypotheses about microscopic mechanism, was combined by Maxwell with the Scottish interest in alternative models as aids to discovery. Lagrange proposed a purely algebraic approach to calculus and an analytical mechanics that was purely mathematical and indifferent to atomic hypotheses. Lagrange had bragged that his analytical mechanics contained no pictures, but Maxwell could combine Lagrangian formulations with a variety of pictorial models employed opportunistically for their usefulness in exploration and discovery. The Lagrangian formulation of a mechanical system is compatible with an infinite number of particular physical mechanisms. Thus, the investigator could explore several mechanical models of a system described by the same Lagrangian.

The antiempiricism of Kant or W. R. Hamilton is a separate issue from the belief in extended spherical atoms versus Boscovichian point atoms or in forces and energy as opposed to mass and matter as the ultimate reality. Maxwell's implicit denial of physical atomism with respect to electric charges would place him in the antimechanical camp if atomism is considered an intrinsic component of mechanism. Siegel, however, in his recent detailed examination of Maxwell's concept of the displacement currents, argues that Maxwell was a mechanist in the tradition of continuum mechanics.[23]

It is certainly true that Maxwell's equations were partial differential equations and were developed and interpreted by Thomson and Maxwell himself as equations of classical continuum mechanics. In eighteenth-century continuum mechanics, although the field is subsumed under "mechanics," there are aspects that lead to a very different metaphysical worldview from that of atomistic mechanism. The great figures of eighteenth-century continuum mechanics are more often than not quite ambiguous with respect to the extent of their commitment to the reality of continua as opposed to atoms. They clearly were committed to mathematical models involving continua in hydrodynamics and elasticity. Their metaphysical commitments are often quite guarded and implicit. There do seem to be several roots for a commitment to real continua. Leibniz's metaphysical theory of the continuum, which was the philosophical basis of Continental physics, is one. The development of a phenomenological physics that bases itself on the observable solidity and continuity of medium-sized physical bodies is another commitment not wholly consistent with the first. The belief in the real continuum is sometimes belief in an unobservable Platonic continuum of mathematical points, but commitment to the priority of the phenomenal continuum bases itself on the direct deliverance of sense. Both, however, conclude with the rejection of physical atoms with size and shape.

The further stage of abstraction Maxwell and many later physicists pursued was the formulation of the theory in Lagrangian or Hamiltonian form. Since Lagrange's equations and Hamiltonian mechanics arose as generalizations or abstractions from classical mechanics, it is often thought that to formulate a theory in terms of Lagrange's or Hamilton's equations is to give it a mechanical formulation, or a mechanical model, but this involves several confusions, to which Maxwell was less prone than many later physicists. Lagrangian (or Hamiltonian) formalism is compatible with a multiplicity of physical models, or with having no physical model at all. To identify the equations or formalism of Lagrange or Hamilton with the commitment to a mechanical model is to confuse syntax and semantics. To use the Lagrangian notation involves no further interpretation of that notation in terms of mechanical point particles. One can draw analogies (as did Maxwell and later workers in electromagnetism) between quantities in the Lagrangian formulation and quantities in a mechanical model (for instance, between the variables that in mechanics would be positions and quantities of electricity whose positions are not given in the theory of electrical circuits) without claiming a literal identity between quantities of electricity and locations. The Lagrangian method introduces generalized coordinates, which need not represent spatial position.[24] Similarly, the position and momentum coordinates of Hamiltonian mechanics yield a multidimensional abstract space for a system in which a single point representing the whole system, not to be identified with any one particle, can move.

The denial that Maxwell was a full-fledged atomist may seem strange in light of his role in the formulation of the kinetic theory of gases (the derivation of thermodynamic quantities from the assumption of atoms); however, Maxwell was careful to qualify his characterization of particles. He even speculates in several passages that the laws and properties of particles and the microrealm may be different from those of the macrorealm.[25] Some readers have suggested that he also hints at the nonmechanical nature of irreversibility, although this is controversial.[26] Maxwell suggests that the particle be defined by an infinitesimal mathematical volume. This latter definition would not distinguish particle theories from continuum theories. It may be for this reason that in Britain there was no major opposition to kinetic theory, while in Europe Ludwig Boltzmann's realistic atomism was opposed by the energeticism of Duhem and Ostwald, and the denial of the reality of atoms by Mach.[27]

The German assimilators of Maxwell, among them Helmholtz and Weber, attempted to compromise between the wholly field-based approach of Faraday and

Maxwell and the charge-based approach of such figures as Ampère. With the development of the theory of the electron (initially treated by such later Maxwellians as Joseph Larmor as a singularity in the field, but later empirically discovered by Joseph J. Thomson as a particle), the need to reconcile an atomic- or particle-based theory of electricity with the continuum or field-based theory of Maxwell became imperative. Figures in the late nineteenth and early twentieth centuries worked out the compromise that is to be found in the modern exposition of Maxwell's equations involving electrons and charged particles that were not considered to be fundamental entities by Maxwell himself. Einstein's papers of 1905 included attempts to reconcile the experimental existence of electrical charges with the wholly field-based approach of Maxwell. On the one hand, Einstein argued for the existence of particles of light, or photons, made manifest in the photoelectric effect. This idea was so radical that it was not accepted even by Niels Bohr, the later formulator of the wave-particle duality, until 1923, eighteen years later and four years before Bohr's codification of that duality.

The same year that Einstein presented the light-particle hypothesis, he also published his famous paper on special relativity theory, which modified Newton's equations to fit Maxwell's. Insofar as Einstein was able to give up Newton to fit Maxwell, he was asserting the primacy of the field point of view. To the extreme proponents of the electromagnetic worldview, however, Einstein's own special relativity was considered too mechanical. In one respect, Einstein removed a mechanical residue from the electromagnetic worldview itself by eliminating the mechanical ether. Einstein's later (1907) derivation of $E = mc^2$ from special relativity theory grounded one aspect of the looser speculations of the energeticists by making mass equivalent to energy, but unlike the theory of the energeticists it did not eliminate the notion of mass.

Maxwell carefully balanced the dynamical method of the Romantics and the developers of abstract, analytical mechanics with the ingenious use of mechanical models and analogies as means to hypothetical theories of electromagnetic phenomena, while maintaining the detachment of the Scottish school toward the reality of these analogies.

CHAPTER 19

Conclusion: Metaphysics of Science—Creation and Justification

———

$\mathcal{T}$he history of the "three traditions" needs to be set against the background of some general philosophical considerations concerning the nature of science and scientific methods.

Science as Formalism versus Science as Interpreted with Models and Guiding Principles

A common view of physics is as consisting solely of equations and the experiments or observations that confirm the equations. The theory appears not to have assumptions or commitments outside of the equations themselves. This can occur because an area of the discipline has settled into an established routine and become noncontroversial and can appear to have no presuppostions. The presuppositions are still present, but they have become ignored and forgotten once the assumptions are no longer controversial. The presuppositions are so agreed upon and assimilated that they function only tacitly.

What is the moral of our story? The contemporary physical scientist who identifies science solely with formalism and predictions or the follower of traditional logical empiricism may say: Absolutely nothing. You have told some interesting historical tales, which are quaint at best, strange at worst. But they have *nothing* to do with genuine science, philosophy, or with the *logic* of electromagnetic theory. Heinrich Hertz, who said "Maxwell's theory is Maxwell's system of equations," had finally reached the truly scientific stage.[1] The earlier figures were only partly scientific, and for the rest, well, they are mired in meaningless prescientific metaphysics, and offensive metaphysics at that. For instance, Steven Weinberg claims, "Whatever cultural influences went into the discovery of Maxwell's equations and other laws of nature have been refined away like slag from ore."[2]

Yet it was the Chinese Taoists and correlative cosmologists who discovered and refined the magnet, Faraday who conceptualized fields of force, and Ørsted who discovered the fundamental electromagnetic interaction. Hertz contributed central, confirming experiments for Maxwell's theory, but it was Ørsted and Faraday who contributed the conceptual substance of the theory. The skeptic may reply, however, "But that is only the 'context of discovery,' where your metaphysical nonsense plays a role. It plays no role in the abstract calculus of the completed theory or the experimental confirmation of the theory—that is the 'context of justification.'"

The position of that skeptic is not quite correct. Faraday's metaphysics guided his careful, inductive confirming experiments. Ørsted's metaphysics allowed him to describe and record the fundamental phenomenon discovered. Scholastic metaphysics prevented the West from understanding the compass for centuries, while in China it seemed a very natural and easily assimilated discovery, considered within the context of correlative cosmology. Not just the context of discovery, but the interpretation and models of the theory were metaphysical, until, of course, Hertz's puritanical iconoclasm. Hertz himself was not a formalist and made his famous statement because he accepted Maxwell's equations empirically but could not reconcile them with his own extreme particle mechanism.[3]

We may again encounter the physicist-formalist or the logical empiricist who would answer: Models, too, are inessential to the pure logic of the theory. All you need are the uninterpreted equations and some definitions to tie their deductive consequences to measurements. The models merely help us learn and teach the theory. They have merely psychological or heuristic functions and, at best, communicative functions. They are, however, irrelevant to the truth or falsehood of the theory.

The model of abstract, formalistic deduction that the logical empiricists assume is questionable. At the purely syntactical formal level an infinity of trivial deductions can be made from any set of formal axioms. Only a tiny, finite subset of them are of interest to the scientist, or even to the working mathematician. Theoretical science and mathematics are not just a matter of deducing consequences at random, but of deducing relevant, significant, illuminating consequences, and making fruitful modifications of one's original hypothesis.[4] This process is guided by models and metaphysics in our sense.

Theories are in principle "underdetermined" by the empirical evidence. Willard V. O. Quine and others have pointed out that any finite set of observation sentences is deducible from an infinity of different axioms, yet we do not choose our theories solely by their compatibility with our observations or by their yielding empirically confirmed consequences. The fruitfulness of our theories in making further predictions narrows the set of formal alternatives compatible with the evidence to a very few. The deployment of the theory to pursue further predictions involves the choice of auxiliary assumptions guided by models and interpretive considerations. Taken into account is the coherence of the theory with other scientific theories, and also with other nonscientific theories, such as metaphysical ones.

The successive development and modification of scientific theories is important for the understanding of science, so the "logic of discovery," better characterized as "the logic of pursuit," has a place alongside the "logic of confirmation."[5] The metaphysical presuppositions and frameworks of science become part of the subject matter of the logic of science. Metaphysical views, such as those recounted here, are relevant to the logic as well as the history of science. Even Karl Popper who, despite titling his book *Logic of Scientific Discovery*,[6] rejects the existence of a logic of discovery, does admit criteria of "acceptance" of theories prior to testing. This fits into the rubric of the "logic of discovery" in the sense of justifications for the proposal of hypotheses and the serious consideration of conjectures.[7]

Much recent philosophy of science has undermined the sharp distinction between the logic of scientific justification and the logic of scientific discovery. No one claims that there is an automatic method for creating new hypotheses (such an algorithm for creation would be a contradiction in terms); however, there are at least two levels at which forms of justification are involved in the development and deployment of scientific hypotheses that are different from the role of justification in the traditional, hypothetico-deductive model of science.

In this model of science a conjecture is formed whose logical status is totally arbitrary or random with respect to the data, at least in some Popperian versions of the method. Consequences are deduced and hypotheses are rejected or tentatively maintained. This model of method neglects that the worthiness of the hypothesis for consideration or for further testing itself involves logical considerations. Popper and Lakatos talk about this in terms of the excess testability of a hypothesis. Thus, falsifiability is a criterion for worthiness of consideration, although it is not the only one.

The notions that Norwood Russell Hanson discussed in his logic of discovery fall into this area rather than into that of the literal creation of hypotheses. Hanson adds the initial success of a hypothesis in explaining some fact or other as his primary model for "abduction," basing himself on American philosopher and scientist Charles Peirce. Hanson, however, adds several other sorts of factors that may make a hypothesis worthy of further consideration, including analogy with another successful hypothesis. A hypothesis successful in hydrodynamics might be a plausible candidate for use in electrodynamics or quantum mechanics. Another consideration supporting worthiness of further testing in Hanson's logic of discovery is symmetry of the equations.[8] Einstein appealed to formal symmetry in the opening paragraphs of his first paper on special relativity theory.

Both these factors of abductive, explanatory capacity and analogy that Hanson considers under the logic of discovery could be seen as forms of induction in the broadest sense. Insofar as the abductive evidence of the ability of a theory to treat a single case is not treated as one piece of inductive support for the theory itself but as evidence that the theory is at least plausible, this case could be said to be part of the logic-of-hypothesis consideration rather than the logic-of-confirmation of hypothesis.

Some recent philosophers, starting with Gilbert Harman, have replaced the notion of abduction as well as the notion of hypothetical inference with the notion of "inference to the best explanation."[9] It adds little, if anything, to the logic of combined abduction, induction in the broadest sense, or hypothetico-deductive method; however, it avoids a number of confusing connotations of previous terminology. It avoids oversimple enumerative notions of induction, identification of abduction with logic of discovery in the literal sense of creation, and separation of the hypothetico-deductive inference as a form of after-the-fact justification from the inference to hypotheses.

A second area in which considerations that are not part of the traditional logic of justification play a role is in the deployment of auxiliary hypotheses in the development of research programs. In the traditional method, a prediction or description is deduced from a hypothesis, and if the prediction fails the hypothesis is rejected. Hypotheses are generally used in groups, however, to derive a prediction (the Duhem thesis). If the prediction fails, it is not determined which of the members of the group of hypotheses that was used to derive it has failed. Thus, it is possible to modify one or a few of the hypotheses to retain the others in successfully deducing the new result that is found (the Duhemian argument).[10]

The logic of modifying or replacing some of the hypotheses in one's arsenal of hypotheses in order to save and successfully use the others is a different logic from that of simple conjecture and refutation or disconfirmation. Lakatos, among others, has talked about "scientific research programmes" that have a "positive heuristic" that serves as a guideline for the formation of new auxiliary hypotheses when a theory runs into trouble. The addition of epicycles in Ptolemaic theory, or the addition of ad hoc properties to the ether in late-nineteenth-century electrodynamics, or the addition of modifier genes to classical genetics are some standard examples of these patterns of hypotheses modification.

The way that hypotheses are modified may be guided by a general conception of nature and of of what constitutes a good theory. Imre Lakatos's "Methodology of Scientific Research Programmes" provided a description and terminology for hypothesis modification.[11] Which hypotheses are retained and protected from modification by the "protective belt" of auxiliary hypotheses may depend on one's conception of the metaphysical structure of reality. Lakatos makes the matter of which hypotheses are shielded from modification or rejection simply a matter of convention;[12] however, one can also look at the injunction not to modify the hard core of central hypotheses as a matter of adherence to a metaphysical view in which the structure pictured by the core hypothesis is considered to be ultimate reality. Thus, metaphysical views may guide the direction of hypothesis modification.

What philosophers of science sometimes call metaphysical presuppositions of science are often cosmological versions of ideologies or social conceptions of the world. (Since this connection is at the metaphysical, not the empirical, level, admitting it does not reduce science to ideology as the consequences of the theory may fit more or less well with observation and experiment.) Larry Laudan, in opposing the sociological account of science, objects to the move from the underdetermination of theory by evidence to the claim that hypothesis formation is socially caused. Laudan notes correctly that the logical point about underdetermination has no particular causal consequences.[13] If, however, one moves from underdetermination not to claims about social causality but to claims about the role of ideological or metaphysical presuppositions in scientific-theory acceptance, one is not jumping the gap between the normative and causal (a gap that pragmatist defenders of scientific method ought not make too wide). If considerations other than empirical evidence are admitted in the evaluation of scientific theories, then consistency with other theories, including nonscientific ones, can be admitted. Similarly, if metaphysical presuppositions play a logical role in relation to scientific theories, then the move to theories of metphysical and social cosmology in logical relation to the assessment of theories need not move out of the realm of logical or normative considerations. Laudan thinks that it is a novelty of his "research tradition" account of scientific progress that he considers "conceptual problems" as well as empirical problems (including among "external conceptual problems" both worldview problems and "extrascientific beliefs") among the issues of logical coherence to be dealt with by a scientific theory.[14] This leaves open an "internalist" consideration of issues of social ideology and cosmology in the consideration of the assessment and deployment of scientific theories.

Some of the problems of dealing with the logic and status of metaphysical presuppositions of science can be solved or avoided by treating it in terms of Bas van Fraassen's logic of presupposition, in which the truth or falsehood of a statement necessitates the truth of a presupposition. Presuppositions do not entail or imply the statements that presuppose them.[15]

Previous Approaches to the Metaphysical Presuppositions of Science

Over the last century there have been several claims about the status of metaphysical presuppositions of science.[16] The claim that there is no relation between science and metaphysics may be based on the claim that there is no such thing as metaphysics, or on the view that science is purely instrumental and technological. One can also believe that metaphysics is legitimate but deny that science can have presuppositions. Roman Catholic physicist Duhem held this latter position. To know about the structure of reality, consult a Catholic theologian, but to make quantitative predictions about observable events, learn physics.

Much British ordinary language analysis and some European existentialism treats science as instrumental and as lacking in propositional truth-content.[17] On the "instrumentalist" view, science is a tool, not a picture of reality. (Indeed, the tool-versus-picture metaphor goes back to Edgar Zilsel's craftspeople and painters of the Renaissance as groups whose activities and knowledge fused in giving birth to modern science.) The no-relation views depend upon the definitions of metaphysics and science. If science has no truth-content and is purely instrumental, there is no scientific representation of the world that could have presuppositions. If metaphysics is sheer nonsense and totally impractical, giving no hints or directives to other fields, then metaphysics has no connection with science. Even the nineteenth century positivists, however, allowed that metaphysics could have a bad psychological effect on science. The no-relation view can also claim that metaphysics and science are actual enterprises, but deal with totally different subject matters and thus do not influence one another.

Logical positivism, for almost a century the most influential view of the relations of metaphysics and science, denied metaphysics' cognitive meaningfulness, thereby denying a logical role for metaphysics in science, but allowed it a psychological role, usually pernicious.[18] Logical empiricists sharply distinguished the "context of discovery" from the "context of justification," and allowed an irrational, psychological role in discovery for metaphysics.

"Fictionalist" philosophers of science, such as Duhem, allowed metaphysics a cognitive role. In contrast to logical positivists, who tacitly rested their case for the intellectual supremacy of science on the prestige of science, fictionalists deny that science is the highest form of knowledge. Duhem maintained the logical separation between science and metaphysics, writing, "Physics proceeds by an autonomous method, absolutely independent of any metaphysical opinion."[19]

Various logical difficulties arose for the positivists' verificationist criterion demarcation between science and metaphysics, distinguishing meaningful from meaningless statements (the criterion allowed in metaphysical statements when logically conjoined to testable ones, and barred high-level theoretical scientific statements that were not directly testable). As awareness of these difficulties grew, Popper's falsifiability criterion became the leading means of demarcation; however, Duhem's earlier arguments against conclusive falsifiability of scientific theories blurred the dichotomy between falsifiable science and unfalsifiable metaphysics and were admitted in various forms by Popperians.[20] Also, the problem of unfalsifiable existential statements (such as "atoms exist") led to difficulties for falsifability as a means of distingishing science and metaphysics.[21]

In the wake of the retreat of the verification criterion, there was widened recognition of the positive role of metaphysics in the development of science. Historical work of Emile Meyerson, Pierre Duhem, Alexandre Koyré, Edwin Burtt, and Ernst Cassirer, written before the heyday of logical empiricism, became influential among philosophers.[22] The heuristic role of metaphysics, however, became central to discussions. The later Karl Popper, Stephan Koerner, and Marx Wartofsky discussed metaphysical "directives" and "heuristics" for science.[23]

More recently a stronger logical role for metaphysics has been suggested. Popper's prime examples of science are the grand theories, such as Newton's or Einstein's, that, interpreted realistically, can be taken as metaphysical views of the world.[24] Popper's student Joseph Agassi has gone even further by asserting that metaphysical theories are refutable, and thus have a kind of entailment relationship to scientific theories. Agassi says, "Some of the greatest single experiments in the history of modern physics are experiments related to metaphysics" and claims that the refutation by the Russian revolution of Marx's prediction of revolution in advanced capitalist countries "amount[s] to" a refutation of materialism.[25]

Popperians have not wholly jettisoned the heuristic approach because of their difficulty with the entailment relation from metaphysics to science. Marx Wartofsky also turned to the "heuristic" interpretation, but ends, in his discussion of models, by collapsing the distinction between science and metaphysics.[26]

The entailment relation and the heuristic approach leave a dilemma: either identify science with metaphysics (via Wartofsky's "models," or Agassi's entailment relationship, or Lakatos's explicit "blurring" of science and metaphysics in his "research programmes")[27] or admit that the role of metaphysics is merely "heuristic" and not logical.

If simple refutationism is used as the logical account of the relation of refutation of a scientific theory to refutation of a metaphysical theory, then metaphysical statements turn out to have the logical status of higher-level scientific hypotheses. If "sophisticated" refutationism is appealed to (later Popper or Quine's Duhemian argument), then the line between science and metaphysics is blurred. If the "heuristic" approach is used, then either metaphysics is *logically* irrelevant to science (Koerner, positivists) or else it is *identified* with science (Wartofsky). The relationship between metaphysics and science is more than a merely psychological or historical one; however, it is false that refutation of a scientific hypothesis overthrows metaphysics by a simple modus tollens argument.

The Logic of Presupposition and the Metaphysics of Science

The use of Fraassen's logic of presupposition to account for the relation of science and metaphysics prevents this blurring of the difference between them, allowing for a relative distinction between scientific and metaphysical assumptions at any state of research, and allowing both metaphysical and scientific statements to have truth-values under certain conditions.[28] Thus, it does not preserve the cognitive status of metaphysics by depriving science of truth content, and does not preserve the cognitive status of science by treating metaphysics as merely instrumental or heuristic. It also allows for progress both in science and in metaphysical cosmology. (Ironically, Fraassen, who developed the formal logic of presuppositions rejects scientific realism and the notion of metaphysical presuppositions of science, and so does not apply his formalism to this issue.)

A statement such as "The chemical in the jar is red" can be wrong in two different ways. In the first sense, the chemical in the jar is not red but green. In the second sense, there is no chemical in the jar. The assumption that there is a chemical in the jar is a presupposition. It does not imply that the chemical is red or that the chemical is not red; however the truth that the chemical is red or the truth that the chemical is not red (say, green) necessitates that the chemical exists. If the presupposition fails (there is no chemical), then it is not the case that the chemical in the jar is red and it is not the case that the chemical in the jar is not red.[29] If a double negation is equivalent to an affirmation, as in standard negation, then the previous statement is contradictory. If, however, the last appearance of "not" is a nonclassical negation different from the first "not," then the statement is noncontradictory. Fraasen's logic of presupposition utilizes two kind of negation (choice and exclusion) and a nonclassical necessitation by which a sentence necessitates its presupposition. It makes sense to say, "The chemical in the jar is neither red nor not red. There is no chemical in the jar."

One common characterization of metaphysical statements, is that they are so general as to be confirmed by any happening whatsoever. On the presupposition relation of metaphysics and science, we can see why this would appear to be the case because the metaphysical presuppositions are the presuppositions of the truth or falsity of the statement of which they are a presupposition. Thus, if M is a metaphysical presupposition of a scientific statement, S, then whether S is confirmed or S is refuted, M is confirmed. For M is a (nonclassical) consequent of S; that is, S necessitates M and not-S necessitates M. The only case where M fails to be confirmed is where S can neither be confirmed nor refuted. If we take the "can" here in a strong sense, we mean S cannot, in principle, be confirmed or refuted. That is, if S is experimentally neither verifiable nor falsifiable or if S is positivistically meaningless (that is, cannot be shown confirmed or disconfirmed by experimental methods in any acceptable conception of science), then M fails to be confirmed.

This presupposition schema also shows that the Agassi notion that metaphysical statements can be disconfirmed by a simple modus tollens is erroneous. One can see, in terms of presupposition logic, why someone might be mistakenly led to such a view. The connections between metaphysical presuppositions and a scientific statement is not classical entailment. Indeed, classical entailment does not relate the metaphysical presupposition to the scientific statement of which it is a presupposition.

Agassi suggests that the relation between metaphysical presuppositions and scientific statements is one of entailment and that the great experiments in science have been crucial *refutations* of metaphysical presuppositions.[30] Agassi's schema would be M implies S, not $-S$, therefore, not $-M$. Metaphysical statements do not yield scientific statements by any simple, classical entailment, however, as Agassi recognizes but does not resolve.[31]

On the presupposition-logic model of the relation of metaphysical presuppositions to scientific theories, metaphysical statements are never simply disconfirmed by the refutation of the scientific statement that presupposes them. Indeed, if a metaphysical presupposition functions as the presupposition of a false scientific statement, not only is the metaphysical presupposition heuristically fruitful (as Popper, Agassi, Koerner, Wartofsky, and others recognize),[32] but the metaphysical presupposition is confirmed as well. That is why metaphysical presuppositions of falsified scientific statements are nonetheless good metaphysical statements. Aristotelian metaphysics is not overthrown by the replacement of Aristotelian science by Galilean science. Elementarist metaphysics is not overthrown by the replacement of Newtonian science with quantum mechanics or relativity theory. Thus, metaphysical statements may linger on, later functioning as presuppositions for new scientific theories.

Unlike Imre Lakatos's "research programme" model of scientific change, the presupposition logic account separates the presuppositions from the scientific theories and distinguishes between science and metaphysics without treating either instrumentally. Lakatos's instrumentalization of science and his purely heuristic treatment of research directives eliminates the distinction between science and metaphysical presuppositions;[33] however, the model presented here preserves the cognitive status of both metaphysics and science. Metaphysics plays a role as a base for research directives, but it does not become identified with them. Our model also allows one to make positive, rational sense of Lakatos's "negative heuristic" and to avoid the collapse of his distinctions of which both he and his followers himself were sometimes guilty.

Lakatos's account of a "research programme" originally had two components. One was the "series of theories," the sequences of common main hypotheses with successively modified auxiliary hypotheses. The series of theories, possessing a common "hard core" and succession of modified auxiliary hypotheses constituting the protective belt, are sets of *statements*. A second component is the "heuristics." The "positive heuristic" is the contentful set of rules for the modification of the auxiliary hypotheses. The positive heuristic, as a set of rules, not of theories, is a partial reconstruction of the nonstatement components of Thomas Kuhn's "paradigm."

The other nonempiricist aspect of Kuhn's paradigm is reconstructed by Lakatos, however, as the nonfalsifiable hard core of the research programme. This hard core is nonfalsifiable because of the negative heuristic, which states, "Don't reject the hard core." Unlike the positive heuristic, which contains a wealth of guides to hypothesis modification, the negative heuristic contains merely a single negative heuristic. This has led Laudan to mistakenly identify the hard core with the negative heuristic.[34] Lakatos himself in lectures tended to collapse the distinction between hard core and negative heuristic and, even occasionally, between the protective belt and positive heuristic, thereby instrumentalizing the entire programme.

If one gives more positive content to the negative heuristic, however, one can claim that the decision to hold on to the hard core is not based upon a conventionalist fiat but on a belief that its metaphysical presuppositions are true. The hard core is not a set of unfalsifiable statements with status of definitions. Rather, the hard core is falsifiable but is relinquished last precisely because it has metaphysical presuppositions that are believed to be true. Lakatos goes further than Popper's students Joseph Agassi and John Watkins "in blurring the distinction between Popper's science and Popper's metaphysics" and "does not even use the term 'metaphysical' any more."[35] The presupposition-logic approach, though, makes a distinction between science and metaphysics while treating both as statements. An account of the structure of scientific worldviews, something like that by Kuhn and Lakatos, can be maintained while holding to both scientific and metaphysical realism.

The treatment of scientific theories themselves as presuppositions, as in Harold I. Brown's "new image of science,"[36] leads to extremely vague characterizations of science and its "dialectic," or, if one uses the presupposition logic, to incoherence, for no true or false observation statement can serve as basis for a theory different from the one it originally presupposed.[37] Also, the falsehood of a prediction cannot be used to reject a theory. This is problematical when scientific theories are taken to function as presuppositions, but it is illuminating when metaphysical assumptions are taken to be presuppositions.

This model replaces the positivistic two-tier account not with a Kuhn-Feyerabend one-tier account but with a three-tier account. Observations can refute theories but cannot refute metaphysical frameworks. Rejection of theories as false (though scientific because falsifiable) confirms the metaphysical frameworks.

Starting with Durkheim, many have observed the correspondences between a culture's view of social structure and its models of cosmology. We have noted such connections between atomism and individualism in ancient Greece and seventeenth-century England, and between holistic, organic thought and the centralized, family-based society of traditional China, as well as between holistic and analogical thought and peasant societies in general and formalistic thought and elite cultures.

Some neoinductivist historians of science have added to their valuable emphasis on instrumentation and practice an unjustifiable rejection of the importance of conceptual influences on science. For instance, an able historian of electricity rejects the investigation of the metaphysical assumptions of Faraday and others as "a search for what is literally a philosophers' stone."[38] Besides trading on the popular underestimation of the scientific significance of alchemy (referring to "alchemical analysts" of history), this approach trades on a popular dismissal of metaphysics as worthless (an odd position for a historian of scholastic physics). Also, cultural materialists, again with a salutary emphasis on physical equipment, sometimes overreact in their rejection of the impact of ideas as such.[39] Such a reaction verges on the science-history equivalent of vulgar Marxism if it neglects the qualification (to paraphrase Marx for the current context) that theory also becomes a material force when embodied in science and technology.[40]

Apparently abstract metaphysical ideas can often be traced to everyday social viewpoints and movements of a time. Often the conceptual struggle of metaphysical perspectives expresses the oppositions and movements in society. It is hoped that this work, in tracing the social bases of three major holistic thought-complexes, has contributed to articulating this point within the history of physical theory. In particular, major incidents in the rise of the concepts of forces and fields in physical science owe their inspiration precisely to these modes of thought, although the results are often restated and the sources of discovery suppressed as more mainstream scientists cover their tracks.

NOTES

Abbreviations of Periodical and Series Titles in Notes

AHES	*Archives for the History of the Exact Sciences*
AS	*Annals of Science*
BJHS	*The British Journal of the History of Science*
BSPS	Boston Studies in the Philosophy of Science
DSB	*Dictionary of Scientific Biography*
EC	*Early China*
HJAS	*Harvard Journal of Asiatic Studies*
HS	*History of Science*
HSPS	*Historical Studies in the Physical Sciences*
HR	*History of Religions*
JHA	*Journal of the History of Astronomy*
JHI	*Journal of the History of Ideas*
SHPS	*Studies in the History and Philosophy of Science*
SR	*Studies in Romanticism*
TP	*T'oung Pau*

CHAPTER 1 *Introduction*

1. Christopher Hill, *The World Turned Upside Down* (New York: Penguin, 1991), 295, correcting his *Intellectual Origins of the English Revolution* (London: Panther Books, 1972). But see J. A. Bennett, "The Mechanics' Philosophy and the Mechanical Philosophy," *HS* 24 (1986): 1–28.

2. Karl Marx, *Capital*, vol. 1, trans. Samuel Moore and Edward Aveling (New York: International Publishers, 1967), 368n. See Franz Borkenau, "The Sociology of the Mechanistic World-Picture," trans. Richard W. Hadden, *Science in Context* 1, no. 1 (March 1987): 109–127.

3. Karl Marx, *The Eighteenth Brumaire of Louis Bonaparte* (New York: International Publishers, 1963), 124. See, for a contentiously negative portrayal of this strand, David Mitrany, *Marx against the Peasant: A Study of Social Dogmatism* (1951; reprint, New York: Collier, 1961).

4. E. J. Dijksterhuis, *The Mechanization of the World Picture*, trans. C. Dikshoorn (Oxford: Oxford University Press, 1961).

5. See Robert Crease and Charles C. Mann, *The Second Creation: Makers of the Revolution in Twentieth Century Physics* (New York: Macmillan, 1986), 306–307, for Murray Gell-Mann's rejection of what he called Feynman's "put-ons" as entities distinct from quarks.

6. Silvan S. Schweber, *QED and the Men Who Made It: Dyson, Feynman, Schwinger and Tomonaga* (Princeton: Princeton University Press, 1994). James Gleich, *Genius: The Life and Science of Richard Feynman* (New York: Pantheon, 1992); and Jagdish Mehra, *The Beat of a Different Drum: The Life and Science of Richard Feynman* (Oxford: Oxford University Press, 1994). Steven Weinberg, in *Dreams of a Final Theory* (New York: Vintage, 1994), 172, emphatically describes the present theory as one of particles as "bundles of the energy of the field" and "the physicists' recipe . . . no longer included particles." In a chapter titled "Against Philosophy," Weinberg also oddly blames the doctrine of mechanism on philosophy.

7. Oliver Darrigol, "The Origin of Quantitized Matter Waves," *HSPS* 7 (1976): 252–253.

8. Schweber, *QED*, 6. B. Norton Wise, "Pascual Jordan: Quantum Mechanics, Psychology, National Socialism," in *Science, Technology and National Socialism*, ed. Monika Renneberg and Mark Walker (Cambridge: Cambridge University Press, 1994), 224–254, 391–396.

9. See Gad Freudenthal, *Atom and Individual in the Age of Newton* (Dordrecht: D. Reidel, 1986).

10. Fritjof Capra, The Tao of Physics (1976; reprint, New York: Bantam, 1977). After Capra, the most influential was Gary Zukav's *Dancing Wu-li Masters* (New York: William Morrow, 1979). Neither Capra nor Zukav focuses on the relation of Chinese Taoism to modern physics.

11. Richard Bernstein, "A Cosmic Flow," *American Scholar* 48 (1978/1979): 6–9. Ed Gerrish and Val Dusek, "The Reader Replies," *American Scholar* 48 (1978/1979): 572. See also Leon Lederman and Dick Teresi, "The Dancing Mu-shu Masters," in *The God Particle* (New York: Houghton Mifflin, 1993), 189–198.

12. Werner Heisenberg, *Physics and Philosophy* (1958; reprint, New York: Harper Torchbooks, 1962), chap. 4, 59–75; Erwin Schrödinger, *Nature and the Greeks* (1954; reprinted in *Nature and the Greeks and Science and Humanism*, Cambridge: Cambridge University Press, 1996).

13. L. Pearce Williams, *Michael Faraday* (New York: Basic Books, 1965).

14. These new forms of holism include Niels Bohr's holism of the measurement situation (there is an unanalyzable interaction between measuring instrument and object being measured, which must be treated as a whole) and David Bohm's or John Bell's holism of reality (the universe must be treated as an undivided whole in which all parts simultaneously are influenced by all other parts).

15. C. B. Macpherson, *The Political Theory of Possessive Individualism* (Oxford: Oxford University Press, 1962).

16. Eric R. Wolf, *Europe and the People without History* (Berkeley: University of California Press, 1982).

17. Sheila Rowbotham, *Hidden from History* (New York: Vintage, 1976).

18. Jesse Lemisch, "History from the Bottom Up," in *Towards a New Past: Dissenting Essays in American History*, ed. Barton J. Bernstein (New York: Pantheon, 1968).

19. Mikhail Bakhtin, Rabelais and his World (1968; reprinted Bloomington: Indiana University Press, 1984).

20. Carlo Ginzburg, *The Cheese and the Worms*, trans. John Tedeschi and Anne Tedeschi (Harmondsworth, Eng.: Penguin, 1982); idem, *Night Battles*, trans. John Tedeschi and

Anne Tedeschi (Baltimore: Johns Hopkins University Press, 1983); idem, *Ecstasies* (1991; reprint, New York: Penguin, 1992).

21. Antonio Gramsci, *Selections from the Prison Notebooks of Antonio Gramsci* (New York: International Publishers, 1977).

22. Hans Mayer, *Outsiders* (Cambridge. Mass.: MIT Press, 1982).

23. Marc Ferro and the Staff of *Annales*, eds., *Social Historians in Contemporary France* (New York: Harper and Row, 1972). Jacques Revel and Lynn Hunt, eds., *Histories: French Constructions of the Past*, (New York: New Press, 1995).

24. Antonio Gramsci, *Selections from the Political Writings, 1921–1926* (New York: International Publishers, 1978); idem, *Selections from the Prison Notebooks*; idem, *Selections from the Cultural Writings* (London: Lawrence and Wisehart, (1985; reprint, Cambridge, Mass.: Harvard University Press, 1991.)

25. Simone de Beauvoir, *The Second Sex*, trans. H. M. Parshley (New York: Bantam, 1952).

26. Friedrich Heer, *The Intellectual History of Europe* [1953], trans. Jonathan Steinberg (London: Weidenfeld and Nicolson, 1966).

27. Lucien Febvre, *The Problem of Unbelief in the Sixteenth Century*, trans. Beatrice Gottlieb (Cambridge, Mass.: Harvard University Press, 1982).

28. Friedrich Heer, *Europe: Mother of Revolutions*, trans. Charles Kessler and Jennetta Adcock (London: Weidenfeld and Nicolson, 1972).

29. Friedrich Heer, *The Medieval World: Europe, 1100–1350*, trans. Janet Sondheimer (New York: New American Library of World Literature, 1963), 183, 283, 320; idem, *Intellectual History of Europe*, 87–89. Hildegard of Bingen, *Hildegard of Bingen's Book of Divine Works*, ed. Matthew Fox (Santa Fe, N.M.: Bear Books, 1987).

30. Heer, *Intellectual History of Europe*, 50.

31. Edgar Zilsel, "The Sociological Roots of Science," *American Journal of Sociology*, 47 (1941–1942): 544–562.

32. Brian P. Copenhaver and Charles B. Schmitt, *Renaissance Philosophy* (Oxford: Oxford University Press, 1992), 306.

33. Allen G. Debus, *Man and Nature in the Renaissance* (Cambridge: Cambridge University Press, 1978), 10.

34. Walter Pagel, *Paracelsus*, 2nd ed. (Basel: Karger, 1982), 296.

35. Henry Cornelius Agrippa, *Declamation on the Nobility and Preeminence of the Female Sex*, trans., ed. Albert Rabil, Jr. (Chicago: University of Chicago Press, 1996).

36. Frank E. Manuel, *The Prophets of Paris* (New York: Harper and Row, 1964).

37. G.E.R. Lloyd, *Polarity and Analogy: Two Types of Argumentation in Early Greek Thought* (1966; reprint, Indianapolis: Hackett, 1992).

38. Lucien Lévy-Bruhl, *Primitive Mentality*, trans. L. A. Clare (London: George Allen and Unwin, 1923); and idem, *How Natives Think*, trans. L. A. Clare (London: George Allen and Unwin, 1926).

39. F. M. Cornford, *From Religion to Philosophy* (Princeton, N.J.: Princeton University Press, 1992).

40. Claude Lévi-Strauss, *Structural Anthropology*, trans. Claire Jacobson (New York: Basic Books, 1963).

41. Marshall Sahlins, *Culture and Practical Reason* (Chicago: University of Chicago Press, 1976).

42. Daniel Kahneman, Paul Slovic, and Amos Tversky, *Judgment under Uncertainty* (Cambridge: Cambridge University Press, 1983).

43. Karl Mannheim, *Ideology and Utopia*, trans. Louis Wirth and Edward Shils (London: International Library of Philosophy, Psychology and Scientific Method, 1936).

44. Bruno Latour, *Science in Action* (Cambridge, Mass.: Harvard University Press, 1987). David Bloor, *Knowledge and Social Imagery* (Chicago: University of Chicago Press, 1976).

45. Mary Hesse, *Models and Analogies in Science* (Notre Dame, Ind.: University of Notre Dame Press, 1966). Rom Harré, *Principles of Scientific Thinking* (Chicago: University of Chicago Press, 1970). Max Black, *Models and Metaphors* (Ithaca: Cornell University Press, 1964). Marx Wartofsky, *Models* (Dordrecht: D. Reidel, 1979).

46. Imre Lakatos, "Methodology of Scientific Research Programmes," in *Philosophical Papers*, ed. John Worrall and Gregory Currie (Cambridge: Cambridge University Press, 1978), 1:8–101.

47. Larry Laudan, *Progress and Its Problems* (Berkeley: University of California Press, 1977), 70–120.

48. Frances A.Yates, "The Hermetic Tradition in Renaissance Science," in *Art, Science and History in the Renaissance*, ed. Charles S. Singleton (Baltimore: Johns Hopkins University Press, 1968), 270.

49. Arnold Thackray, "Comment" (on Mary Hesse), in *Historical and Philosophical Perspectives of Science*, ed. Roger H. Steuwer (New York: Gordon and Breach, 1989), 162.

50. See Aristides Baltas, "On the Harmful Effects of Excessive Anti-Whiggism," in *Trends in the Historiography of Science*, ed. Kostas Gavroglu et al. (Dordrecht: Kluwer, 1994), 107–120.

51. Frances A. Yates, *Giordano Bruno and the Hermetic Tradition* (New York: Random House, 1964), 447.

52. C. B. Wilde, "Whig History," in *Dictionary of the History of Science*, ed. W. F. Bynum, E. J. Browne, and Roy Porter (Princeton: Princeton University Press, 1981) concludes, "'Whig criteria are still generally applied by historians of science in choosing subjects of study" (446), although he naively writes that Alexandre Koyré "interprets the past on its own terms" (445).

53. Lakatos, *Philosophical Papers*, 2:257, 222 n. 1.

54. Jan Smuts, *Holism and Evolution* (New York: Viking, 1961). Anne Harrington, in *Reenchanting Science* (Princeton: Princeton University Press, 1996), 269 n. 94, notes that most German holists avoided the English-derived term.

55. Immanuel Kant, *Critique of Judgement*, trans. Werner Pluhar; foreword by Mary Gregor (Indianapolis: Hackett, 1987).

56. James Gleich, *Chaos: Making a New Science* (New York: Viking, 1987), 163–165.

57. Democritus, fragment B125, in Jonathan Barnes, *Early Greek Philosophy* (New York: Penguin, 1987), 253.

58. Protagoras, fragment B1, in *The Philosophers of Greece*, trans. R. S. Brumbaugh (New York: Crowell, 1964), 117.

59. Cynthia Farrar, *The Origins of Democratic Thinking: The Invention of Politics in Classical Athens* (Cambridge: Cambridge University Press, 1988).

60. Derk Bodde, *Chinese Thought, Society, and Science: The Intellectual and Social Background of Science and Technology in Pre-Modern China* (Honolulu: University of Hawaii Press, 1991), 292–308.

61. David Bloor, "Sociology of (Scientific) Knowledge," in *Dictionary of the History of Science*, ed. W. F. Bynum et al., 392, attempts to make a move that "outflanks standard 'refutations' which cite cases where the 'same' theory is advocated in widely different social circumstances, e.g. atomism in Ancient Greece and 17th century England." Such "outflanking" is unneeded in this case, as both societies share a social individualism which is the base of atomism.

62. Isaac Newton, *Opticks*, with foreword by Albert Einstein; introduction by Edmund Whittaker; preface by I. Bernard Cohen (New York: Dover, 1952), 400.

63. James Mill, *Analysis of the Phenomena of the Human Mind* [1829] (Indianapolis: Bobbs-Merrill, Library of the Liberal Arts, 1962).

64. Edward Chance Tolman was a self-proclaimed follower of logical positivist Rudolf Carnap and brother of Richard Tolman, a leading figure in statistical mechanics.

65. The term *therblig* was coined by the Frederick Taylor's disciple Frank Gilbreth; see Arthur Danto, "Basic Actions," in *Philosophy of Action*, ed. Alan R. White, (Oxford: Oxford University Press, 1968), 43–58.

66. Ludwig Wittgenstein, *Tractatus Logico-Philosophicus*, trans. C. K. Ogden; introduction by Bertrand Russell (London: Routledge, 1922; new trans. D. F. Pears and B. F. McGuiness, London: Routledge, 1961).

67. Bertrand Russell, "The Philosophy of Logical Atomism," *Monist* 28 (1918): 495–525; 29 (1919): 32–63, 190–222, 345–380; reprinted in Bertrand Russell, *Logic and Knowledge*, ed., R. C. Marsh, (London: George Allen and Unwin, 1956).

68. See Julius R. Weinberg, *Nicholas of Autrecourt: A Medieval Hume* (Princeton: Princeton University Press, 1948). Discontinuity of process and occasionalism also in al-Ghazzali, *Al-Ghazzali's "The Incoherence of the Philosophers,"* trans. S. A. Kamali (Lahore, 1958).

69. Walter Pagel, *The Smiling Spleen: Paracelsianism in Storm and Stress* (Basel: Karger, 1984), 1.

70. Justus von Liebig, "Ueber das Studium der Naturwissenschaften und ueber den Zustand der Chemie in Preussen" [1840], in *Reden und Abhandlungen* (Leipzig and Heidelberg, 1874), 24.

71. See Edward Manier, *The Young Darwin and His Cultural Circle* (Dordrecht: D. Reidel, 1976); Robert J. Richards, *The Meaning of Evolution: The Morphological Construction and Ideological Reconstruction of Darwin's Theory* (Chicago: University of Chicago Press, 1992).

72. See, for instance, J. S. Haldane, *Mechanism, Life, and Personality*, 2nd ed. (New York; 1923).

73. Ludwig von Bertalanffy, *Problems of Life* (1952; reprint, New York: Harper and Brothers, 1960); and idem, *Modern Theories of Development*, trans. J. H. Woodger (1933; reprint, New York: Harper and Brothers, 1960).

74. Ralph S. Lilly, *General Biology and Philosophy of Organism* (Chicago: University of Chicago Press, 1945).

75. E. S. Russell, *Form and Function* (1916; reprint, Chicago: University of Chicago Press, 1982).

76. Val Dusek, "The Bukharin Delegation on Science and Society: Action and Reaction in British Science Studies," in *Nikolai Ivanovich Bukharin*, ed. Nicholas N. Kozlov and Eric D. Weitz (New York: Praeger, 1990), 129–147. Gary Werskey, *The Visible College* (New York: Holt, Rinehart and Winston, 1978).

77. Richard Levins and Richard Lewontin, *The Dialectical Biologist* (Cambridge, Mass.: Harvard University Press, 1985).

78. C. H. Waddington, *The Evolution of an Evolutionist* (Ithaca, N.Y.: Cornell University Press, 1975), chap. 1.

79. Jacques Monod, in *Chance and Necessity* (New York: Random House, 1972), opposes dialectical materialism and Catholicism to an "existential" conception of the accidental origin of life, but he earlier was a Marxist in the French resistance. François Jacob, *The Logic of Life*, trans. Betty E. Spillmann (New York: Random House, 1973),

presents a more emergentist conception of biology. He borrows the term *holon* from Arthur Koestler.

80. John Graves, *The Conceptual Foundations of Contemporary Relativity Theory* (Cambridge: MIT Press, 1971), 216, 314.
81. Poincaré studied not the particular numerical solutions of differential equations, but the general features of whole families of curves. Poincaré's work as forerunner of chaos theory did not become a popular topic until some seven decades after his work.
82. The biology of the noted "modern synthesis" evolutionist Ernst Mayr exemplifies this position. He criticizes "beanbag genetics," emphasizes population thinking, rejects group selection, and does not accept the punctuated equilibrium of Stephen Jay Gould. His position on species is more holistic. He invented the "biological definition of species" as interbreeding systems, and rejects the species selection of 1970s paleontologists.
83. John Burnet, *Early Greek Philosophy* (New York: Meridian Books, 1957), 259. G. S. Kirk and J. E. Raven, *The Presocratic Philosophers* (Cambridge: Cambridge University Press, 1957), 381.
84. Daniel Garber, *Descartes' Metaphysical Physics* (Chicago: University of Chicago Press, 1992), 117–155.
85. Charles Hartshorne, "Continuity as the Form of Forms in Charles Peirce," *Monist* 39 (1929): 521–534. R. Valentine Dusek, "Peirce as Philosophical Topologist," in *Charles S. Peirce and the Philosophy of Science*, ed. Edward C. Moore (Tuscaloosa: University of Alabama Press, 1993), 49–59.
86. G. Spencer Brown, *The Laws of Form* (London: George Allen and Unwin, 1969).
87. *Oxford English Dictionary*, s.v. "holism."
88. F[rancis] H[erbert] Bradley, *Appearance and Reality* (London, 1893).
89. "It is not possible for what is nothing to be" (Burnet, *Early Greek Philosophy*,174). "It can be, and nothing can not" (Jonathan Barnes, ed., *Early Greek Philosophy* [London: Penguin, 1987], 133).
90. Hippasus of Metapontum, who proved or made public the irrationality of the square root of 2, was drowned in punishment, either by fellow Pythagoreans or by the gods, and was said to have been a follower of the process philosopher Heraclitus (Barnes, *Early Greek Philosophy*, 214).
91. Rom Harré, *The Philosophies of Science*, 2nd ed. (Oxford: Oxford University Press, 1984), 105, 107.
92. Emile Meyerson, *Identity and Reality*, trans. Kate Loewenberg (1930; reprint, New York: Dover, 1962).
93. Olivier Costa de Beauregard, "Relativity: Arguments for a Philosophy of Being," in *The Voices of Time: A Cooperative Survey of Man's View of Time as Expressed by the Sciences and by the Humanities*, 2nd ed., ed. J. T. Fraser (Amherst: University of Massachusetts Press, 1981), 417–433.
94. Kurt Gödel, "A Remark about the Relationship between Relativity Theory and Idealistic Philosophy," in *Albert Einstein: Philosopher-Scientist*, ed. Paul Schillp (New York: Harper and Row, 1959), 557.
95. Rudy Rucker, *Infinity and the Mind* (Boston: Birkhauser, 1982),170–171. Rucker is a direct descendent of Hegel.
96. Hermann Weyl, *Philosophy of Mathematics and Natural Science*, trans. Olaf Helmer; rev., aug. English ed. (Princeton, N.J.: Princeton University Press, 1949), 116.

CHAPTER 2 *Discovery of the Compass and Magnetic Declination*

1. Francis Bacon, *Novum organum* (1620), in *The Works of Francis Bacon*, 7 vols., ed. James Speding, Robert Leslie Eillis, and Douglas Dennon Heath, new ed. (London: Longmans, 1870), 4: 114, quoted in Debus, *Man and Nature*, (1978), 1.

2. Joseph Needham, *Science and Civilization in China*, 7 vols. (Cambridge: Cambridge University Press,1962), vol. 4, pt. 1, xxv.

3. Colin Ronan, ed., *The Shorter Science and Civilization in China*, 5 vols. (Cambridge: Cambridge University Press, 1978), 1:37, 64.

4. Ibid., 1:74.

5. Edward H. Schafer, *The Golden Peaches of Samarkand* (Berkeley: University of California Press, 1962), discusses numerous imports to China and their cultural role.

6. Lynn White, *Medieval Technology and Social Change* (Oxford: Oxford University Press, 1966), 10–12, 14–15, 20–22.

7. E. L. Eisenstein, *The Printing Press as an Agent of Change*, 2 vols. (Cambridge: Cambridge University Press, 1979), abridged as *The Printing Revolution in Early Modern Europe* (Cambridge: Cambridge University Press, 1983).

8. Ronan, *Shorter Science*, 3: 217–218.

9. Dusek, "Bukharin Delegation," 129–147. Helena Sheehan, *Marxism and the Philosophy of Science: A Critical History*, vol. 1, *The First Hundred Years* (Atlantic Highlands, N.J.: Humanities Press, 1985), 306–308.

10. Henry Holorenshaw [Joseph Needham], "The Making of an Honorary Taoist," in *Changing Perspectives in the History of Science: Essays in Honor of Joseph Needham*, ed. Mikulás Teich and Robert Young (London: Heineman Educational Books, 1973), 1–20.

11. Joseph Needham, *Science in Traditional China* (Cambridge: Harvard University Press, 1981), 3.

12. Ibid., 13–14.

13. H. Floris Cohen, in *The Scientific Revolution* (Chicago: University of Chicago Press, 1994), 424, shows his ignorance of the similarities of process philosophy (including Whitehead's philosophy of organism, which interested Needham) with Marxist dialectics of nature, writing that Needham's reflections "find their nadir in how he managed to weave the verbal contortions of dialectical materialism into his philosophy of integrative levels."

14. Holorenshaw, "Honorary Taoist," 20.

15. Charles Coulston Gillispie, "Perspectives," *American Scientist* 45 (1957): 173, 176.

16. David Joravsky, *The Lysenko Affair* (New York: Columbia University Press, 1970). In a letter to the editor, Gillispie explicitly frames his criticisms of Needham in terms the dangers of Lysenkoism. Gillispie writes that to admit that views of science vary with sciences (physics vs. biology) or with cultures "is to give the game away. . . . It is to agree that the Soviet state was justified when it imposed Lysenkoism" ("Perspectives," 272A). Next to Gillispie's defense of the purity of Western science is an advertisement for Lockheed Missile Systems, recruiting "missile systems scientists" (ibid., 270A).

17. Joseph Needham, *The Grand Titration* (Toronto: University of Toronto Press, 1969), 43, 54.

18. Review in *Publishers Weekly* of Alan Cromer, *Uncommon Sense: The Heretical Nature of Science* (Oxford: Oxford University Press, 1993), quoted in *Physics: New and Recent Books, 1995–96,* (Oxford University Press catalog), 41.

19. Cromer, *Uncommon Sense*, 117.

20. Jay A. Levenson, ed., *Circa 1492* (New Haven: Yale University Press, 1991), 328–332.
21. Schafer, *Golden Peaches*.
22. Cromer, *Uncommon Sense*, citing Daniel Boorstin, *The Discoverers* (New York: Random House, 1985). Boorstin excludes women and nonwhites from his pantheon of inventors in world history.
23. Cromer, *Uncommon Sense*, 119.
24. Francesca Bray, "Joseph Needham, 9 December 1900–24 March 1995," *Isis* 87 (1996): 316.
25. Zaiuddin Sardar, "Conventional Wisdoms," *Nature* 360 (1992): 713–714; idem, "Logic and Laws," *Nature* 368 (1994): 378.
26. Bray, "Joseph Needham."
27. Needham, *Science and Civilization*, vol. 4, pt. 1, 245–246.
28. Ibid., 250. On Shen Kua, see Nathan Sivin, "Shen Kua," in *DSB*, vol. 12, revised in *Science in Ancient China* (Aldershot, Eng.: Ashgate Variorum, 1995), art. 3, 1–53.
29. Needham, *Science and Civilization.*, vol. 4, pt. 1, 315, 333.
30. Ibid., 319–323.
31. Ibid., chart on 331.
32. Wang Ch'ung, *"Lun-Heng": Philosophical Essays of Wang Chhung*, 2 vols., ed. A. Forke (1907–1911; reprint, New York: Paragon, 1962), chap. 47.
33. Donald J. Harper, "The Han Cosmic Board (*Shih*)," *EC* 4 (1979–1980): 1–10. Christopher Cullen, "Some Further Points on the *Shih*," *EC* 6 (1980–1981): 37–38, suggests "cosmic model" or "cosmographic model."
34. John S. Major, *Heaven and Earth in Early Han Thought* (Albany: State University of New York Press, 1993), 13, 39–43.
35. Harper, in "Cosmic Board" (9 n. 56, 10 n. 57), notes the connections of the Dipper to Taoist meditation and magic mirrors. Cullen, in "Further Points," criticizes Harper on the incorporation of metaphors "left over from defunct schemes of cosmography" and of incorporating later T'ang dynasty material (such as pacing the stars) with earlier Han dynasty material (41 n. 1). Harper replies that Cullen's interest in predictive, scientific astronomy leads him to dismiss its embedding in "the subtleties of thought which characterize the ancient mind" ("The Han Cosmic Board: A Response to Christopher Cullen," *EC* 6 [1980–1981] 50).
36. Michael Loewe, *Ways to Paradise* (London: George Allen and Unwin, 1979), 74–78. Edward H. Schafer, "A Tang Taoist Mirror," *EC* 4 (1979–1980): 56–59.
37. Isabelle Robinet, *Taoist Meditation* (Albany: State University of New York Press, 1993), 201–202, 205–212.
38. The limping step of Yü the Great resembles the worldwide mythology of the limp or the single sandal from Achilles to Cinderella. See Ginzburg, *Ecstasies*, 230–231, 243, though Ginzburg misses Yü in his survey.
39. Edward H. Schafer, *Pacing the Void: T'ang Approaches to the Stars* (Berkeley: University of California Press, 1977), 45–50 on the Dipper, 236–238 on space travel by immortals and shamans, 238–242 on walking the pattern of the Dipper on earth. Robinet, *Taoist Meditation*, also discusses the "Step of Yü" and the pacing of the Dipper (221–222).
40. Needham, *Science and Civilization*, vol. 4, pt. 1, 251–252.
41. Ibid., 273; Ronan, *Shorter Science*, 2: 25–26.
42. Loewe, *Ways to Paradise*, 85.
43. Nathan Sivin, "The Myth of the Naturalists," in *Medicine, Philosophy and Religion*

in Ancient China (Aldershot, U.K.: Ashgate Variorum, 1995), art. 4; 19–29, debunks the Chi-Hsia Academy as a kind of think tank or school, but does grant that the state of C'hi financed buildings 26 C.E.–400 C.E. (ibid., 25), and that among those patronized were "a handful of remarkable philosophers among their crowd of weapons consultants, experts on treachery, etc." (ibid., 29).

44. Needham, *Science and Civilization*, vol. 4, pt. 1, 240, passage translated in ibid., 2: 42–45; Ronan, *Shorter Science*, 3: 7. For a modern translation of the work (unfortunately vol. 2, which contains the chapter in question, is not yet published), see *Guanzi*, ed., trans. W. Allyn Rickett (Princeton: Princeton University Press, 1985–).

45. Richard J. Smith, *Fortune-Tellers and Philosophers* (Boulder, Colo.: Westview, 1991), 148–171. Ernest J. Eitel, *Feng-Shui: The Science of Sacred Landscape in Old China* (1873; reprint, Oracle, Ariz.: Synergetic Press, 1984).

46. Susan Hornick, "How to Get That Extra Edge on Health and Wealth," *Smithsonian*, 24, no. 5 (August 1993): 70–75; also "Feng Shui," *USAir Magazine* (November 1993).

47. Needham, *Science and Civilization*, vol. 4, pt. 1, 239. Needham gives little coverage to geomancy and does not compare it sympathetically to contemporary environmental views. He considers it a pseudoscience and does not give it the sympathetic interpretation and comparison to modern views that he gives to alchemy, though he says that geomancy as a protoscience of the environment is parallel to alchemy as protochemistry and to astrology as protoastronomy.

48. Hornick, "Health and Wealth," 74.

49. Jan J. M. de Groot, *The Religious System of China*, 3: 938, cited in Needham, *Science and Civilization*, vol. 4, pt.1, 293.

50. Needham, *Science and Civilization*, vol. 4, pt. 1, 309.

51. Ibid., 299. The twelfth and seventeenth rings, in contrast with the fifth, are "staggered," the result of the fact that the circles added in the Tang (603–906) and Sung dynasties accommodated the shifted magnetic declinations of those times.

52. Ibid., 297; Ronan, *Shorter Science*, 3: 37.

53. Needham, *Science and Civilization*, vol. 4, pt. 1, 299.

54. Ibid., 302.

55. Ibid., 303.

56. Ibid., 305.

57. Ibid., 313.

58. Ibid., 286.

59. Ibid., 249.

60. Thomas Young, *A Course of Lectures on Natural Philosophy and the Mechanical Arts*, (1807), 1: 743, cited in Alfred Still, *Soul of the Lodestone* (New York: Murray Hill Books, 1946).

61. Hans Breuer, *Columbus Was Chinese* (New York: McGraw-Hill, 1972), 102.

CHAPTER 3   *Major Movements of Chinese Thought*

1. Nathan Sivin, "On the Word 'Taoist' as a Source of Perplexity, with Special Reference to the Relations of Science and Religion in Traditional China," *HR* 17 (1978): 303–329.

2. Bodde, *Chinese Thought,* 329–330. Sivin, "Word 'Taoist.'"

3. A. C. Graham, *Disputers of the Tao* (La Salle, Ill.: Open Court, 1989), 160–170; idem, *Later Mohist Logic, Ethics and Science* (Hong Kong: Chinese University Press, 1978).

4. Sivin, "Shen Kua," 369–393.

5. Bodde, *Chinese Thought*, 190–191, 364–367.
6. Recounted in *The History of the Former Han*, chap. 56, quoted by Wing-Tsit Chan, *A Source Book in Chinese Philosophy* (Princeton: Princeton University Press, 1963), 272 n. 3.
7. Graham, *Disputers*, 239–241.
8. *Hsün Tzu*, book 17, trans. in Bodde, *Chinese Thought*, 312; Graham, *Disputers*, 241.
9. Hsün Tzu, *Hsün Tzu: Basic Writings*, trans. B. Watson (New York: Columbia University Press, 1963), chap. 17, 44–46, 86.
10. Graham, *Disputers*, 271–273.
11. Ibid., 245 (Hsün Tzu); Benjamin Schwartz, *The World of Thought in Ancient China* (Cambridge: Harvard University Press, 1985), 335 (legalists and Skinner).
12. Schwartz, *World of Thought*, 328.
13. Ibid., 333.
14. Wang Ch'ung, "*Lun-Heng.*"
15. John B. Henderson, *The Development and Decline of Chinese Cosmology* (New York: Columbia University Press, 1984), 149–224.
16. Joseph R. Levenson, *Modern China and Its Confucian Past* (Garden City, N.Y.: Doubleday Anchor, 1964), 6–10.
17. Benjamin Schwartz, "The Absence of Reductionism in Chinese Thought," in *China and Other Matters* (Cambridge, Mass.: Harvard University Press, 1996), 81–97.
18. Nathan Sivin, *Traditional Medicine in Contemporary China* (Ann Arbor: University of Michigan, Center for Chinese Studies, 1987), 73, 75.
18. John S. Major, "A Note on Two Technical Terms in Chinese Science: *Wu-hsing* and *Hsiu*," *EC* 2 (1976): 1–3.
20. John Major, "The Five Phases, Magic Squares and Schematic Cosmography," in *Explorations in Early Chinese Cosmology*, ed. Henry Rosemont, Jr., *Journal of the American Academy of Religion* 50, no. 2 (1984): 133–166.
21. Iulien K. Shchutskii, *Researches on the "I Ching*," trans. William L. MacDonald, Tsuyoshi Hasegawa, and Hellmut Wilhelm (Princeton: Princeton University Press, 1979).
22. Needham, *Science and Civilization*, 2: 334–335; Ronan, *Shorter Science*, 1: 186–187.
23. Needham, *Science and Civilization*, 2: 330–333.
24. Graham, *Disputers*, 368–370.
25. Needham, *Science and Civilization*, 2: 311; Ronan, *Shorter Science*, 1: 184.
26. Graham, *Disputers,* 369.
27. Ronan, *Shorter Science*, 1: 187–88.
28. Ibid., 188. Needham, *Science and Civilization*, 2: 338.
29. Ronan, *Shorter Science*, 1: 189.
30. I owe this point to Lynda Shaffer.
31. Kirk and Raven, *Presocratic Philosophers*, 238; Aristotle, *Metaphysics*, A5. 985b23.
32. Graham, *Disputers*, 323–324, bases himself on Jacques Derrida, *Of Grammatology*, trans. G. C. Spivak (Baltimore: Johns Hopkins University Press, 1976).
33. Gilbert Ryle, *The Concept of Mind* (1949; reprint, New York: Barnes and Noble, 1968).
34. Thomas Kuhn, *The Structure of Scientific Revolutions*, 2nd ed. (Chicago: University of Chicago Press, 1970).
35. Sivin, "The Myth of the Naturalists," in *Medicine, Philosophy and Religion*, 4:2.
36. Ibid., 14.
37. Ronan, *Shorter Science*, 1: 144.
38. Needham, *Science and Civilization*, 2: 325.

39. Sivin, "Myth," 16, quotes the passage from the *Shih chi* which recounts the changes in symbolism instituted by the First Emperor. Sivin suggests ("Myth," 18) that Tsou was forgotten in the official writing of the Han dynasty because of his association with the First Emperor, a usurper who was remembered as tyrant, burner of books, and slaughterer of the literati.

40. Graham emphasizes that Tsou was not a philosopher, but Sivin ("Myth," 8–9) argues that although Tsou's concepts had background in the diviner tradition, he should not be excluded from philosophy by means of too narrow a conception. Graham notes that yin/yang thinking was around before Tsou, and that Tsou was not founder of a "Yin Yang school" or "School of Naturalists" as historian of philosophy Fung Yu-Lan and Joseph Needham, respectively, designate him. Sivin agrees with these claims of Graham, but says this need not lead us to deny that he was a philosopher in some broad sense.

41. Kurt von Fritz, *Pythagorean Politics in Southern Italy* (1940; reprint, New York: Octagon, 1970).

42. Mary Ellen Waithe, ed., *A History of Women Philosophers*, vol. 1 (Dordrecht: D. Reidel, 1987), 11–74.

43. Hans Joachim Krämer, *Plato and the Foundations of Metaphysics*, trans. John R. Caton (Albany: State University of New York Press, 1990). J. N. Findlay, *Plato: The Written and Unwritten Doctrines* (Atlantic Highlands, N.J.: Humanities Press, 1974).

44. Graham, *Disputers*, 349.

45. Robert Temple, *The Genius of China* (New York: Simon and Schuster, 1986), 94–96.

46. Needham, *Science and Civilization*, vol. 5, pt. 5, 301–337.

47. Ibid., 132–137.

48. Ronan, *Shorter Science*, 1: 165–166.

49. Ibid., 166.

50. Needham, *Science and Civilization* 2: 298.

51. Ronan, *Shorter Science*, 1: 167.

52. Schwartz, *World of Thought*, 371.

53. Needham, *Science and Civilization*, 2: 302.

54. Schwartz, *World of Thought*, 372.

55. Needham, *Science and Civilization*, 2: 302.

56. Jack Lindsay, Introduction to Giordano Bruno, *Cause, Principle and Unity* (New York: International Publishers, 1964), 23.

57. Graham, *Disputers*, 320.

58. Ibid., 349.

59. Ibid., 318.

60. The first major twentieth-century Chinese scholar sympathetic yet textually critical of Taoist religion, Ch'en Kuo-fu, was an MIT-trained chemist. In Communist China studies of Taoism in relation to science were tolerated even when studies of Taoist religion were suppressed. See T. H. Barrett, "Taoism: History of Study," in *Encyclopedia of Religion*, Mircea Eliade, gen. ed. (New York: Macmillan, 1987), 331.

61. Needham, *Science and Civilization*, 2: 35.

62. Sivin, "Word 'Taoist,'" 303–329.

63. Michel Strickmann, "On the Alchemy of T'ao Hung-Ching," in *Facets of Taoism*, ed. Holmes Welch and Anna Seidel (New Haven: Yale University Press, 1979), 165.

64. Ibid., 167.

65. Herrlee G. Creel, *What Is Taoism?* (Chicago: University of Chicago Press, 1970), 24.

66. Ofuchi Ninji and Kubo Noritada, cited in T. H. Barrett, Introduction to *Taoism and*

Chinese Religion, by Henri Maspero (Amherst: University of Massachusetts Press, 1981), xxi.

67. Graham, *Disputers*, 172, notes, "However, Strickmann himself acknowledges that he cannot quite dispense wn4h the term 'philosophical Taoism.'"

68. Maspero, *Taoism*, 376.

69. Ofuchi Ninji, "The Formation of the Taoist Canon," in Welch and Seidel, *Facets*, 265.

70. Michel Strickmann, *Le Taoism du Mao Chan* (Paris: Collège de France, Institute des Haute Etudes Chinoise, 1981), 4, trans. Laurence Thompson in "Taoism: Classic and Canon," in *The Holy Book in Comparative Perspective*, by Frederick M. Denny and Rodney L. Taylor (Columbia: University of South Carolina Press, 1985), 208.

71. A. C. Graham, "The Origins of the Legend of Lao Tan," in *Studies in Chinese Philosophy and Philosophical Literature* (Albany: State University of New York Press, 1990), 111–124.

72. Graham, *Disputers*, 220.

73. John Herman Randall, *Plato: Dramatist of the Life of Reason* (New York: Columbia University Press, 1970).

74. Chuang Tzu, *Zhuangzi Speaks: The Music of Nature*, trans. Brian Bruya; adapted and illustrated by Tsai Chih Chung; afterword by Donald J. Munro (Princeton: Princeton University Press, 1992).

75. *The Book of Lieh Tzu: A Classic of the Tao*, trans. A. C. Graham (1960; reprint, New York: Columbia University Press, 1990).

76. Needham, *Science and Civilization*, vol. 5, pt. 2, 83 n. b.

77. Lao Tzu, *Tao Te Ching*, trans. Victor Mair (New York: Bantam, 1990), chap. 10, p. 69 (breathing like baby); Chuang Tzu, *The Wanderer and the Way*, trans. Victor Mair (New York: Bantam, 1994), chap. 2, p. 11 (trance state), chap. 15, p. 145 (stretching).

78. Livia Knaul Kohn, "Lost Chuang-Tzu Passages," *Journal of Chinese Religions* 10 (1982): 53–79.

79. Graham, "How Much of Chuang Tzu Did Chuang Tzu Write?" in *Studies in Chinese Philosophy*, 283–321.

80. Hisayuki Miyakawa, "Local Cults around Mt. Lu at the Time of Sun En's Rebellion," in Welch and Seidel, *Facets*, 83–102. B. J. Mansvelt Beck mentions a score of rebel emperors in the latter two-thirds of the second century C.E. ("The Fall of Han," in *The Cambridge History of China*, ed. Denis Twitchett and Michael Loewe [Cambridge: Cambridge University Press, 1986]), 1: 337.

81. Sivin, "Word 'Taoist,'" 323.

82. Livia Knaul Kohn, *Early Chinese Mysticism* (Princeton, N.J.: Princeton University Press, 1992), 125, 127.

83. Richard B. Mather, "K'ou Ch'ien-chih and the Taoist Theocracy at the Northern Wei Court, 425–451," in Welch and Seidel, *Facets,* 103, 122.

84. Creel, *What Is Taoism?*, 48–78.

85. John Major, review of *Guanzi*, trans. W. Allyn Rickett, *EC* 14 (1988): 243. See also R. P. Peerenboom, *Law and Morality in Ancient China: The Silk Manuscripts of Huang-Lao* (Albany: State University of New York Press, 1993), 234–241.

86. Maspero, *Taoism,* 258–259.

87. Wing-Tsit Chan, *Source Book*, 314. Neither Maspero nor Wing-Tsit Chan, when he made this judgment, had any access to Huang-Lao texts, which were not discovered until 1973.

88. Maspero, *Taoism*; Barrett, Introduction, xxi–xxii.

89. Peerenboom, *Law and Morality*, 2, mentions two books in Chinese and none in Japanese prior to his, although numerous articles had appeared in Asian languages. There were only eight articles in English (five by the same author) prior to Peerenboom's book.

90. Ibid., 250.

91. Ibid., 261.

92. Strickmann, "On the Alchemy," 167, citing E. Zürcher, *The Buddhist Conquest of China*, 2 vols. (Leiden: Brill, 1959), 1: 87. See also Zürcher, "Buddhist Influence on Early Taoism," *TP* 66 (1980): 120.

93. Graham, *Disputers*, 171.

94. Liu I-ch'ing, *"Shih-shuo Hsin-yü": A New Account of Tales of the World*, trans. Richard B. Mather (Minneapolis: University of Minnesota Press, 1976), 54–55, 60–62, 98–99, 122, 259, 272 –273, 326.

95. Wang Bi, "Commentary on the *Book of Changes*," in *The Classic of Changes: A New Translation of the "I Ching" as Interpreted by Wang Bi*, trans. Richard Lynn John (New York: Columbia University Press, 1994); idem, *Commentary on the "Lao Tzu*," trans. Ariane Rump (Honolulu: University of Hawaii Press, 1979). T'ang Yung T'ung, "Wang Pi's New Intepretation of the *I Ching* and *Lun-Yu*," trans. Walter Liebenthal, *HJAS* 10 (1947): 124–161.

96. Alan K. L. Chan, *Two Visions of the Way: A Study of the Wang Pi and the Ho-Shang Kung Commentaries on the Lao-Tzu* (Albany: State University of New York Press, 1991).

97. Zürcher, *Buddhist Conquest*, vol 1: 87.

98. The claim that Wang Pi died by execution is made by Richard Mather, but it has been questioned as unjustified by reviewers of his work.

99. Etienne Balazs, "Nihilistic Revolt or Mystical Escapism," in *Chinese Civilization and Bureaucracy*, trans. by H. M. Wright; ed. Arthur F. Wright (New Haven: Yale University Press, 1964), 234–236.

100. "Two Songs by Ts'ao Ts'ao," in Balazs, *Chinese Civilization*, 173–178.

101. Kristofer Schipper, *The Taoist Body* (Berkeley: University of California Press, 1993), 13–14, 192–194.

102. Wing-Tsit Chan, *Source Book,* 316, 323. See also Wang Bi, *Commentary.*

103. Fung Yu-lan, *A Short History of Chinese Philosophy*, trans. Derk Bodde (New York: Macmillan, 1948), 217, 231.

104. Liu I-ch'ing, *"Shih-shuo Hsin-yü*," chap. 23, sections 1, 2, 5, 7, 8–10, 23.

105. Ibid., chap. 23, sec. 6, 374. Benjamin Schwartz implicitly notes that the imagery which the skeptic Wang Ch'ung uses portraying humans as lice in the garments of the universe is also to be found in Juan Chi and the Neo-Taoists, although Schwartz does not mention Juan Chi by name (Schwartz, "On the Absence of Reductionism," in *China and Other Matters*, 93).

106. Kohn, *Early Chinese Mysticism*, 106.

107. Hsi Kang, *Philosophy and Argumentation in Third-Century China*, trans. Robert G. Henricks (Princeton: Princeton University Press, 1983).

108. Maspero, *Taoism*, 308. Liu I-ch'ing, *"Shih-shuo Hsin-yü*," chap. 4, sec. 21 n. 2; chap. 6, sec. 2 n. 1.

109. Liu I-ch'ing, *"Shih-shuo Hsin-yü*," 393. Needham, *Science and Civilization*, 2: 434.

110. Max Kaltenmark, "The Ideology of T'ai-Ping ching," in Welch and Seidel, *Facets*, 23, 41.

111. Sivin, "Word 'Taoist,'" 319.

112. David Noble, *A World without Women* (New York: Knopf, 1992).

113. Wing-Tsit Chan, *Source Book*, 47: "From Confucius on down, Confucianists have always considered women inferior."

114. Kaltenmark, "Ideology," 38.

115. Maspero, *Taoism*, 378.

116. Michael Loewe, *Chinese Ideas of Life and Death* (London: George Allen and Unwin, 1982), 104.

117. Edward H. Schafer, *Ancient China* (New York: Time-Life Books, 1967), 59.

118. Maspero, *Taoism*, 533–539.

119. Livia Knaul Kohn, *Laughing at the Tao Debates among Buddhists and Taoists in Medieval China* (Princeton: Princeton University Press, 1995), 149–150.

120. Kristofer Schipper, "The Taoist Body," *HR* 17 (1978): 377 n. 122.

121. Maspero, *Taoism*, 540; Needham, *Science and Civilizaton*, vol. 5, pt. 5, 184–218. Ling Li and Keith McMahon, "The Contents and Terminology of the Mawangdui Texts on the Arts of the Bedchamber," *EC* 17 (1992): 145–185.

122. Schipper, *Taoist Body*, 126.

123. Ibid., 125.

124. Ibid., 16.

125. Strickmann, "On the Alchemy," 164 .

126. Rolf A. Stein, "Religious Taoism and Popular Religion from the Second to the Seventh Centuries," in Welch and Seidel, *Facets*, 53–81.

127. Kimura Eichi, quoted in Sivin, "Word 'Taoist,'" 314.

128. Sivin, "Word 'Taoist,'" 319.

129. Ibid., 316.

130. Wolfgang Bauer, *China and the Search for Happiness* (New York: Continuum, 1976), 116–119.

131. Ibid., 110–111.

132. Sivin, "Word 'Taoist,'" 320.

133. Chan Wing-Tsit, *Source Book*, 425, quoting D. T. Suzuki and Hu Shih.

134. Robert M. Hartwell, "Demographic, Political, and Social Transformations of China, 750–1550,." *H JAS*. 41 (1982): 365–427.

135. Sivin, "Shen Kua," 1: 1–53.

136. Ssu-ma Kuang, *The Chronicle of the Three Kingdoms (220–265)*, 2 vols., ed. Glen W. Baxter; trans. Achilles Fang (Cambridge: Harvard University Press, 1952, 1965). Robert M. Hartwell, "Historical Analogism, Public Policy and Social Science in Eleventh- and Twelfth-Century China," *American Historical Review* 76 (1971): 690–727.

137. Daniel K. Gardner, Introduction to Chu Hsi, *Learning to Be a Sage* (Berkeley: University of California Press, 1990), 61–65; Wing-Tsit Chan, *Chu Hsi: Life and Thought* (Hong Kong: Chinese University Press, 1987), 181–182.

138. On the Ch'eng brothers, see A. C. Graham, *Two Chinese Philosophers: The Metaphysics of the Brothers Ch'eng*, 2nd ed. (La Salle, Ill.: Open Court, 1992).

139. Wing-Tsit Chan, *Chu Hsi: Life and Thought*, 7–8, 178–179.

140. Mao Hauixin, "The School of Chu Hsi and Its Propagation," in *Chu Hsi and Neo-Confucianism*, ed. Wing-Tsit Chan (Honolulu: University of Hawaii Press, 1986), 513.

141. Bodde, *Chinese Thought*, 191. Needham, *Science and Civilization*, 3: 193–194.

142. Wing-Tsit Chan, *Chu Hsi: New Studies* (Honolulu: University of Hawaii Press, 1989), 564.

143. Wing-Tsit Chan, *Chu Hsi: Life and Thought*, 52, 114.

144. Lao Tzu, *Tao Te Ching*, trans. Victor M. Mair (New York: Bantam, 1990), chap. 42.

145. Wing-tsit Chan, Introduction to Wang Bi, *Commentary*, xii; and idem, *Source Book*, 323, on another pair of terms: substance and function.
146. Kirk and Raven, *Presocratic Philosophers*, 238 n, from Aristotle, *Metaphysics*, A5. 986a. 15–16. G.E.R. Lloyd, *Adversaries and Authorities* (Cambridge: Cambridge University Press, 1996), 118–139, emphasizes that Chinese opposites are complementary while Greek ones are antagonistic.
147. Jonathan Bennett, *A Study of Spinoza's Ethics* (Indianapolis: Hackett, 1984).

CHAPTER 4 *Social Background of Organismic Thought*

1. Karl August Wittfogel, *Oriental Despotism* (New York: Random House, 1981).
2. Rudolf Bahro, *Die Alternative: Zur Kritik des real existierenden Sozialismus* (Reinbek bei Hamburg: Rowohlt Taschenbuch Verlag, 1977).
3. George Lichtheim, "Oriental Despotism," in *The Concept of Ideology* (New York: Vintage, 1967), 67, 76.
4. François Quesnay, "Le Despotisme de la Chine," *Éphémérides du Citoyen* (Paris: March–June 1767), trans. in *China: A Model for Europe*, by Lewis A. Maverick (San Antonio, Tex.: Paul Anderson, 1946), 139–304.
5. Christian Wolff, *The Real Happiness of a People under a Philosophical King Demonstrated* (London, 1750), quoted in *Essays in the History of Ideas*, by Arthur O. Lovejoy (New York: George Braziller, 1955), 108–109.
6. Karl August Wittfogel, *Wirtschaft und Gesellschaft Chinas, Part I, Produktivkräfte, Produktions und Zirkulationsprozess* (Economy and society of China: Part 1, Productive forces, production, and circulation process) (Leipzig, 1931); idem, "The Stages of Development of Chinese Social and Economic History," *Zeitschrift für Sozialforschung* 4 (1935): 35–58, reprinted in *The Asiatic Mode of Production: Science and Politics,* ed. Anne M. Bailey and Josep R. Llobera (London: Routledge and Kegan Paul, 1981), 113–140.
7. Matthias Greffrath and Fritz J. Raddatz,, "Conversations with Wittfogel," trans. David J. Parrent, *Telos*, no. 43 (spring 1980): 146.
8. Marvin Harris, *The Rise of Anthropological Theory* (New York: Columbia University Press, 1968), 686. Jeremy A. Sabloff, *The New Archaeology and the Ancient Maya* (New York: Freeman, 1990), 81–84, 180–182.
9. Some Chinese Marxist reaction against the theory began earlier, with the break with the Nationalists after the 1927 slaughter of Shanghai Communist workers by Chiang Kai-shek. See Wolfram Eberhard, *Conquerors and Rulers* (Leiden: Brill, 1965), 67.
10. Wittfogel, *Oriental Despotism*, 408.
11. Karl Marx, *The Grundrisse*, ed. Martin Niklaus (Harmondsworth, U.K.: Penguin, 1973).
12. Karl Marx, *Precapitalist Economic Formations*, ed. E. J. Hobsbawm (London: Lawrence and Wisehart, 1964).
13. Jürgen Habermas's use of the Engels unilinear evolutionary schema in his "Reconstruction of Historical Materialism," in *Communication and the Evolution of Society* (Boston: Beacon Press, 1975), is odd for a supposedly "sophisticated" student of Marx. Habermas, here and in his accounts of labor in Marx, tends to use the more "vulgar" versions of Marxism, perhaps thereby necessitating his own "revisions."
14. Ellen W. Schrecker, *No Ivory Tower* (Oxford: Oxford University Press, 1986), 165–166. Greffrath and Raddatz, "Conversations," 143–174, 170.
15. Harris, *Anthropological Theory*; idem, *Cultural Materialism* (New York: Random House, 1979).

16. Bahro, *Die Alternative*, 70–99.
17. Jürgen Rühle quoted in Greffrath and Raddatz, "Conversations," 143–144.
18. Steven Shapin, "Needham Thesis," in *Dictionary of the History of Science*, ed. W. F. Bynum et al., 295.
19. Friedrich Engels, *The Origin of the Family, Private Property, and the State*, intro. Eleanor Burke Leacock (New York: International Publishers, 1972), 179.
20. Schwartz, *World of Thought*.
21. Kwang-chih Chang, *Shang Civilization* (New Haven: Yale University Press, 1980), 245–248, cited by Schwartz, *World of Thought*, 14.
22. Paul Wheatley, *The Pivot of the Four Quarters* (Chicago: Aldine, 1971). David N. Kneightley, "The Origin of the Ancient Chinese City: A Comment," *EC* 1 (1975).
23. Schwartz, *World of Thought*, 23, cites R. E. Bradbury, a student of Edo religion, on the limitations to imaginative storytelling imposed by taking ones' own ancestors rather than nature gods as the object of religion.
24. Sarah Allen, *The Shape of the Turtle* (Albany: State University of New York Press, 1991), 57, 63.
25. Schwartz, *World of Thought*, 417, based on Donald Munro, *The Concept of Man in Early China* (Stanford, Calif.: Stanford University Press, 1969).
26. Max Weber, *The City*, trans. Don Martindale and Gertrud Neuwirth (New York: Collier, 1962), 89.
27. Ibid., 103–104.
28. Reinhard Bendix, *Max Weber: An Intellectual Portrait* (Garden City, N.Y.: Doubleday, 1962), 100. Max Weber, *The Religion of China* (New York: Free Press, 1951), 14.
29. An exception showing resemblences to Greek cities is discussed by Vitaly Rubin, "Tzu-Ch'an and the City State of Ancient China," *TP* 52 (1965–1966): 8–34.
30. Lichtheim, *Concept of Ideology*, 83.
31. Schwartz, *World of Thought*, 27.
32. Emile Durkheim, *The Elementary Forms of the Religious Life* (New York: Free Press, 1965), 52–53.
33. Ibid., 236–239.
34. Emile Durkheim and Marcel Mauss, *Primitive Classification* (Chicago: University of Chicago Press, 1963).
35. Marcel Granet, *Festivals and Songs of Ancient China* (New York: E. P. Dutton, 1932).
36. Marcel Granet, *The Religion of the Chinese People*, trans. Maurice Freedman (New York: Harper and Row, 1977), 43.
37. Ibid., 49.
38. Quoted in Granet, *Religion*, 165 n. 24.
39. Ibid., 66.
40. A later, classic, English-language study of this topic is William Soothill, *The Hall of Light: A Study of Early Chinese Kingship* (New York: Philosophical Library, 1952).
41. Marcel Granet, *La Pensée Chinoise*, quoted, trans. in C. Wright Mills, *Power, Politics and People*, ed. Irving Louis Horowitz (New York: Ballantine, 1963), 479.
42. Graham, *Disputers*, 404; Hansen, *Daoist*, 45. See also Bodde, *Chinese Thought* 42–55; and Rudolph G. Wagner, "Interlocking Parallel Style: Laozi and Wang Bi," *Asiatische Studien/Etudes Asiatiques* 34 (1980): 18–58.
43. Granet, *La Pensée Chinoise*, 473–479, excerpted and translated by Alfred Bloom, *The Linguistic Shaping of Thought: A Study in the Impact of Language on Thinking in China and the West* (Hillsdale, N.J.: Lawrence Erlbaum, 1981), 58.
44. D. C. Twitchet, Introduction to Henri Maspero, *China in Antiquity*, trans. Frank A. Kierman, Jr. (Amherst: University of Massachusetts Press, 1978), xix.

45. Mills, "Language and Thought in Ancient China," in *Power, Politics and People.*
46. Claude Lévi-Strauss, *The Elementary Structure of Kinship*, trans. J. S. Bell, J. H. von Sturmer, and Rodney Needham (Boston: Beacon Press, 1968), 311–404.
47. Norman J. Girardot, *Myth and Meaning in Early Taoism: The Theme of Chaos (Huntun*) (Berkeley: University of California Press, 1983), 17.
48. Rodney Needham, Introduction to Durkheim and Mauss, *Primitive Classification*, xi. Similarly, Needham notes that a major midcentury symposium on the topic, *Language and Culture*, ed. Harry Hoijer (Chicago: University of Chicago Press, 1954), made no mention of Durkheim.
49. Benjamin Lee Whorf, *Language, Thought and Reality* (1956; reprint, Cambridge, Mass: MIT Press, 1964).
50. Edward Sapir, *Language: An Introduction to the Study of Speech* (New York: Harcourt, Brace, 1949).
51. Henry Rosemont, Jr., "On Representing Abstractions in Archaic Chinese," *Philosophy East and West* 24 (1974): 71–88, and idem, "Gathering Evidence for Linguistic Innateness," *Synthese* 38 (1978): 127–148.
52. Steve Pinker, *The Language Instinct* (New York: Morrow, 1994), 57–67. Christoph Harbsmeier similarly attacks Chad Hansen for claiming that classical Chinese does not have a word for *truth*, as correspondence, by claiming that Hansen is making the racist claim that the Chinese are liars. See Harbsmeier "Marginalia Sino-Logica," in *Understanding the Chinese Mind*, ed. R. E. Allinson (Oxford: Oxford University Press, 1989), 147. A Deweyan notion of warranted assertability might be more sensible than the traditional Platonic or Scholastic conception of absolute truth, which hardly makes Deweyans liars.
53. Pinker uses G. K. Pullum, *The Great Eskimo Vocabulary Hoax and Other Irreverent Essays on the Study of Language* (Chicago: University of Chicago Press, 1991).
54. Pinker, *Language Instinct*, 67.
55. Willard van Orman Quine, *Word and Object* (Cambridge: MIT Press, 1960), 52–58.
56. Sapir, *Language.*
57. Donald Davidson, "On the Very Idea of a Conceptual Scheme," in *Post-Analytic Philosophy*, ed. John Rajchman and Cornel West (New York: Columbia University Press, 1985), 139.
58. A. C. Graham, "Conceptual Schemes and Linguistic Relativism in Relation to Chinese," in *Unreason within Reason* (La Salle, Ill.: Open Court, 1992), 68.
59. See Joscelyn Godwin, *Athanasius Kircher* (London: Thames and Hudson, 1979).
60. There is an English translation of Kircher's *China Illustrated* by Charles D. Van Tuyl from the 1677 original Latin edition (Muskogee, Okla.: Indian University Press, 1987).
61. Michael D. Coe, *Breaking the Maya Code* (New York: Thames and Hudson), 1992.
62. Coe, *Code*, 30.
63. Ibid., 38, 90, 106.
64. Erik Iverson, *The Myth of Egypt* (Princeton, N.J.: Princeton University Press, 1993), 94.
65. Iverson, *Myth*, 96–97.
66. Umberto Eco, *The Search for a Universal Language* (Oxford: Blackwell, 1996), 148–149. Eco bases himself upon Serge Sauneron, *Les prêtres de l'ancien Egypte* (Paris: Seuil, 1957); and Jean Yoxotee, "Jeux d'écriture. Sur une statuette de la XIXè dynastie," *Revue d'égyptologie* 10 (1955): 84–89.
67. Jean Jacques Rousseau, *Essay on the Origin of Language*, cited in Derrida, *Grammatology*, 3.

68. See the quotation from Hegel's *Philosophy of Fine Art* (London, 1920), 15–16, in Derrida, *Grammatology*, 12.

69. Gregory Ulmer, *Applied Grammatology* (Baltimore: Johns Hopkins University Press, 1985), 17. Ulmer (16–18) emphasizes the importance of hieroglyphics to Derrida's reorientation of emphasis in grammatology.

70. Ibid., 18, quoting Derrida, *Grammatology*, 25.

71. Raymond Chang and Margaret Sero, *Speaking of Chinese* (New York: Norton, xxxx), 129.

72. Hansen, *Daoist*, 25–26, 39–40.

73. Coe, *Code*, 25.

74. John DeFrancis, *The Chinese Language: Fact and Fantasy* (Honolulu: University of Hawaii Press, 1984), 74.

75. Ibid., 128–129; and Hansen, *Daoist*, 36 n.

76. Rosemont, "On Representing Abstractions," 71–88.

77. Hansen, *Daoist*, 34 n.

78. Graham, *Disputers*, 390.

79. DeFrancis, *Chinese Language*, 1–22.

80. Hansen, *Daoist*, 36 n.

81. Derrida, *Grammatology*, 10. Christopher Norris, *Deconstruction: Theory and Practice* (London: Methuen, 1982), 29.

82. Graham, *Disputers,* 331; and idem, *Reason and Spontaneity* (Totowa, N.J.: Barnes and Noble, 1985).

83. David L. Hall and R. T. Ames, *Thinking Through Confucius* (Albany: State University of New York Press, 1987).

84. Jacques Derrida, "Plato's Pharmacy," in *Dissemination*, trans. Barbara Johnson (Chicago: University of Chicago Press, 1981), 16–162.

85. Leonard Bloomfield, "Linguistic Aspects of Science," in *International Encyclopedia of Unified Science* (Chicago: University of Chicago Press, 1939), 6. Hansen, *Daoist*, 34–35.

86. Alexander Marshack, *The Roots of Civilization*, rev. ed. (Mount Kisco, N.Y.: Moyer Bell, 1991).

87. Hansen, *Daoist*, 41–42.

88. Bloom, *Linguistic Shaping,* 42–44.

89. Bodde, *Chinese Thought*, 42–55; and Wagner, "Interlocking Parallel Style," 18–58.

90. Hansen, *Language and Logic*, 30–54.

91. An example of atomistic bias is the Western translation in the early twentieth century by the Frenchman S. Le Gall of *ch'i* as "atoms" (Graham, "Conceptual Schemes and Linguistic Relativism in Relation to Chinese," in *Reason within Unreason*, 60, 74).

92. Christoph Harbsmeier, "The Mass Noun Hypothesis," in *Chinese Texts and Philosophical Contexts*, ed. Henry Rosemont, Jr., (La Salle, Ill.: Open Court, 1991), 49–66.

93. Graham, *Disputers*, 83, 262, 286, 402.

94. Graham, "Reflections and Replies," in Rosemont, *Chinese Texts*, 276. Graham, "Conceptual Schemes," 74.

95. Davidson, "Conceptual Scheme," 137.

96. Graham, "Conceptual Schemes," 74–75.

CHAPTER 5   *Interlude: Peter the Wayfarer and the Western Reception of the Compass*

1. Alfred Still, *Soul of the Lodestone* (New York: Murray Hill, 1946), 54.
2. Ibid., 55.
3. Ronan, *Shorter Science*, 3: 9.
4. Still, *Lodestone*, 55.
5. A. C. Crombie, *Medieval and Early Modern Science*, 2 vols. (Garden City, N.Y.: Doubleday Anchor, 1959), 2: 189.
6. Carolyn Eisele, ed., *Historical Perspectives on Peirce's Logic of Science: A History of Science*, 2 vols. (Berlin: Walter de Gruyter, 1985), 2: 112.
7. Ibid., 99.
8. Roger Bacon, *Opus Tertium*, chap. 13, quoted in Edward Grant, "Peter Peregrinus," in *DSB*, 533–534.

CHAPTER 6   *Occultism and the Rise of Early Modern Science*

1. Theophrastus Bombastus Paracelsus, *Paracelsus: Selected Writings*, ed. Jolande Jacobi (1952; reprint, Princeton: Princeton University Press, 1988), 79–80.
2. Copenhaver and Schmitt, *Renaissance Philosophy* 306.
3. Debus, *Man and Nature,* 7
4. Andrew Weeks, *Paracelsus* (Albany: State University of New York Press, 1977), 104.
5. Henry Pachter, *Paracelsus: Magic into Science* (New York: Henry Schuman, 1951), 313 n. 7; Jolande Jacobi, *Paracelsus: Selected Writings* (Princeton, N.J.: Princeton University Press, 1988), 168, 176–178.
6. Brian Easlea, *Witch Hunting, Magic and the New Philosophy* (Atlantic Highlands, N.J.: Humanities Press, 1980), 102–103.
7. Heer, *Intellectual History of Europe*, 236.
8. Ibid., 237, citing K. Goldhammer, ed., *Paracelsus, Politische Schriften* (Tübingen, 1952), 44, 53, 57.
9. Piyo M. Rattansi, "Paracelsus and the Puritan Revolution," *Ambix* 11 (1963): 24–32, 25.
10. Ibid., 27–28.
11. William W. Lowrance, *Modern Science and Human Values* (Oxford: Oxford University Press,1986), 4, citing Henry Lyons, *The Royal Society*, 1660–1940 (Cambridge: Cambridge University Press, 1944), 41.
12. Debus, *Man and Nature*, 10.
13. Pachter, *Magic into Science*, 95.
14. Easlea, *Witch Hunting*. Carolyn Merchant, *The Death of Nature* (San Franciso: Harper and Row, 1980).
15. Evelyn Fox Keller, *Reflections on Gender and Science* (New Haven, Conn.: Yale University Press, 1985), 48–53.
16. Sandra Harding, *The Science Question in Feminism* (Ithaca: Cornell University Press, 1986), 116.
17. Keller, *Gender and Science*, 190.
18. Weeks, *Paracelsus*, 79–83, 92–93.
19. Debus, *Man and Nature*, 27.
20. Charles Webster, review of *Resultate und Desiderate der Paraceusu-Forschung*, ed. Peter Dilg and Harmut Rudolph, *Isis*, 87 (1996): 164.
21. Peuckert, *Gabalia*, 271, quoted in Joscelyn Godwin, *Robert Fludd: Hermetic Philosopher and Surveyor of Two Worlds* (Boulder, Colo.: Shambhala, 1979), 9.

22. Walter Pagel, *Joan Baptista van Helmont* (Cambridge: Cambridge University Press, 1982), 10–11, 32–33.

23. Pagel, *Paracelsus*,142–143.

24. Debus, *Man and Nature*, 31.

25. Easlea, *Witch Hunting*, 101.

26. Pagel, *Paracelsus*, 67.

27. Ibid., 66–71.

28. J.L.E. Dreyer, *Tycho Brahe: A Picture of Scientific Life and Work in the Sixteenth Century* (1890; reprint, New York: Dover, 1963), 25.

29. Victor E. Thoren, *The Lord of Uraniborg: A Biography of Tycho Brahe* (Cambridge: Cambridge University Press, 1991), 52–53.

30. Charles Webster, *From Paracelsus to Newton: Magic and the Making of Modern Science* (Cambridge: Cambridge University Press, 1982), 29, citing J. R. Christianson, "Tycho Brahe's Observations on the Comet of 1577," *Isis* 70 (1979): 110–140.

31. Dreyer, *Tycho Brahe*, 236.

32. Ibid., 128–129, 232.

33. Ibid., 259.

34. Thoren, *Lord of Uraniborg*, 84.

35. Ibid., 79–84.

36. William H. Donahue, "The Solid Planetary Spheres in Post-Copernican Natural Philosophy," in *The Copernican Achievement*, ed. Robert S. Westman (Berkeley: University of California Press, 1975), 254.

37. Tycho Brahe, *Learned Tycho Brahe His Astronomical Conjectur of the New and Much Admired * Which Appeared in the Year 1572*, trans. V.V.S. (1632; facsimile reprint, New York: Da Capo, 1969), 19, 21.

38. Webster, *Paracelsus to Newton,* 30.

39. Max Caspar, *Kepler*, trans., ed. C. Doris Hellman (New York: Dover, 1993), 183. J. V. Field, "Kepler's Rejection of Numerology," in *Occult and Scientific Mentalities of the Renaissance*, ed. Brian Vickers (Cambridge: Cambridge University Press, 1978), 291.

40. Webster, *Paracelsus to Newton,* 30.

41. Doris Hellman, *The Great Comet of 1577: Its Place in the History of Astronomy* (New York: Columbia University Press, 1944), 92–93, cited in Donahue, "Planetary Spheres."

42. Webster, *Paracelsus to Newton* 29.

43. Yates, *Bruno*; and idem, "The Hermetic Tradition."

44. John Herman Randall, *The Career of Philosophy*, 2 vols. (New York: Columbia University Press, 1962), 1: 327.

45. Yates, *Bruno*, 354.

46. Hans Blumenberg, *The Genesis of the Copernican World* (Cambridge: MIT Press, 1987), 372.

47. Yates, "Hermetic Tradition." Idem, *The Rosicrucian Enlightenment* (Boulder, Colo.: Shambhala, 1972).

48. John Maynard Keynes, "Newton the Man," in *Essays in Biography* (New York: Norton, 1963), 311.

49. Alexandre Koyré, *From the Closed World to the Infinite Universe* (Baltimore: Johns Hopkins University Press, 1957), 46, 54.

50. Frances Yates, *The Occult Philosophy in the Elizabethan Age* (London: Routledge and Kegan Paul, 1979), 1.

51. Brian R. Copenhaver, "Natural Magic, Hermeticism and Occultism in Early Modern

Science," in *Reappraisals of the Scientific Revolution*, ed. David C. Lindberg and Robert S. Westman, (Cambridge: Cambridge University Press, 1990), 261–301.

52. Yates, *Bruno*, 13.

53. Ibid., 14.

54. Ibid., 398–402. Mary Leftkowitz, *Not Out of Africa: How Afrocentrism Became an Excuse to Teach Myth as History* (New York: Basic Books, 1996), 57, 100.

55. Meric had privileges and properties taken away in 1644 and apparently wished to discredit certain foes in the Puritan Revolution who claimed divine inspiration. See Peter French, *John Dee* (New York: Dorset Press, 1989), 11–13.

56. Yates, *Bruno*, 431, citing J. Doresse, *The Secret Books of the Egyptian Gnostics* (London, 1960), 255–257.

57. James M. Robinson, ed., *The Nag Hammadi Library*, rev. ed. (San Franciso: Harper and Row, 1988), 330–338. Also, the *Art of Eudoxus*, the "earliest illustrated scientific work to have survived," is subtitled "Oracles of Hermes" (D. J. Thompson, *Memphis under the Ptolemies* [Princeton: Princeton University Press, 1988], 252–254).

58. A.-J. Festugière, *Hermes Trismegiste*, 4 vols. (Paris: J. Gabalda, 1950–1954), 1: 427, quoted in Garth Fowden, *The Egyptian Hermes* (Princeton: Princeton University Press, 1993), xxiii.

59. A.-J. Festugière, *Hermes Trismegiste*, 1: 85, quoted in Brian P. Copenhaver, Introduction to *Hermetica* (Cambridge: Cambridge University Press, 1992), lv.

60. J.-P. Mahe, cited in Copenhaver, Introduction to *Hermetica*, lvii.

61. Fowden, *Egyptian Hermes*, 73, 116–120, 172.

62. James Henry Breasted, *Ancient Times: A History of the Early World*, 2nd ed. (Boston: Ginn, 1944), 131, 133.

63. Martin Bernal, *Black Athena*, 2 vols. (New Brunswick, N.J.: Rutgers University Press, 1987), 1: 140–141.

64. Ibid., 1: 136–137.

65. Ibid., 1: 138.

66. Derrida, "Plato's Pharmacy" in *Dissemination*.

67. Fowden, *Egyptian Hermes*, 167, 122, 125.

68. Raphael Patai, *The Jewish Alchemists: A History and Source Book* (Princeton, N.J.: Princeton University Press, 1994), 60–91.

69. Frank J. Yurco, "How to Teach about Ancient History: A Multicultural Model," in *Alternatives to Afro-Centrism* (New York: Manhattan Institute, 1994), 70–77.

70. Joyce L. Haynes, *Nubia: Ancient Kingdoms of Africa* (Boston: Museum of Fine Arts, 1992), 20–28.

71. Fowden, *Egyptian Hermes*, 65–70.

72. Much of this discussion of the debate concerning the Black Athena thesis is based on Christy Hammer, "A Sociological Study of the Manifestation of Multi-Culturalism in Public Schools" (Ph.D. diss., University of New Hampshire, 1994), 279–296, and conversations with the author.

73. Joseph A. A. Ben Jochannon, *Black Man of the Nile and His Family* (Baltimore: Black Classic Press, 1972), 328.

74. Mary Lefkowitz, "Not Out of Africa," *New Republic* 206 (February 10, 1992): 29–36; and Letters to the Editor, *New Republic* 206 (March 9, 1992): 4–5. Idem, *Not Out of Africa*.

75. At the conference, Civil Rights, Civil Liberties and Civic Responsibility, at the University of New Hampshire, October 16, 1994, Lefkowitz admitted that her criticisms do not touch later, Hellenistic, Egypt.

76. Lefkowitz, *Not Out of Africa*, 100–101.

77. Roger S. Bagnall, *Egypt in Late Antiquity* (Princeton: Princeton University Press, 1993), 230–260.

78. Robert Palter, "*Black Athena*, Afro-Centrism and the History of Science," *HS* 31 (1993): 227–287, reprinted in *Black Athena Revisited*, ed. Mary R. Lefkowitz and Guy MacLean Rogers (Chapel Hill: University of North Carolina Press, 1996), 209–266.

79. John Pappademos, "The Newtonian Synthesis in Physical Science and Its Roots in the Nile Valley," *Journal of African Civilizations* 6 (1984): 84–101.

80. Palter, "Afro-Centrism," 244, citing Betty Jo Teeter Dobbs, *The Janus Faces of Genius* (Cambridge: Cambridge University Press, 1991), 193–212.

81. Edward Rosen, "Was Copernicus a Hermeticist?" in *Minnesota Studies in the Philosophy of Science*, vol. 5, *Historical and Philosophical Perspectives of Science*, ed. Roger H. Steuwer (New York: Gordon and Breach, 1989). 171.

82. Ibid., 169.

83. Fernand Hallyn, *The Poetic Structure of the World: Copernicus and Kepler* (New York: Zone Books, 1990), 313.

84. Herbert Butterfield, *The Origins of Modern Science, 1300–1800*, rev. ed. (New York: Macmillan, 1965), 141. Brian Vickers writes, "Such comments may have raised a laugh among undergraduates in the Cambridge history faculty . . . but they seem unworthy of a serious historian. . . . When we consider that the authors who could been have aimed at include Walter Pagel, . . . Joseph Needham and J. R. Partington, it reveals a sadly closed mind" (Vickers, *Occult and Scientific Mentalities*, 2–3).

85. Paolo Rossi, "Hermeticism, Rationality, and the Scientific Revolution," in *Reason, Experiment, and Mysticism in the Scientific Revolution*, ed. M. L. Righini Bonelli and William R. Shea (New York: Science History Publications, 1975), 272–273; idem, *Francis Bacon: From Magic to Science* (London: Routledge, and Kegan Paul, 1968); and idem, *Philosophy, Technology and the Arts in the Early Modern Era* (New York: Harper & Row, 1970). Rossi's only reference to Zilsel's social theories: "Zilsel draws conclusions from this contrast [between cooperative labor of the workshop and solitary study of the monk or humanist] which are not wholly acceptable" (*Philosophy, Technology and the Arts*, 70).

86. A. Rupert Hall, "Magic, Metaphysics and Mysticism in the Scientific Revolution," in *Reason, Experiment, and Mysticism*, eds. Righini Bonelli and Shea, 277.

87. Larry Laudan, *Between Positivism and Relativism: Theory, Method and Evidence* (Boulder, Colo.: Westview, 1996), 95.

88. For instance, Mary Hesse, "Hermeticism and Historiography: An Apology for the Internal History of Science," in Steuwer, *Historical and Philosophical Perspectives*, 134–162. This volume originally appeared in *Minnesota Studies in the Philosophy of Science*, in 1970, at the height of the counterculture and the interest among philosophers in historical approaches to science. It is, however, the one volume of the Minnesota Studies that was allowed to go out of print by the University of Minnesota Press. It was farmed out, to the series Classics in the History and Philosophy of Science. The exile of this volume by the publisher from the rest of those issued by the Minnesota Center for the Philosophy of Science is significant.

89. Hesse, "Hermeticism," 146: "No general specification of rationality seems to be forthcoming from the philosophical analyses discussed, and therefore we cannot expect an answer to the question of whether there is a relatively autonomous internal history."

90. W. H. Newton-Smith, "Science, Rationality and Newton," in *Newton's Dream*, ed. Marcia Sweet Stayer (Montreal: McGill-Queen's University Press, 1988), 19–35.

91. Richard Rorty, "Is Natural Science a Natural Kind?" in *Construction and Constraint*, ed. Edwin McMullin, (Notre Dame, Ind.: University of Notre Dame Press, 1988), 61.

92. John Wilkins, *Mathematical Magick; or, The Wonders That May Be Performed by Mechanicall Geometry* (London, 1648), cited in Frances Yates, *The Theater of the World* (Chicago: University of Chicago Press, 1969), 51, n. 19. Wilkins cited hermeticists John Dee and Robert Fludd as authorities.

93. Desmond M. Clark, *Occult Powers and Hypotheses: Cartesian Natural Philosophy under Louis XIV* (Oxford: Oxford University Press, 1989), 73, refers to Keith Hutchinson, "What Happened to Occult Qualities in the Scientific Revolution?" *Isis* 73 (1982): 233–253.

94. For instance, Henri Poincaré's famous account of mathematical discovery or Bertrand Russell's description of his own creative process.

95. Allan Sandage, in *The Carnegie Atlas of Galaxies* (1994), vol. 1, chap. 1, quoted by Timothy Ferris, "Minds and Matter," *New Yorker* 71, no. 12 (May 15, 1995): 50.

96. William R. Shea, *The Magic of Numbers and Motion* (Canton, Mass.: Science History Publications, 1991), 98–100. Shea also discusses Descartes's early interest in Rosicrucianism (ibid., 109).

97. See Larry Laudan, "Why Was the Logic of Discovery Abandoned?" in *Scientific Discovery, Logic and Rationality* ed. Thomas Nickles (Dordrecht: D. Reidel, 1980).

98. Edgar Zilsel, "Origins of Gilbert's Scientific Method," in *The Roots of Scientific Thought*, ed. Philip Wiener and Aaron Noland (New York: Basic Books, 1957), 243 n. 35, cites Leonard Olschki on the scientific literature in the vernacular, *Geschichte der Neusprachlichen Wissenschaftlichen Literatur*, vol. 1 (Heidelberg, 1918); vol. 2, (Leipzig, 1922); vol. 3, (Halle, 1927).

99. Jan Golinski, "The Secret Life of an Alchemist," in *Let Newton Be!* ed. John Fauvel et al. (Oxford: Oxford University Press, 1988), 160.

100. Karl Popper, *Conjectures and Refutations*, 2d ed. (New York: Harper Torchbooks, 1968). Robert K. Merton, "The Normative Structure of Science," in *The Sociology of Science* (Chicago: University of Chicago Press, 1973). Both Popper's *Open Society and Its Enemies* and Merton's article were written during World War II, when Nazi irrationalism was seen as the main enemy and science was identified with Western democracy, despite the fact that "totalitarian" nations (Nazi Germany and the Soviet Union) were supporting physics.

101. Robert K. Merton, "Science and the Economy in Seventeenth Century England," *Science and Society*, 3 (1939): 37–61, reprinted in Merton, *Social Theory and Social Structure* (Glencoe, Ill.: Free Press, 1957), 607–627.

102. See William Broad and Nicholas Wade, *Betrayers of the Truth: Fraud and Deceit in the Halls of Science* (New York: Simon and Schuster, 1982); and Alexander Kohn, *False Prophets: Fraud and Error in Science and Medicine* (Oxford: Basil Blackwell, 1986). Broad and Wade move from the well-established cases of fraud to cases in past science in which standards of measurement were looser than they are today.

103. See Lowrance, *Modern Science*, 71, 77, 112–114.

104. Here notable are the criticisms of studies of the heritability of IQ by evolutionists Stephen J. Gould and Richard Lewontin, linguist Noam Chomsky, and astrophysicist David Layzer. Not surprisingly, psychometricians who claim heritability of IQ and heritable racial differences in IQ claim that IQ is largely inherited and dismiss these criticisms by leading scientists as incompetent because they do not come from within the psychometric profession. See Ned J. Block and Gerald Dworkin, eds., *The IQ Controversy* (New York: Pantheon, 1976), for a sample of the criticisms; and M. Snyderman

and S. Rothman, *The IQ Controversy: Media and Public Policy* (New Brunswick, N.J., Transaction,1988), for a survey showing that the "experts," members of the hereditarian psychometrics community, reject such criticisms. The primary author of the second work is a collaborator of Richard Herrnstein, a Harvard hereditarian psychologist.

105. L. Kowarski of CERN, Geneva, at the Boston Colloquium for the Philosophy of Science, October 26, 1971.

106. Fred Kaplan, *The Wizards of Armageddon* (New York: Simon and Schuster, 1983).

107. Michael Polanyi, *Personal Knowledge: Towards a Post-Critical Philosophy* (Chicago: University of Chicago Press, 1958), 53. Kuhn, *Scientific Revolutions*, 44–47.

108. Rustrum Roy has been one of the most vocal critics of peer review of grant proposals. See Rustrum Roy, "An Alternative Funding Mechanism," *Science* 211 (March 21, 1981): 1377; and idem, "Alternatives to Review by Peers: A Contribution to the Theory of Scientific Choice," *Minerva* 22, nos. 3/4 (Autumn/Winter 1984): 316–328. Daryl E. Chubin and Edward Hackett, *Peerless Science: Peer Review and US Science Policy* (Albany: State University of New York Press, 1990), review arguments about the peer review system. Criticisms of scientists who report their results to the general media prior to having had their results published or even accepted for publication, as in June Goodfield, *Reflections on Science and the Media* (Washington, D.C.: AAAS, 1981), and the editorial "Gene Cloning by Press Conference," *New England Journal of Medicine*, March 27, 1980, 743, cut both ways. On the one hand, announcement of scientific results to the media prior to publication bypasses evaluation by the scientific community. On the other hand, the threats by leading journals in physics and medicine to refuse to publish articles that receive prior publicity in the newspapers totally oppose the appeal to the broader public by scientists.

109. Brian Goodwin, "Science and Alchemy," in *Rules of the Game: Cross-Disciplinary Essays on Models of Scholarly Thought*, ed. T. Sharin (London: Tavistock, 1972), 360; and in C. H. Waddington, ed., *Towards a Theoretical Biology*, 4 vols. (Chicago: Aldine, 1968–1972).

110. Quoted favorably by Needham, *Science and Civilization*, vol. 5, pt. 5, 20.

111. Waddington, *Evolution of an Evolutionist*, 2–3.

112. Jerome Ravetz, *Scientific Knowledge and Its Social Problems* (Oxford: Oxford University Press, 1971), 424, 432–434.

113. Marx Wartofsky, "Is Science Rational?," in *Science, Technology, and Freedom*, ed. Willis H. Truitt and T. W. Graham Solomons (Boston: Houghton Mifflin, 1974), 202–210.

CHAPTER 7   *Occultism and the Social Background of Science*

1. Zilsel, "Sociological Roots of Science," 544–562.

2. Robert S. Brumbaugh, *The Philosophers of Greece* (New York: Crowell, 1964), 16, 69.

3. Aristotle, *"De Partibus Animalium I"* and *"De Generatione Animalium I,"* trans., ed. D. M. Balme (Oxford: Oxford University Press, 1972), 1. 5. 17–18.

4. On the economic situation, see Robert Lopez, "Hard Times and the Investment in Culture," in *The Renaissance: Six Essays*, by Wallace K. Ferguson et al. (New York: Harper and Row, 1962).

5. Edgar Zilsel, "The Genesis of the Concept of Progress," in Wiener and Noland, *Roots of Scientific Thought*, 251–275.

6. Zilsel, "Gilbert's Scientific Method," 224.

7. Robert K. Merton, *Science, Technology, and Society in Seventeenth-Century England* (New York: Harper and Row, 1970).

8. I. Bernard Cohen, ed., *Puritanism and the Rise of Modern Science: The Merton Thesis* (New Brunswick, N.J.: Rutgers University Press, 1990), 51.

9. Ibid., 73.

10. Margaret C. Jacob, *The Cultural Meaning of the Scientific Revolution* (Philadelphia: Temple University Press, 1988), 77.

11. For instance, see Frederick B. Tolles, *Meeting House and Counting House* (Chapel Hill: University of North Carolina Press, 1948); and Arthur Raistrick, *Quakers in Science and Industry* (London: Bannisdale, 1950).

12. George Rosen, "Left-Wing Puritanism and Science," *Bulletin of the Institute of the History of Medicine* 15 (1944): 375–380, reprinted in Cohen, *Puritanism*, 171–177. Interestingly, Joseph Needham, in his favorable review of Merton (*Science and Society* 2 [1938]: 566–571) took the latter to task solely for not discussing the Levelers.

13. P. M. Rattansi, "The Social Interpretation of Science in the Seventeenth Century," in *Science and Society, 1600–1900*, ed. Peter Mathius (Cambridge: Cambridge University Press, 1972); and idem, "The Intellectual Origins of the Royal Society," *Notes and Records of the Royal Society of London* 23 (1968): 129–143.

14. Merton interestingly contrasts Presbyterianism with earlier Calvinism by noting that in Calvinism hard work was to justify conviction of already predestined salvation, while in Presbyterianism hard work actually helped one to be saved (Merton, *Science, Technology*, 57, noted by Cohen, introduction to *Puritanism*, 65).

15. Harold J. Cook, "Charles Webster on Puritanism and Science," in Cohen, *Puritanism*, 267, following Charles Webster, *The Great Instauration* (London: Duckworth, 1975), 505.

16. Christopher Hill, *Antichrist in Seventeenth Century England* (London: Verso, 1990), 119n. The Fifth Monarchists were the most extreme and bizarre of the political millenarians. They entered Parliament chanting and their disruptions helped Cromwell to become dictator. See Austin Woolrych, "Oliver Cromwell and the Rule of the Saints," in *The English Civil War*, ed. R. H. Parry (Berkeley: University of California Press, 1970), 73–76.

17. Beale to Samuel Hartlib, March 22, 1658/9, in Webster, *Great Instauration*, 12 (spelling modernized).

18. A. Rupert Hall, "Science, Technology and Utopia," in Mathias, *Science and Society*, cited in Webster, *Great Instauration*, 496.

19. See, for instance the Leveler proposals for legal reform, and Gabriel Plattes, author of the utopian *Macaria*, with his proposal for medical reform, via training of parish priests in medicine (Webster, *Great Instauration*, 257–262).

20. Ibid., 504.

21. On John Graunt, Matthew Hale, and William Petty as forerunners of mathematical population biology, see G. Evelyn Hutchinson, *An Introduction to Population Ecology* (New Haven: Yale University Press, 1978), 5–10, 176.

22. Jacob, *Cultural Meaning*, 94–95.

23. Richard Westfall, *Never at Rest* (Cambridge: Cambridge University Press, 1980), 143–144, raises questions about the traditional characterization of Newton's *annus mirabilis* as 1666 and titles his chapter "Anni Mirabiles" to emphasize that three years of young Newton's work were equally extraordinary.

24. Anne Llewellyn Barstow, *Witchcraze: A New History of the European Witch Hunts*

(San Francisco: HarperCollins, Pandora, 1994); Frances Hill, *The Delusion of Satan: The Full Story of the Salem Witch Trials* (New York: Doubleday, 1995); Selma R.Williams and Pamela Williams Adelman, *Riding the Nightmare: Women and Witchcraft from the Old World to Colonial Salem* (New York: HarperCollins, 1978); Carol F. Karlsen, *The Devil in the Shape of a Woman* (New York: Vintage, 1987).

25. Keller, *Gender and Science*, 63, citing Ruth Perry.
26. Ibid., 62.
27. Barbara Ehrenreich, and Deirdre English, *Witches, Midwives and Nurses: A History of Women Healers* (Old Westbury, N.Y.: The Feminist Press, 1973).
28. M. J. Hughes, *Women Healers in Medieval Life and Literature* (n.p.: Books for College Libraries Press, 1968), nn. 80, 86. J. Donnison, *Midwives and Medical Men* (London: Heinemann, 1977), 1–15, cited in Easlea, *Witch Hunting*, 39.
29. Easlea, *Witch Hunting*, 214.
30. Sandra Harding, *Whose Science, Whose Knowledge* (Ithaca, N.Y.: Cornell University Press, 1991), 43; Merchant, *Death of Nature*, 168; Keller, *Gender and Science*, 37. Alan Soble notes that this quotation does not unambiguously refer to rape or to inquisition in the sense of torture of witches. Soble also attempts to mitigate the sexism of the claim by noting even if Bacon's model is rape, it is marital rape! (Alan Soble, "In Defense of Bacon," *Philosophy of the Social Sciences* 25 [1995]: 193). Soble's defense of Bacon does not eliminate the association of science with "manliness" and the exclusion of the feminine in Glanvill, Oldenburg, Sprat, and others.
31. Keller, *Gender and Science*, 54.
32. Ibid., 56. Easlea, *Witch Hunting,* 214.
33. H. R. Trevor-Roper, *The European Witch-Craze of the Sixteenth and Seventeenth Centuries and Other Essays* (New York: Harper and Row, 1967), 116n, 243.
34. Ginzburg, *Ecstasies*, 8–9.
35. Ginzburg, *Cheese and Worms*, 57–58.
36. Ibid., 50.
37. Hans Peter Dürr, *Dreamtime: Concerning the Boundary between Wilderness and Civilization* (Oxford: Basil Blackwell, 1985).
38. Wilhelm Burkert, *Lore and Science in Ancient Pythagoreanism* (Cambridge: Harvard University Press, 1972).
39. Agrippa, *Declamation*.
40. Yates, *Occult Philosophy*, 116.
41. Trevor-Roper, *Witch-Craze*, 146–147. Easlea, *Witch Hunting*, 13. Johann Weyer, *Witches, Devils, and Doctors in the Renaissance: Johann Weyer, de Praestigiis Daemonum* (Binghamton, N.Y.: Medieval and Renaissance Texts and Studies, 1991).
42. Trevor-Roper, *Witch-Craze*, 122.
43. Easlea, *Witch Hunting*, 44.
44. Valerie I. J. Flint, *The Rise of Magic in Early Medieval Europe* (Princeton, N.J.: Princeton University Press, 1991), discusses the ways in which the Church consciously absorbed and transformed various forms of magic in order to draw on the energy of pagan views.
45. Bakhtin, *Rabelais*.
46. Febvre, *Problem of Unbelief.*
47. Ginzburg, *Cheese and Worms*, xviii.
48. Ibid., 126.
49. Ibid., 53, 57.

50. Ibid., fn. 58, 154.
51. Allen G. Debus, *The English Paracelsians* (London: Oldbourne, 1965), 25.
52. Bernal, *Black Athena*, 1: 132, citing Elaine Pagels, *The Gnostic Gospels* (1979; reprint, New York: Vintage, 1989), xix.
53. Keith Thomas, *Religion and the Decline of Magic* (New York: Charles Scribner's Sons, 1971), 579, cited in Easlea, *Witch Hunting*, 197, 271 n. 1.
54. Easlea, *Witch Hunting*, 218.
55. Webster, *Paracelsus to Newton*.
56. Ibid., 92–93.
57. Ibid., 96.
58. Weyer, *Witches, Devils*, introduction, lxxvii–lxxviii. Weyer was first presented as a precursor of the psychoanalytical interpretation in G. Zilboorg, *The Medical Man and the Witch* (Baltimore: Johns Hopkins University Press, 1935).
59. John C. Greene, *The Death of Adam* (New York: New American Library of World Literature, 1961), 27–34.
60. Webster, *Paracelsus to Newton*, 98.
61. Steven Weinberg, "Sokal's Hoax," *New York Review of Books* 43, no. 13 (August 1996): 15.
62. Mandrou's views are briefly summarized in part in *From Humanism to Science, 1480–1700* (New York: Penguin, 1978), 114–117, 143–149, 247–252.
63. Suggested by Harry Marks.
64. C. V. Wedgewood, *Thirty Years War* (New York: Book-of-the-Month Club, 1995), 64.
65. Ibid., 96.
66. Ibid., 78–80.
67. Ibid., 119.
68. Ibid., 120.
69. Johann Valentin Andreä, *Christianopolis: An Ideal State of the Seventeenth Century*, trans. F. E. Held (Oxford: Oxford University Press, 1916).
70. Wedgewood, *Thirty Years War*, 130. Yates, *Rosicrucian*, 23, 54, plate 9.
71. *The Hermetic Romance; or, Chemical Wedding, Written in High Dutch by C. R.*, trans. Ezechiel, 1690; reprinted in A. E. Waite, *The Real History of the Rosicrucians* (1887, reprint, Klima, Mont.: Kessinger, n.d.), and in John Warwick Montgomery, *Cross and Crucible: Johann Valentin Andreae (1586–1654): Phoenix of the Theologians*, vol. 2, *The "Chymische Hochzeit," with Notes and Commentary* (The Hague: Martinus Nijhoff, 1972).
72. Yates, *Rosicrucian's*, 66–67.
73. See, for instance, John Dee, *General and Rare Memorials Pertayning to the Perfect Arte of Navigation* (1577; reprint, Amsterdam: Da Capo, 1968).
74. Yates, *Rosicrucian*, 70.
75. Ibid., 36.
76. Ibid., 81.
77. Brian Vickers, "Frances Yates and the Writing of History," *Journal of Modern History* 51 (June 1979): 287–316. A criticism of Vickers, mainly on the basis of an ahistorical view of the difference between science and magic, is Patrick Curry, "Revisions of Science and Magic," *HS* 23 (1985): 299–325.
78. Vickers, "Frances Yates," 296.
79. Yates, *Rosicrucian*, 115.
80. Shea, *Magic of Numbers*, 97, 102–104, 109.

CHAPTER 8   *Gilbert and Early Modern Theories of the Magnet*

1. Gerrit L. Verschuur, *Hidden Attraction: The Mystery and History of Magnetism* (New York: Oxford University Press, 1993), 19, 236–237. Verschuur presents a history of science that, though influenced by Kuhn, accepts a naive conception of clearing away superstition and access to absolute truth. He writes, "We are well on the way to a significant and possibly complete understanding of the nature of the physical universe" (ibid., 241).

2. Ibid., 20–21.

3. Suzanne Kelly, *The "De Mundo" of William Gilbert* (Amsterdam: Menno Hertzberger, 1965).

4. Verschuur recounts several articles in the *Journal of the History of Astronomy* that discuss Kircher and the "magnetical philosophy": J. A. Bennett, "Cosmology and the Magnetical Philosophy, 1640–1680," *JHA* 12 (1981), and M. R. Baldwin, "Magnetism and the Anti-Copernican Polemic,"*JHA* 16 (1985), in Verschuur, *Hidden Attraction* 35–42. Verschuur also cites Robert C. Staufer's "Speculation and Experiment in the Background of Oersted's Discovery of Electromagnetism," *Isis* 48 (1957): 33–50, and Pearce Williams's *DSB* article on Ørsted, but does not mention the Romantic, panpsychist side of Ørsted in his own chapter on him.

5. William Gilbert, *De Magnete*, trans. P. Fleury Mottelay (New York: Dover, 1958), xlvii.

6. Ibid., 5, 103.

7. Ibid., xlvii.

8. Ibid., xlviii. The phrase "the common mother of all" appears frequently, as Zilsel says, "whenever the earth is mentioned": 12, 26, 38 (twice), 41, 117, 152, 210 (Zilsel, "Gilbert's Scientific Method," 223).

9. William Gilbert, *De Mundo* (1651), 115, quoted in Pierre Duhem, *The Aim and Structure of Physical Theory*, trans. Philip P. Wiener (Princeton, N.J.: Princeton University Press, 1954), 230.

10. Merchant, *Death of Nature*, chap. 1, "Nature as Female." The term *coition* is introduced by Gilbert, *De Magnete*, 97, and is emphasized in Easlea, *Witch Hunting*, 92.

11. James Lovelock, *The Gaia Hypothesis: A New Look at Life on Earth* (Oxford: Oxford University Press, 1979); and idem, *The Ages of Gaia: A Biography of Our Living Earth* (New York: Norton, 1988).

12. Gilbert, *De Magnete*, 309.

13. Ibid., 310.

14. Aristotle, *De Anima*, A2. 405a.19.

15. Gilbert, *De Magnete*, 311.

16. Gilbert, *De Mundo*, 307, quoted in Duhem, *Aim and Structure* 236.

17. Gilbert, *De Magnete*, 305.

18. Ibid., 19.

19. Ibid., 24.

20. Ibid., 12.

21. Ibid., 11. Giambattista della Porta, *Natural Magick* (New York: Basic Books, 1957). See William Eamon, *Science and the Secrets of Nature* (Princeton, N.J.: Princeton University Press, 1994), chap. 6 on Porta, who was not truly an experimentalist, but collected most of his "secrets" from classical texts. Eamon notes that for Porta, the model of attraction and repulsion—opposing forces—was a model for all phenomena (211). Oddly, Porta did not treat the magnet extensively until the second edition of his book.

22. Marie Boas, *The Scientific Renaissance, 1450–1630* (New York: Harper and Row, 1962), 190.

23. Keller, *Gender and Science*, chaps. 2, 3.

24. Mary Hesse, "Gilbert and the Historians," *BJPS* 11 (1960): 9.

25. Galileo Galilei, *Dialogue Concerning the Two Chief World Systems*, trans. Stillman Drake (1953; reprint, Berkeley: University of California Press, 1962), 400–411.

26. Ibid.

27. Galilei, *Dialogue,* cited in Hesse, "Gilbert and the Historians," 4.

28. Hesse, "Gilbert and the Historians," 141–142.

29. *Novum Organum* 1: liv, cited in Hesse, ibid.

30. Gilbert, *De Magnete*, 358.

31. Ibid., 347.

32. Ibid., 353.

33. Thomas Kuhn, *The Copernican Revolution* (Cambridge: Harvard University Press, 1957), 138.

34. Peter Urbach, *Francis Bacon's Philosophy of Science* (La Salle, Ill.: Open Court, 1987).

35. Gilbert, *De Mundo*, in *De Magnete*, 346n.

36. Zilsel, "Gilbert's Scientific Method."

37. Stephen Pumfrey, "O Tempora, O Magnes!" *BJHS* 22 (1989): 181–214. Idem, "Magnetical Philosophy and Astronomy, 1600–1650," in *The General History of Astronomy*, ed. Michael Hoskin, vol. 2, *Planetary Astronomy from the Renaissance to the Rise of Astrophysics*, Part A: *Tycho Brahe to Newton*, ed. René Taton and Curtis Wilson (Cambridge: Cambridge University Press, 1989), 45–53.

38. Stephen Jay Gould, "The Reverend Thomas's Dirty Little Planet," in *Ever since Darwin* (New York: Norton, 1977), 141–146.

39. Stephen Pumfrey, "William Gilbert's Magnetic Philosophy, 1580–1684: The Creation and Dissolution of a Discipline" (Ph.D. diss., University of London, 1987), 60–66.

40. Henry Power, *Experimental Philosophy: In Three Books Containing New Experiments, Microscopical, Mercurial, Magnetical* (1664; reprint, New York: Johnson, 1966).

41. Stephen Pumfrey, "Mechanizing Magnetism in Restoration England: The Decline of Magnetic Philosophy," *AS* 44 (1987): 5.

42. Ibid., 5–7.

43. Robert Palter, "The Earliest Measurements of Magnetic Force," *Isis* 63 (1972): 554–558.

44. Pumfrey, "Mechanizing Magnetism," 9–15.

45. Athanasius Kircher, *Magnes: Sive de Arte Magnetica,* 3 parts (Coloniae Agrippinae apud Iodocum Kalcoven, 1643).

46. Paula Findlen, *Possessing Nature: Museums, Collecting, and Scientific Culture in Early Modern Italy* (Berkeley: University of California Press, 1994), 85.

47. Athanasius Kircher, *Mundus Subterraneus*, 3rd ed. (Amstelodami: Joannem Jansnssonium a Waesberg & Filios, 1678).

48. Martha Baldwin, "Alchemy and the Society of Jesus in the Seventeenth Century: Strange Bedfellows?" *Ambix* 40, no. 2 (July 1992): 47–54.

49. Ibid., 47.

50. Mario Biagioli, *Galileo, Courtier: The Practice of Science in the Culture of Absolutism* (Chicago: University of Chicago Press, 1993), 347–348.

51. Baldwin, "Anti-Copernican Polemic," 155.

52. Findlen, *Possessing Nature*, 381, 96, 237, 131.

CHAPTER 9   ***Kepler and the Magnetic Force Model of the Solar System***

1. Caspar, *Kepler*, 35–36.
2. Johannes Kepler, *Mysterium Cosmographicum: The Secret of the Universe*, trans. A. M. Duncan (New York: Abaris, 1981).
3. Johannes Kepler, *The Six-Cornered Snowflake*, trans. Colin Hardie; intro. Lancelot Law Whyte (Oxford: Oxford University Press, 1966.)
4. V. I. Arnold, *Theory of Singularities and Its Applications* (Cambridge: Cambridge University Press, 1991), 27–29.
5. J. V. Field, *Kepler's Geometrical Cosmology* (Chicago: University of Chicago Press, 1988), 38.
6. Robert S. Westman's phrase.
7. Owen Gingerich, ed., *The Eye of Heaven: Ptolemy, Copernicus, Kepler* (New York: American Institute of Physics, 1993), 289.
8. Blumenberg, *Genesis of Copernican World*, 296–297.
9. Caspar, *Kepler*, 63.
10. Ibid., 62.
11. Ibid., 63.
12. Arthur Koestler, *The Watershed* (Garden City, N.Y.: Doubleday, 1960), 48–49.
13. Rosen, who, throughout his career polemically attempted to play down as much as possible any "unscientific" hermetic influences on Copernicus or Newton, collects many passages of negative comments by Kepler concerning astrology and mysticism, but is forced to admit Kepler's belief in a special form of astrology. See the apposite remark concerning Rosen on Copernicus in Hallyn, *Poetic Structure of the World*, 313: "One series of Rosen's articles, all presented in the form of a question, 'Was Copernicus an X?' and all leading to a negative conclusion, illustrates the will to reduce the speaking subject to an empty zero state."
14. Edward Rosen, "Kepler's Attitude toward Astrology and Mysticism," in Vickers, *Occult and Scientific Mentalities*, 253–272, quote on 265.
15. Ernst Cassirer, "Giovanni Pico della Mirandola (Part II)," *JHI* 3 (1942): 343 n. 73.
16. Galileo Galilei, *Sidereus Nuncius; or, The Sidereal Messenger* (Chicago: University of Chicago Press, 1989), 31–32.
17. J. V. Field, "A Lutheran Astrologer, Johannes Kepler," *AHES* 31 (1984): 223.
18. Caspar, *Kepler*, 342.
19. Rosen, "Mysticism," 261–262.
20. Quoted in ibid., 263–264.
21. Gérard Simon, "Kepler's Astrology: The Direction of Reform," in Arthur Beer and Peter Beer, eds., *Kepler: Vistas in Astronomy* 18 (1975): 439–448.
22. Field, "Lutheran Astrologer," 220.
23. Otto Neugebauer, *The Exact Sciences in Antiquity* (New York: Harper, 1969), 171, cited in J. Bruce Brackenridge, Foreword to Johannes Kepler, "Johannes Kepler on the More Certain Fundamentals of Astrology," *Proceedings of the American Philosophical Society* 123 (1979): 85.
24. Simon, "Astrology," 448.
25. Rosen, "Mysticism," 266, 271.
26. Kepler, "Fundamentals of Astrology," 97.
27. Field, "Lutheran Astrologer," 224.
28. Job Kozhamthadam, *The Discovery of Kepler's Laws: The Interaction of Science, Philosophy and Religion* (Notre Dame, Ind.: University of Notre Dame Press, 1994), 20.
29. Simon, "Astrology," 446.

30. Field, "Lutheran Astrologer," 225; idem, *Kepler's Geometrical Cosmology*, 165.
31. Tycho's skeleton showed traces of green on the nasel area of his skull. Of course, copper or brass might have been used along with silver and gold, or a brass nose might have been exaggerated into a gold one in later tales. See C. Doris Hellman "Brahe, Tycho," in *DSB*, 402.
32. Koestler, *Watershed*, 101.
33. Letter to Maestlin, February 26, 1599 (Kozhamthadam, *Kepler's Laws*, 88).
34. R.J.W. Evans, *Rudolf II and His World* (Oxford: Oxford University Press, 1973).
35. On the culture of collectors and cabinets, see Findlen, *Possessing Nature*; and Kenseth Joy, ed., *The Age of the Marvelous* (Hanover, N.H.: Hood Museum of Art, 1991).
36. Edward Kelley, *The Alchemical Writings of Edward Kelley* (1676; reprint, Klima, Mont.: Kessenger, n.d.), xxxviii–xxxix. Evans, *Rudof II*, 227.
37. Evans, *Rudolf II,* 154–156.
38. Johannes Kepler, *Kepler's "Somnium: The Dream; or, Posthumous Work on Lunar Astronomy,"* trans. Edward Rosen (Madison: University of Wisconsin Press, 1967).
39. Carl Sagan, *Cosmos* (New York: Harper and Row, 1980), 65–67.
40. Trans. in Koestler, *Watershed.*, 220. Johannes Kepler, *The Harmonies of the World, Book 5*, trans. Charles Glenn Wallis, vol. 16 of *Great Books of the Western World*, ed. Robert Maynard Hutchins (Chicago: Encyclopaedia Britannica, 1952), 1010.
41. Several historians claim that Newton did not have to wade through Kepler's musical fantasies to find the third law, but that Newton read an English summary of Kepler's results. Westfall writes that in 1664 Newton took notes from Thomas Streete, *Astronomia carolina* (1661) (Westfall, *Never at Rest*, 94). Curtis Wilson doubts that Newton ever read Kepler's *Astronomia Nova* with its elaborate calculations (Curtis Wilson, "The Newtonian Achievement in Astronomy," in Taton and Wilson, *Planetary Astronomy*, 238). On Jeremiah Horrocks, who developed Kepler's most technical work in England, see Curtis Wilson, "Predictive Astronomy in the Century after Kepler," in Taton and Wilson, *Planetary Astronomy*, 166–171.
42. Gingerich, *Eye of Heaven*, 51.
43. Field, "Lutheran Astrologer," 205.
44. Field, "Kepler's Rejection of Numerology," 275–290; and idem, *Kepler's Geometrical Cosmology.*
45. Field, "Kepler's Rejection of Numerology," 275–290, 279.
46. "Many students of music theory, from Mersenne to Euler, Padre Martini and Marpurg, were cognizant of his musical learning," as was astronomer and composer William Herschel (Eric Werner, "The Last Pythagorean Musician: Johannes Kepler," in *Aspects of Medieval and Renaissance Music: A Birthday Offering to Gustave Reese*, ed. Jan La Rue [New York: Norton, 1966], 867–882).
47. Field, "Lutheran Astrologer," 209.
48. J. V. Field, "Kepler's Star Polyhedra," *Vistas in Astronomy* 23 (1980): 109–141, contains a translation of book 2 of *Harmonies of the World*. Branko Grünbaum and G. C. Shepard, *Tilings and Patterns* (New York: W. H. Freeman, 1977).
49. J.L.E. Dreyer, *A History of Astronomy from Thales to Kepler*, 2nd ed. (New York: Dover, 1953), 406–409.
50. Field, *Kepler's Geometrical Cosmology*, 150.
51. Victor G. Szebehely, *Adventures in Celestial Mechanics* (Austin: University of Texas Press, 1989), 135. Also see E. L. Stiefel, and G. Scheifele, *Linear and Regular Celestial Mechanics* (New York: Springer Verlag, 1971), 265.
52. Laurence G. Taff, *Celestial Mechanics: A Comprehensive Guide for the Practitioner* (New York: John Wiley, 1985), 299.

53. Charles Yvon Le Blanc, "The Idea of Resonance (Kan-Ying) in the Huai nan Tzu," Ph.D. diss., University of Pennsylvania, 1978; published as *Huai-nan-tzu: A Philosophical Synthesis of Early Han Thought* (Hong Kong: Hong Kong University Press, 1985).

54. Johannes Kepler, "Preface to the Rudolphine Tables," trans. Owen Gingerich and William Walderman, *Quarterly Journal of the Royal Astronomical Society* 13 (1972): 360–373. Koestler, *Watershed*, 236–241; Caspar, *Kepler*, 319–326.

55. Ernst Cassirer, "Mathematical Mysticism and Mathematical Science," in *Galileo, Man of Science*, ed. Ernan McMullin (Princeton: Scholar's Bookshelf, 1988), 338–351.

56. Debus, *Man and Nature*, 123–124.

57. Kozhamthadam, *Kepler's Laws*, 53. Easlea, *Witch Hunting*, 102. Caspar, *Kepler*, 292. Wolfgang Pauli, "The Influence of the Archetypal Ideas on the Scientific Theories of Kepler," in *Essays: Writings on Physics and Philosophy*, ed. Charles Enz and Karl von Meyenn (Berlin: Springer Verlag, 1994), 251, 253.

58. Cassirer, "Mathematical Mysticism," 349.

59. Werner Heisenberg, *Physics and Beyond: Encounters and Conversations* (New York: Harper and Row, 1971), 27.

60. Arthur Koestler, *The Roots of Coincidence* (New York: Vintage, 1973), 89–91, 98–100.

61. Robert S. Westman, "Nature, Art and Psyche: Jung, Pauli, and the Kepler-Fludd Polemic," in Vickers, *Occult and Scientific Mentalities*, 207–220.

62. Pierre Duhem, *To Save the Phenomena* (Chicago: University of Chicago Press, 1969), 5.

63. A. A. Long, "Ptolemy on the Criterion: An Epistemology for the Practising Scientist," in *The Question of Eclecticism: Studies in Later Greek Philosophy*, ed. John Dillon and A. A. Long (Berkeley: University of California Press, 1989). Surprisingly, Liba Chaia Taub, in *Ptolemy's Universe* (La Salle, Ill.: Open Court, 1993), a work wholly devoted to Ptolemy's philosophy of astronomy, does not discuss the skeptical essay attributed to Ptolemy.

64. See Duhem, *To Save*, although this distinction has come in for criticism with regard to several medieval astronomers.

65. Robert Palter, "Interpreting the History of Ancient Astronomy," *SHPS* 1, no. 1 (1970): 93–133, esp. 113–114. Palter traces the confusion back to Berry's *Short History of Astronomy* (1898), followed by E. A. Burtt in 1925 and by Herbert Butterfield in 1949.

66. Nicholas Copernicus, *On the Revolutions*, trans. Edward Rosen (Baltimore: Johns Hopkins University Press, 1978), 22.

67. See Palter, "Interpreting." Kuhn, *Copernican Revolution*.

68. Copernicus, *Revolutions*, bk. 1, chap. 9, cited in Duhem, *Aim and Structure*, 226.

69. Galilei, *Dialogue*, 461–462n.

70. Duhem, *Aim and Structure*, 240.

71. Ibid.

72. The *Echeneis* is the remora, or sucker-fish. The *Torpedo* (an electric fish) and the magnet, two classical examples of occult powers, were both electromagnetic. See Copenhaver, "Natural Magic," 278.

73. Duhem, *Aim and Structure*, 225.

74. Owen Gingerich, "Kepler," in *DSB*, 310. Robert Small, *An Account of the Astronomical Discoveries of Johannes Kepler* (1820; reprint, Madison: University of Wisconsin Press, 1963).

75. Bruce Stephenson, *Kepler's Physical Astronomy* (New York: Springer Verlag, 1987), 2, and passim.

76. Johannes Kepler, *New Astronomy*, trans. William H. Donahue (Cambridge: Cambridge University Press, 1992), 169.

77. Norwood Russell Hanson, "The Copernican Disturbance and the Keplerian Revolution," *JHI* 22 (1961): 169–184.

78. Crombie, *Medieval*, 2: 191; Dijksterhuis, *Mechanization*, 310.

79. Koestler, *Watershed*, 158; Kepler, *Mysterium Cosmographicum*, chap. 20.

80. Dijksterhuis, *Mechanization*, 310; Richard Westfall, *The Construction of Modern Science: Mechanisms and Mechanics* (Cambridge: Cambridge University Press, 1977), 12.

81. Koestler, *Watershed*, 158; Kepler, *New Astronomy*, 170.

82. Kepler, *New Astronomy*, 169.

83. Ibid., 170.

84. Ibid. Koestler, *Watershed*, 158, runs this quotation together with a striking one that is many pages distant in the text.

85. Stephenson, *Physical Astronomy*, 120. Kepler, *New Astronomy*, 560–567.

86. Kepler, *New Astronomy*, 570.

87. Ibid., 569.

88. Gerald Holton, "Kepler's Universe: Its Physics and Metaphysics," in *Thematic Origins of Scientific Thought: Kepler to Einstein* (Cambridge: Harvard University Press, 1973), 72.

89. Kepler, *New Astronomy*, 379.

90. Ibid., 386.

91. Stephenson, *Physical Astronomy*, 72–73; Kepler, *New Astronomy*, 398. Stephenson suggests that this passage misled commentators to think that the moving force is in a *single* plane, and thus decreases as the inverse, rather than as the inverse square as in Newton, but it occupies a stack of parallel planes between Earth's poles.

92. Dreyer, *History of Astronomy*, 398 n. 4.

93. Kepler, *New Astronomy*, 549–550.

94. Eric J. Aiton, *The Vortex Theory of Planetary Motions* (New York: American Elsevier, 1972).

95. David C. Lindberg, "The Genesis of Kepler's Theory of Light: Light Metaphysics from Plotinus to Kepler," *Osiris*, 2nd ser., 2 (1986): 5–42.

96. Koestler, *Watershed*, 156–157.

97. This account is given by Koestler, following Jean Delambre's early-nineteenth-century praise of Kepler's account of gravity.

98. Dijksterhuis, *Mechanization*, 315.

99. Karin Figala, "Kepler and Alchemy," in Beer and Beer, *Kepler*, 547–469, 462.

100. Dijksterhuis, *Mechanization*, 315.

101. Kepler, *New Astronomy*, 55.

102. Ibid., 56.

103. Westfall, *Construction*, 9.

104. Stephenson, *Physical Astronomy*, 4–7.

105. Fritz Krafft, "Kepler's Contribution to Celestial Physics," in Beer and Beer, *Kepler*, 571–572.

106. Dreyer, *History of Astronomy*, 399–400; Kepler, *Somnium*, 123 n. 202.

107. Kozhamthadam, *Kepler's Laws*, 85; Johannes Kepler, *Epitome of Copernican Astronomy, Book 4*, in Hutchins, *Great Books of the Western World*, 16: 850.

108. Kozhamthadam, *Kepler's Laws*, 84.

109. Ibid., 16.

110. Bennett, "Cosmology," 165–177.

111. Peter Medawar, "Is the Scientific Paper Fraudulent?" *Saturday Review*, August 1, 1964.

112. Ibid.; and idem, *The Art of the Soluble* (New York: Barnes and Noble, 1968), 7.

113. Charles S. Peirce, *Collected Papers*, 6 vols., ed. Charles Hartshorne and Paul Weiss (Cambridge: Harvard University Press, 1931–1935), 1: 73; Norwood Russell Hanson, *Patterns of Discovery* (Cambridge: Cambridge University Press, 1958), 70–85.

114. Robert Newton, *The Crime of Claudius Ptolemy* (Baltimore: Johns Hopkins University Press, 1977).

115. A. Aaboe and D. J. Price, "Qualitative Measurement in Antiquity," in *L'Aventure de la Science*, vol. 2 of *Mélanges Alexandre Koyré* (Paris: Hermann, 1964), 1–20; Palter, "Interpreting," 123–126; Gingerich, *Eye of Heaven*, 55–73, 74–80.

116. Heinrich von Staden, University of Texas, Austin, April 15, 1971.

117. Hanson, *Patterns*, 74.

118. Marjorie Hope Nicolson, *The Breaking of the Circle: Studies in the Effect of the "New Science" on Seventeenth Century Poetry* (New York: Columbia University Press, 1960).

119. John Donne, "First Anniversary," ll. 205–214.

120. Art historian Abby Warburg recovered from a mental breakdown when philosopher Ernst Cassirer reassured him that an idea that obsessed Warburg—that the transition from circle to ellipse had affected all modern thought—was legitimate. Cassirer quoted from memory in Latin a lengthy passage from Kepler on the ellipse, and Warburg became confident of his sanity (H. Saxl, "Ernst Cassirer," in *The Philosophy of Ernst Cassirer*, ed. Paul Arthur Schilpp [Evanston, Ill.: Library of the Living Philosophers, 1949], 49–50).

121. Hanson, *Patterns,* 78.

122. I have not seen this hypothesis in the literature on Kepler, although Koestler mentions the possibility of "some unconscious biological bias," in part because of Kepler's wife's giving birth (Koestler, *Watershed*, 144).

123. Kuno Fladt, "Das Keplerische," *Elemente der Mathematik* 17 (1962): 73–78.

124. Hanson, *Patterns*, 82, has "truth" not nature. Koestler, *Watershed*, 221–222, portrays her as a teasing hussey who gives herself to him just as he was about to give up, as also he quotes Virgil in the *Harmonies of the World*, bk. 5, chap. 3, prop. 8, when Kepler discovers the third law. The poetry quoted with third law has no teasing hussey, but an "unskilled man" (see *Harmonies of the World*, bk. 5,). Koestler mixes the unromantic Virgil quotation here with the Keplerian outburst concerning the third law and the romantic Virgil quotations in the *New Astronomy*.

125. Kepler, *New Astronomy,* 573.

126. Ibid., chap. 58, 575, 576, quoted by Koestler, *Watershed*, 147, and Hanson, *Patterns*, 202, 83. Koestler gives "what a foolish bird I have been" for "ridiculous me."

127. Boas, *Scientific Renaissance*, 287.

128. Rosen, "Mysticism," 267.

129. Edwin Arthur Burtt, *The Metaphysical Foundations of Modern Physical Science*, rev. ed. (Garden City, N.Y.: Doubleday Anchor, 1954), 58, 60. Debus, *Man and Nature*, 92. Gingerich, *Eye of Heaven*, 51.

130. Edward W. Strong, *Procedures and Metaphysics: A Study in the Philosophy of Mathematical-Physical Science in the Sixteenth and Seventeenth Centuries* (Berkeley: University of California Press, 1936).

131. Stephenson, *Physical Astronomy*, Preface.

132. Bruce Stephenson, *The Music of the Heavens: Kepler's Harmonic Astronomy* (Princeton, N.J.: Princeton University Press, 1994), 8–9, 195, 202, 249–252.

133. E. J. Aiton, "Johannes Kepler in the Light of Recent Research," *HS* 14 (1976): 77.

134. Kozhamthadam, *Kepler's Laws*, 13–14.

135. Field, "Kepler's Rejection of Numerology"; and idem, *Kepler's Geometrical Cosmology*, 144, 182.

136. Cassirer, "Mathematical Mysticism," 342–351.

137. Field, *Geometrical Cosmology*, 123.

138. Burtt, *Metaphysical Foundations*, 58.

139. Ibid., 62.

140. Abraham Maslow, *Motivation and Personality*, 2nd ed. (New York: Harper and Row, 1970), 164–165.

CHAPTER 10   *Newton: Alchemy and Active Principles*

1. Derek Gjertsen, *The Newton Handbook* (London: Routledge and Kegan Paul, 1986), 10–11. The low estimate is based upon the 1936 Sotheby's auction of the Portsmouth Collection. The high estimate is made by Richard S. Westfall, "Newton and Alchemy," in Vickers, *Occult and Scientific Mentalities*, 330. Maynard Keynes estimated 100,000 words on Newton's own alchemical experiments, as opposed to copies of or commentary on older alchemists.

2. Richard S. Westfall, "The Role of Alchemy in Newton's Career," in Bonelli and Shea, *Reason, Experiment, and Mysticism*, 189.

3. P. M. Rattansi, "Newton's Alchemical Studies," in *Science, Medicine and Society in the Renaissance: Essays to Honor Walter Pagel*, 2 vols., ed. Allen G. Debus (New York: Neale Watson, 1972), 167. Also cited in Westfall, "Role of Alchemy," 189.

4. Keynes, "Newton the Man," 311.

5. Ibid., 313, 319, 311.

6. Marie Boas and A. R. Hall, "Newton's Alchemical Experiments," *Archives Internationales d'Histoire des Sciences* 9 (1958): 113–152.

7. Betty Jo Teeter Dobbs, *The Foundations of Newton's Alchemy* (Cambridge: Cambridge University Press, 1975); and idem, *Janus Faces*.

8. Karin Figala, "Newton as Alchemist," *HS* 15 (1977): 102–137.

9. Dobbs, *Foundations*, 88. Westfall, *Never at Rest*, 90, 294.

10. But see Peter Spargo, "Newton's Chemical Experiments: An Analysis in the Light of Modern Chemistry," in *Action and Reaction*, ed. Paul Theerman and Adele F. Seef, (Newark: University of Delaware Press, 1993), 123–143, although the latter relates Newton's chemistry solely to modern chemistry, and not to Newton's own alchemical interests.

11. Golinski, "Secret Life of an Alchemist," in Fauvel et al., *Let Newton Be!* 164–166.

12. John L. Brooke, *The Refiner's Fire: The Making of Mormon Cosmology, 1644–1844* (Cambridge: Cambridge University Press, 1994), 107–111.

13. Frank E. Manuel, *A Portrait of Sir Isaac Newton* (Cambridge, Mass.: Harvard University Press, 1968), 234–235.

14. Danton B. Sailor, "Moses and Atomism," *Isis* 25 (1964): 3–16.

15. Lancelot Law Whyte, *The Unconscious before Freud* (New York: Doubleday Anchor, 1960).

16. J. E. McGuire and P. M. Rattansi, "Newton and the 'Pipes of Pan'," *Notes and Records of the Royal Society* 21 (1966): 108–143.

17. Letter, November 23, 1646, in *The Philosophical Writings of Descartes*, vol. 3, *The Correspondence*, trans. John Cottingham, Robert Stoothoff, Dugald Murdoch, and Anthony Kenny (Cambridge: Cambridge University Press, 1991), 304.

18. Brian P. Copenhaver, "Jewish Theologies of Space in the Scientific Revolution: Henry More, Joseph Raphson, Isaac Newton and Their Predecessors," *AS* 37 (1980): 489–548.

19. Max Jammer, *The Concept of Space* (New York: Harper, 1960), 27–29.

20. Copenhaver and Schmitt, *Renaissance Philosophy*, 168–174.

21. Anne Conway, *Principles of the Most Ancient and Modern Philosophy*, trans. Allison Coudert and Taylor Corse (Cambridge: Cambridge University Press, 1996).

22. Carolyn Merchant, in "The Vitalism of Frances Mercury van Helmont: Its Influence on Leibniz," *Ambix* 26 (1979): 170–183, attributes the source of the term *monad* to Anne Conway, but George H. R. Parkinson, in *Leibniz: Philosophical Writings* (London: Dent, 1973), 255, notes a usage of *monas* a year earlier. This is cited by Donald Rutherford, "Metaphysics: The Late Period," in *The Cambridge Companion to Leibniz*, ed. Nicholas Jolley (Cambridge: Cambridge University Press, 1995), 166 n. 24.

23. Merchant, *Death of Nature*, 258–267. Allison P. Coudert, *Leibniz and the Kabbalah* (Boston: Kluwer, 1995).

24. Margaret Jacob, *The Newtonians and the English Revolution, 1689–1720* (Ithaca, N.Y.: Cornell University Press, 1976), 34–35.

25. Westfall, "Newton and Alchemy," 331.

26. John E. McGuire, "Neo-Platonism, Active Principles and the Corpus Hermeticum," in *Hermeticism and the Scientific Revolution: Papers Read at a Clark Library Seminar, March 9, 1974*, ed. Robert S. Westman and J. E. McGuire (Los Angeles: Clark Memorial Library and University of California Press, 1977).

27. Manuel, *Portrait*, 392; and the title of Dobbs, *Janus Faces*.

28. N. R. Hanson, "Hypotheses *Fingo*," Boston Colloquium for the Philosophy of Science, October 10, 1966.

29. Manuel, *Portrait*, 352, 360.

30. Gjertsen, *Handbook*, 87, 90. Westfall, *Never at Rest*, 540.

31. Gjertsen, *Handbook*, 91–92.

32. Burtt, *Metaphysical Foundations*.

33. Robert M. Palter, "Saving Newton's Text: Documents, Readers, and the Ways of the World," *SHPS* 18, no. 4 (1987): 385–439, on disparaging comparisons of Newton to other early modern philosophers.

34. The extreme of this approach can be found in John O. Wisdom, *The Unconscious Origins of Berkeley's Philosophy* (London: Hogarth Press, 1953), where it is argued that Bishop Berkeley's denial of the existence of matter was really motivated by a rejection of his feces.

35. Westfall, *Never at Rest*, 357.

36. Ibid., 529–530.

37. Ibid., 296.

38. Merchant, *Death of Nature*, chap. 8; and Easlea, *Witch Hunting*, chap. 3.

39. Manuel, *Portrait*, 380.

40. Ibid., 390–392.

41. H. G. Alexander, ed., *The Leibniz-Clarke Correspondence* (New York: Barnes and Noble), 1984, 19. On the Socinians and the origins of Unitarianism, see Stanilas Kot, *Socinianism in Poland* (Boston: Starr King, 1957).

42. Alexander, *Leibniz-Clarke* 11, 17–18.

43. Manuel, *Portrait*, 312–313.

44. This latter point was made elegantly in a technical sense much later by Heinrich Hertz, a German physicist. He claimed that physical theories are to be identified with the equations and are to be distinguished from the "gay garment" of models or interpre-

tations. Hertz attempted to replace the dispositional concept of force with purely observable positions and motions of particles. Ironically, he succeeded in eliminating the concept of force only to find that he had to add imaginary "particles" that could never be physically observed. Hertz thought this a triumph of positivist method, but it merely shows that it is impossible to eliminate nonobservable entities from Newtonian mechanics (Heinrich Hertz, *The Principles of Mechanics Presented in a New Form* [1899; reprint, New York: Dover, 1956]).

45. Ernan McMullin, *Newton on Matter and Activity* (Notre Dame, Ind.: University of Notre Dame Press, 1978), 79.
46. Descartes, *Correspondence*, 219, 227–228.
47. Westfall, *Never at Rest*, 308.
48. Ibid., 644.
49. McMullin, *Newton on Matter*, 85.
50. Westfall, *Never at Rest*, 528.
51. Immanuel Kant, *Opus Posthumum*, ed. Eckhart Föster; trans. Eckhart Föster and Michael Rosen (Cambridge: Cambridge University Press, 1993).
52. McMullin, *Newton on Matter*, 100.
53. J. E. McGuire, "Force, Active Principles, and Newton's Invisible Realm," *Ambix* 15 (1968): 184–185. P. M. Heimann, "'Nature Is a Perpetual Worker': Newton's Aether and Eighteenth Century Natural Philosophy," *Ambix* 20 (1973): 7.
54. I owe this suggestion to Willem de Vries.
55. Westfall, *Never at Rest*, 495.
56. Alexander, *Leibniz-Clarke*, 11–12.
57. Freudenthal, *Atom and Individual*, 182.
58. Jacob, *Newtonians in the Revolution* 14.
59. Charles Webster, "New Light on the Invisible College," *Transactions of the Royal Historical Society* 24 (1974): 19–42. Webster, *Great Instauration*, 57–67.
60. Nicholas Clulee, *John Dee's Natural Philosophy: Between Science and Religion* (London: Routledge, 1988), 182. Dee, *Arte of Navigation*.
61. J. B. Bury, *The Idea of Progress* (New York: Dover, 1955), 93–97.
62. Hill, *Intellectual Origins*, 139.
63. Robert Hugh Kargon, *Atomism in England from Hariot to Newton* (Oxford: Clarendon, 1966), 31–32, 18.
64. Freudenthal, *Atom and Individual*, 75.
65. Charles Webster, personal communication.
66. Jacob, *Cultural Meaning*, 94–95. W. R. Albury, "Halley's Ode on the *Principia* of Newton and the Epicurean Revival in England," *JHI* 39 (1978): 24–43.
67. Jacob, *Newtonians in the Revolution*, 116.
68. Gould, *Ever since Darwin*, 145. Idem, *Time's Arrow, Time's Cycle* (Cambridge: Harvard University Press, 1987), chap. 2.
69. Westfall, *Never at Rest*, 652.
70. James L. Axtell, "Thomas Scargill and the Mechanics of Opposition," 1966, manuscript sent to me by author.
71. Jacob, *Cultural Meaning*, 96.
72. Jacob, *Newtonians in the Revolution*, 168.
73. Ibid., 189.
74. J. T. Desaguliers, *The Newtonian System of the World as the Best Model for Government: An Allegorical Poem* (London: A. Campbell for J. Roberts, 1728).
75. Ibid.

CHAPTER 11   *Interlude: Leibniz and China*

1. Joseph Agassi, "Leibniz's Place in the History of Physics," *JHI* 30 (1969): 331–344.
2. Christina Mercer and R. C. Sleigh, Jr., "Metaphysics: The Early Period to the *Discourse on Metaphysics*," in Jolley, *Cambridge Companion*, 67–68. On Leibniz's efforts at religious and political reconciliation, see Leroy E. Loemker, *The Struggle for Synthesis: The Seventeenth Century Background of Leibniz's Synthesis of Order and Freedom* (Cambridge: Harvard University Press, 1972); and R. W. Meyer, *Leibnitz and the Seventeenth Century Revolution* (Cambridge, Eng.: Bowes and Bowes, 1952). Mercer and Sleigh mention neither of these works. Loemker claimed that reconciliation of Catholics and Protestants lay behind Leibniz's metaphysics as developed in the *Discourse*. Leroy Loemker, "A Note on the Origin and Problem of Leibniz's Discourse of 1686," *JHI* 8 (1947): 449–466.
3. Meyer, *Revolution,* 133.
4. Donald F. Lach, "Leibniz and China," *JHI* 6 (1945): 442.
5. Meyer, *Revolution*, 193 n. 168. Leibniz met with the Jesuit Grimaldi.
6. Daniel J. Cook, "Metaphysics, Politics and Ecumenism: Leibniz' *Discourse on the Natural Theology of the Chinese*," *Studia Leibnitiana* 42 (1981): 159.
7. Gottfried Wilhelm Leibniz, "Preface to the *Novissima Sinica*," in *Writings on China*, ed. Daniel J. Cook and Henry Rosemont, Jr. (Chicago: Open Court, 1994), 51.
8. Jonathan Spence, *Emperor of China* (New York: Vintage, 1975), constructed from writings of K'ang Hsi, includes reflections on medicine and natural philosophy as well as on political and autobiographical matters. The chapter "Thinking" includes his reflections on the importance of inquiry and experience (65–68), as well as his studies of Western mathematics (72–73); the chapter "Growing Old" collects reflections on medicine (95–100), and witty and skeptical remarks about Taoist claims concerning the achievement of physical immortality (101–102.).
9. Cook and Rosemont, Introduction, to Leibniz, *Writings on China*, 2.
10. Joseph Needham, *Science and Civilization*, 2: 496–505.
11. Daniel J. Cook and Henry Rosemont, Jr., "The Pre-Established Harmony between Leibniz and Chinese Thought," *JHI* 42 (1981): 253–267, persuasively argue that Leibniz came upon systematic Chinese Neo-Confucian scholasticism only at the end of his career, and only in the forms misinterpreted by Jesuit commentators, such that his own similar system was not derived from Chinese thought.
12. Donald F. Lach, *Asia in the Making of Europe*, vol. 1, *The Century of Discovery*, pt. 3, *China* (Chicago: University of Chicago Press, 1965), 731–738.
13. Matteo Ricci, *China in the Sixteenth Century: The Journals of Matthew Ricci, 1583–1610*, trans. Louis J. Gallagher; ed. Nicholas Trigault (New York: Random House, 1953). Henri Bernard, *Matteo Ricci's Scientific Contribution to China*, trans. Edward Chalmers Werner (1935; reprint, Westport, Conn.: Hyperion, 1973). Jonathan Spence, *The Memory Palace of Matteo Ricci* (New York: Viking Penguin, 1984).
14. Maverick, *China*.
15. On the fascinating ironies of the Jesuit introduction of astronomy to China, see Joseph Needham, *Science and Civilization*, 3: 437–461; and Ronan, *Shorter Science*, 2: 212–221.
16. Henderson, *Development and Decline*, chaps. 8, 9.
17. Stephen Toulmin, *Cosmopolis: The Hidden Agenda of Modernity* (New York: Free Press, 1990).
18. Pasquale D'Elia, *Galileo in China* (Cambridge: Harvard University Press, 1960), emphasized the extent to which ideas of Copernicus, Galileo, and Tycho were introduced

by the Jesuits in China. Nathan Sivin argues persuasively that after the condemnation of Galileo became known to Jesuit missionaries, Copernicanism could only be presented in a confusing and misleading manner, if at all. Although Galileo's telescopic observations were introduced, their significance could not be mentioned. This led to mistaken views by the Chinese of Copernicus's claims, and when Copernicanism was introduced in the eighteenth century, previous misrepresentations reinforced Chinese skepticism about realistic cosmological models. See Sivin, "Copernicus in China," in *Science in Ancient China*, 4: 1–53.

19. Ronan, *Shorter Science*, 2: 205.
20. Ibid., 211–212.
21. On the rites controversy, see Donald Lach and Edwin J. van Kley, *Asia in the Making of Europe*, vol. 3, *A Century of Advance*, bk. 1, *Trade, Missions, and Literature* (Chicago: University of Chicago Press, 1993), 267–269, 385–386, 423–430.
22. Athanasius Kircher, *China Illustrated*, trans. Charles D. Van Tuyl from the 1677 original Latin edition (Muskogee, Okla.: Indian University Press, 1987). Paolo Rossi, "The Twisted Roots of Leibniz' Characteristic," in *The Leibniz Renaissance* (Florence: Leo S. Olschki Editore, 1989), 272–277.
23. Daniel J. Cook and Henry Rosemont, Jr., Introduction to Leibniz, *Writings on China*, 16, referring to Bouvet's letter to Leibniz November 4, 1701, found in *Leibniz korrespondiert mit China* ed. R. Widmaier (Frankfurt: V. Klosterman, 1990), 147–163.
24. For instance, Hellmut Wilhelm, in *Heaven, Earth, and Man in the Book of Changes* (Seattle: University of Washington Press, 1977), 8.
25. Lao Tzu, *Tao Te Ching*, chap. 42, ll. 1–4.
26. David Mungello, *Leibniz and Confucianism: The Search for Accord* (Honolulu: University of Hawaii Press, 1977), 48.
27. See John Wheeler's speculative meditations on the deduction of space from logic, "Pregeometry as a Calculus of Propositions," at the end of Charles W. Misner, Kip S. Thorne, and John Archibald Wheeler, *Gravitation* (San Franciso: Freeman, 1973), 1208–1209, 1211–1212. George Gale noted the similarity to Leibniz.
28. Genesis 2: 19–20.
29. Genesis 11: 1–9. David Rosenberg, trans., *The Book of J* (New York: Vintage, 1990), 73.
30. Paolo Rossi, "Twisted Roots," 273.
31. Rita Widmaier, *Die Rolle der Chinesischien Schrift in Leibniz' Zeichtheorie* (Wiesbaden: Franz Steiner Verlag, 1983), 40–64, discusses the senses of analogy that are appropriate.
32. It turns out that the diagram that Bouvet sent Leibniz, who elaborated his conception of its relation to binary arithmetic, was not truly a product of the ancient sage Fu Hsi, but of an eleventh-century Neo-Confucian thinker, Shao Yung (Shao K'ang-chie) (A. J. Aiton and Eikoh Shimao, "Gorai Kinzo's Study of Leibniz and the *I Ching* Hexagrams," *AS* 36 [1981]: 82). On Shao Yung, see Anne D. Birdwhistell, *Transition to Neo-Confucianism: Shao Yung on Knowledge and Symbols of Reality* (Stanford, Calif.: Stanford University Press, 1989).
33. G. W. Leibniz, *Discourse on the Natural Theology of the Chinese*, in *Writings on China*. My account largely follows that of Cook and Rosemont.
34. Cook and Rosemont,"Pre-Established Harmony," 253–267, claim that Joseph Needham is wrong to say that Neo-Confucianism influenced Leibniz's later organic doctrines, as these were developed before Leibniz encountered Neo-Confucianism.

35. Donald Rutherford, "Metaphysics: The Late Period," in *Cambridge Companion*, 126, 130.
36. Cook and Rosemont, Introduction to Leibniz, *Writings on China*, 2.
37. Ibid., 11, citing Lach, "Leibniz and China," 436.
38. Cook and Rosemont, "Pre-Established Harmony," 260–261.
39. Ibid., 253–267.
40. Lach, "Leibniz and China," 452.
41. These reflections were occasioned by a conversation with Hisa Kuriyama.
42. Wing-Tsit Chan, *Source Book*, 638.
43. Ibid., 412, 639.
44. Nicholas Malebranche, *Dialogue between a Christian Philosopher and a Chinese Philosopher on the Existence and Nature of God*, trans. Dominick A. Iorio (Washington, D.C.: University Press of America, 1980). See David E. Mungello, "Malebranche and Chinese Philosophy," *JHI* 41 (1980): 551–578. Mungello suggests that Malebranche's real target in his criticism of Chinese philosophy is Spinozism. See Lewis A. Maverick, "A Possible Chinese Source of Spinoza's Doctrine," *Revue de Littérature Comparée* 19 (1939): 417–428.
45. Mercer and Sleigh, "Metaphysics: The Early Period," 91. G.H.R. Parkinson, ed., *Leibniz, De Summa Rerum: Metaphysical Papers, 1575–1676* (New Haven: Yale University Press, 1992), 25, 29.

CHAPTER 12   ***Philosophical Background of Romantic Science***

1. Lewis Samuel Feuer, *Spinoza and the Rise of Liberalism* (Boston: Beacon, 1958).
2. Ibid., 43.
3. Ibid., 45.
4. Ibid., 49.
5. Marx, *Eighteenth Brumaire*, 17.
6. Feuer, *Spinoza,* 55.
7. Yirmiyahu Yovel, *Spinoza*, vol. 1, *The Marrano of Reason* (Princeton: Princeton University Press, 1989), 20.
8. Ibid., 91, 26.
9. Ibid., 25.
10. Ibid., 43.
11. Ibid., 27.
12. Ibid., 210n.
13. Ibid., 207.
14. Friedrich Heinrich Jacobi, *The Spinoza Conversations between Lessing and Jacobi*, trans. Gérhard Vallée, J. B. Lawson, and C. G. Chapple (Lanham, Md.: University Press of America, 1988). See also idem, *The Main Philosophical Writings and the Novel "Allwill,"* trans., intro. George di Giovanni (Montreal: McGill-Queens University Press, 1994), 173–251, 339–378. Frederick Beiser, *The Fate of Reason* (Cambridge, Mass.: Harvard University Press, 1987) has a stirring account of the controversy (chaps. 2 and 3, 44–126 and 158–164).
15. Beiser, *Fate*, 74.
16. Johann Gottfried von Herder, *God: Some Conversations*, trans. Frederick H. Burkhardt (Indianapolis: Bobbs-Merrill, 1940).
17. Johann Gottfried von Herder, *Outlines of a Philosophy of the History of Man*, trans. T. O. Churchill, 2nd ed., 2 vols. (London, 1803).

18. Herder, *God*, 103.
19. Ibid., 105.
20. Beiser, *Fate*, 153–158.
21. Friedrich Schelling, *On the History of Modern Philosophy*, trans. Andrew Bowie (Cambridge: Cambridge University Press, 1994), 66. Alan White, *Schelling: An Introduction to the System of Freedom* (New Haven: Yale University Press, 1983), presents Schelling as the counterpoint to Spinoza, substituting freedom for determinism but retaining the latter's monism.
22. Lewis Ford, "The Controversy between Schelling and Jacobi," *Journal of the History of Philosophy* 3 (1965): 75–89.
23. George di Giovanni, "The First Twenty Years of Critique: The Spinoza Connection," in *The Cambridge Companion to Kant*, ed. Paul Guyer (Cambridge: Cambridge University Press, 1992), 417–448. Jacobi, "Open Letter to Fichte," trans. Diana I. Behler, in *The Philosophy of German Idealism: Fichte, Jacobi and Schelling*, ed. Ernst Behler (New York: Continuum, 1987), 119–141.
24. E. H. Gombrich, *The Story of Art*, 15th ed. (Oxford: Phaidon, 1989), 376–377.
25. Beiser, *Enlightenment*, 276.
26. Weeks, *Paracelsus*, 24.
27. Friedrich Wilhelm Joseph von Schelling, *Bruno; or, On the Divine Principle of Things*, trans. Michael Vater (Albany: State University of New York Press, 1984), 169–172. On Kepler via Hegel, see G.W.F. Hegel, "On the Orbits of the Planets," *Graduate Faculty Philosophy Journal* 12 (1987): 278–309.
28. Margaret Jacob, *Living the Enlightenment* (Oxford: Oxford University Press, 1991), 45.
29. Yates, *Rosicrucian*, 210; Jacob, *Living*, 37–38.
30. Jacob, *Living*, 36.
31. Gooch, *Germany*, 303; Saine, *Georg Forster*, 28–30.
32. Beiser, *Enlightenment*, 163, 398 n. 47.
33. Saine, *Georg Forster*, 33, Soemmerring was sufficiently fearful of reprisals to ask permission to leave the Rosicrucians (ibid., 161 n. 2).
34. Rolf Christian Zimmerman, *Das Weltbild des jungen Goethe: Elemente und Fundamente*, vol. 1 (Munich: Fink, 1969), cited in Saine, *Georg Forster*, 160 n. 8.
35. Paul Müller, *Untersuchung des Problems der Freimauerei in Lessing, Herder und Fichte*, Sprache und Dichtung, Neue Folge, vol. 12 (Bern: Verlag Paul Haupt, 1965).
36. Johann Gottlieb Fichte, "The Philosophy of Masonry: Letters to Constant," *Masonic Papers* 2, no. 2 (1945): 26–65, also published as Fichte, *The Philosophy of Masonry: Letters to Constant*, trans. Roscoe Pound (Seattle: Masonic Press, 1945).
37. Jürgen Habermas, "Dialektischer Idealismus im Übergang zum Materialismus: Schellings Idee einer Kontraktion Gottes," in *Theorie und Praxis* (Frankfurt: Suhrkamp Verlag, 1972), 184–185, 198–199.
38. Peter Ackroyd, *Blake: A Biography* (New York: Knopf, 1996), 88–90, 147–149.
39. William Blake, "London," no. 46 in *Songs of Innocence and Experience*, ed. Andrew Lincoln (Princeton: Princeton University Press, 1991).
40. *Newton*, Tate Gallery, London.
41. Donald Ault, *Visionary Physics: Blake's Response to Newton* (Chicago: University of Chicago Press, 1974), 2–4. Ault's book elaborates on the complexity and contradicatory attitudes of Blake toward Newton.
42. William Blake, *The Complete Poems*, ed. Patricia Ostriker (London: Penguin Books, 1977), 494.

43. Quoted in J. Bronowski, *Blake and the Age of Revolution* (London: Routledge, 1965), 137.

44. Letter to Thomas Butts, 1801, quoted in Bronowski, *Blake*, 110.

45. Ackroyd, *Blake*, 25.

46. Quoted by Bronowski, *Blake*, 138.

47. Ackroyd, *Blake*, 193.

48. Friedrich Meinecke, *The German Catastrophe*, trans. Sidney B. Fay (1950; reprint, Boston: Beacon, 1963), 120–121.

49. Karl Loewith, *From Hegel to Nietzsche* (New York: Holt, Rinehart and Winston, 1964), 9, 11.

50. I owe this phrase to Karsten Harries.

51. Translated by Frederick M. Barnard, in "Spinozism," in *Encylopedia of Philosophy*, ed. Paul Edwards, 7: 542.

52. Ernst Cassirer, *Kant's Life and Thought* (New Haven, Conn.: Yale University Press, 1981), 273–274. See Cassirer, "Goethe and the Kantian Philosophy," in *Rousseau, Kant, and Goethe* (New York: Harper and Row, 1963), 61–98.

53. Johann Wolfgang von Goethe, *Goethe's Botanical Writings*, trans. Bertha Mueller (Woodbridge, Conn.: Ox Bow Press, 1989).

54. Lisbet Koerner, "Goethe's Botany: Lessons of a Feminine Science," *Isis* 84 (1993): 470–495.

55. This association is evidently present in Paul R. Gross and Norman Levitt, *The Higher Supersition: The Academic Left and Its Quarrels with Science* (Baltimore: Johns Hopkins University Press, 1994), 137–139. The comparison of feminist analyst of science Evelyn Fox Keller with Goethe is clearly intended to be an insult. They allude to Keller's characterization of the science of Barbara McClintock as identifying with the objects of her research: Keller, *A Feeling for the Organism: The Life and Work of Barbara McClintock* (New York: W. H. Freeman, 1983)—and grant this point about a range of variation.

56. See Koerner, "Botany," 494–495.

57. Johann Wolfgang von Goethe, "An Intermaxillary Bone Is Present in the Upper Jaw of Man as Well as in Animals" (Jena 1786), in *Scientific Studies* (New York: Suhrkamp, 1988), 111–116, and plates I and II.

58. Adrian Desmond, *Archetypes and Ancestors: Paleontology in Victorian London, 1850–1875* (Chicago: University of Chicago Press, 1982), 74–75.

59. Rupert Riedl, *Order in Living Organisms* (New York: Wiley, 1978).

60. Stephen Jay Gould and Richard Lewontin, "The Spandrels of San Marco," *Proceedings of the Royal Society* 205 (1979): 581–598.

61. Johann Wolfgang von Goethe, *Theory of Colors*, trans. Charles Lock Eastlake (Cambridge, Mass.: MIT Press, 1970). An improved translation of the work is in *Scientific Studies*.

62. Miles Jackson, "A Spectrum of Belief: Goethe's 'Republic' versus Newtonian Despotism," *Social Studies of Science* 24 (1994): 673–701.

63. Jackson, "Spectrum," 698 n. 50, credits Albrecht Schöne, *Goethes Farbentheologie* (Munich: C. H. Backe, 1987), 45–67, 76–83, for this insight and its documentation.

64. Hermann von Helmoltz, "On Goethe's Scientific Researches" [1853], in *Popular Scientific Lectures* (New York: Dover, n.d.), 1–21; and Werner Heisenberg, "Goethe and Newton on Color," [1941], in *Philosophical Problems of Quantum Physics* (New Haven: Ox Bow Press, 1979), and idem, "Goethe's View of Nature and the World of Science and Technology" [1967], in *Across the Frontiers* (New York: Harper and Row,

1974). Helmholtz is highly negative concerning Goethe's claims to scientific explanation, while Heisenberg is conciliatory.

65. Dennis L. Sepper, *Goethe contra Newton* (Cambridge: Cambridge University Press, 1988), 88–89.

66. Keld Nielsen, "Another Kind of Light: The Work of T. J. Seebeck and His Collaboration with Goethe," *Historical Studies in the Physical and Biological Sciences* 20 (1989): 107–182.

67. Joseph Agassi, quoted from the roundtable session of the Boston Colloquium on Goethe and the Sciences in "Postscript," by editors Frederick Amrine, Francis J. Zucker, and Harvey Wheeler, *Goethe and the Sciences: A Reappraisal* (Dordrecht: D. Reidel, 1987), 373–374.

68. Sepper, *Goethe contra Newton*, 14–16, has emphasized this similarity to legitimate Goethe's work on color.

69. Edwin Land, "Experiments in Color Vision," *Scientific American* 200 (May 1959): 84–99; and idem, "The Retinex Theory of Color Vision," *Scientific American* 237 (December 1977): 108–128.

70. Johann Wolfgang von Goethe, *Elective Affinities*, trans. James Anthony Froude and R. Dillon Boylan (New York: Frederick Ungar, 1962).

71. Torbern Bergman, *Dissertation on Elective Attractions*, trans., intro. J. A. Schufle (1775; reprint, New York: Johnson, 1967).

72. Hans Gerth and C. Wright Mills, Introduction to *From Max Weber* (Oxford: Oxford University Press, 1946), 62–63.

73. Trans. of Weber by Nancy Metzeland and John Flodstrom, in Maurice Merleau-Ponty, *The Primacy of Perception and Other Essays*, ed., James Edie (London: Routledge and Kegan Paul, 1962), 201. Talcott Parsons, Weber's standard English translator and major American social theorist translates the term as "correlation," turning the relationship into a purely external correspondence. Max Weber, *The Protestant Ethic and the Spirit of Capitalism*, trans. Talcott Parsons (New York: Scribner, 1958),91.

CHAPTER 13   *Social Background to Romanticism*

1. Beiser, *Enlightenment*, 386 n. 1.

2. Ibid., chap. 7; Saine, *Georg Forster*.

3. Examples are George Santayana, *Egotism in German Philosophy* (London: J.M. Dent and Sons, 1916), during World War I, and Bertrand Russell, *A History of Western Philosophy* (New York: Simon and Schuster, 1945), 718, during World War II.

4. Beiser, *Enlightenment*, chap. 3, 57–83; and Reinhold Aris, *History of Political Thought in Germany: From 1789 to 1815* (New York: Russell and Russell, 1965), chap. 33, 106–135, and chap. 12, 345–360.

5. Beiser, *Enlightenment*, chap. 8, 189–221; Aris, *History*, chap. 7, pt. 2, 234–249; Gooch, *Germany*, chap. 6, pt. 1, 160–173.

6. Frederick Beiser, Introduction to *The Early Political Writings of the German Romantics* (Cambridge: Cambridge University Press, 1996), xxiii.

7. Beiser, *Enlightenment* chap. 11, 264–278; Aris, *History*, chap. 9, 266–280; Gooch, *Germany*, chap. 9, pt. 2, 234–238. Novalis, "Christianity or Europe: A Fragment," in *The Political Thought of the German Romantics*, ed. Hans Reiss (Oxford: Blackwell, 1955); Beiser, *Early Political Writings*, xxii, 61–79.

8. Beiser, *Enlightenment*, 252.

9. Ibid., 260.

10. Ibid., 278. Jacques D'Hondt, *Hegel in His Time* (Peterborough, Ont.: Broadview, 1988), documents how the secret police of Berlin considered Hegel a political subversive and supporter of student rebels, even in his maturity.

11. Graham, *Disputers,* 227–229.

12. Robert M. A. Adamson, *Fichte* (Edinburgh: Blackwood, 1881), 70. That this attitude toward the assertiveness of the Romantic women continues into the twentieth century is suggested by unnecessary denial by W. H. Bruford, *Culture and Society in Classical Weimar* (Cambridge: Cambridge University Press, 1962), 347; "Karoline was not slovenly or promiscuous in her habits."

13. Deborah Herz, *Jewish High Society in Old Regime Berlin* (New Haven, Conn.: Yale University Press, 1988), esp. chaps. 6, 7; Steven Lowenstein, *The Berlin Jewish Community: Enlightenment, Family and Crisis, 1770–1830* (Oxford: Oxford University Press, 1994), chaps. 9, 14.

14. Lowenstein, *Jewish Community*, 163.

15. Hannah Arendt, *Rahel Varnhagen* (New York: Harcourt Brace, 1974). Liliane Weissberg, "Turns of Emancipation," in *In the Shadows of Olympus*, ed. Katherine Goodman and Edith Waldstein (Albany: State University of New York Press, 1992), 53–70.

16. Helga Sprung, "Bourgeois Berlin Salons: Meeting Places for Culture and the Sciences," in *World Views and Scientific Discipline Formation*, ed. William R. Woodward and Robert S. Cohen (Boston: Kluwer, 1991), 405, 409.

17. Ilse-Marie Barth, *Literarisches Weimar: Kultur/Literatur/Sozialstruktur im 16–20 Jahrhundert* (Stuttgart: J. B. Metzlersche Verlagsbuchhandlung, 1971), 52–57.

18. Sprung, "Salons," 410–411.

19. Pierce C. Mullen, "The Romantic as Scientist: Lorenz Oken," *SR* 16 (1977): 380–399.

20. Raymond Schwab, *The Oriental Renaissance* (New York: Columbia University Press, 1984), 59.

21. Saine, *Georg Forster*, 9.

22. Beiser, *Enlightenment*, 248–249.

23. Friedrich Schlegel, "Athenaeum Fragments," no. 31, in Beiser, *Early Writings*, 115.

24. Schlegel, "Athenaeum Fragments," no, 420, in Beiser, *Early Writings*, 121. See Werner Weiland, *Der junge Friedrich Schlegel oder Die Revolution in der Frühromantik* (Stuttgart: Kohlhammer Verlag, 1968); and Rudolf Murtfeld, *Caroline Schlegel-Schelling: Moderne Frau in revolutionärer Zeit* (Bonn: Bouvier Verlag Herbert Grundman, 1973), which document the impact of Caroline Schlegel-Schelling on the radical thought of the young Friedrich Schlegel.

25. Gisella Fuhrmann Ritchie, *Caroline Schlegel Schelling in Wahrheit und Dichtung* (Bonn: Bouvier, 196), 8. See also Sara Friedrichsmey, "Caroline Schlegel-Schelling: A Good Woman and No Heroine," in Goodman and Waldstein, *Shadows*, 115–136.

26. *Caroline: Briefe auf der Früromantik*, ed. George Waitz, aug. Erich Schmidt, 2 vols. (Bern: Herbert Lang, 1970). A few occur in translation in *Bitter Healing: German Women Writers, 1700–1830*, ed. Jeanine Blackwell, and Susanne Zantop (Lincoln: University of Nebraska Press, 1990), 285–296.

CHAPTER 14   *Schelling's Philosophy of Nature*

1. Friedrich Wilhelm Joseph von Schelling, *The Unconditional in Human Knowledge: Four Early Essays, 1794–1796*, trans. Fritz Marti (Lewisburg, Penna: Bucknell University Press, 1980).

2. Friedrich Wilhelm Joseph von Schelling, *System of Transcendental Idealism (1800)*, trans. Peter Heath (Charlottesville: University Press of Virginia), 1978.

3. Kant, *Opus Posthumum*, 67–99.

4. Schelling, *Bruno; or, On the Divine Principle of Things*, trans. Michael Vater (Albany: State University of New York Press, 1984).

5. Lloyd, *Polarity and Analogy*.

6. Schelling, *System*, 82–89.

7. E. D. Hirsch, Jr., *Wordsworth and Schelling: A Typological Study of Romanticism* (New Haven: Yale University Press, 1960). Hirsch does not claim any historical connection, but only that there are systematic structural parallels. We do not know what, if anything, Wordsworth absorbed of German philosophy during his visit to Germany.

8. Immanuel Kant, *Critique of Pure Reason*, trans. N. K. Smith (London: Macmillan, 1929), 61: "*sensibility* and *understanding*, which perhaps spring from a common, but to us unknown, root."

9. Schelling, *System*, 219–236.

10. Schelling, *On the History of Modern Philosophy*, trans. Andrew Bowie (Cambridge: Cambridge University Press, 1994), 66.

11. Thomas McFarland, *Coleridge and the Pantheist Tradition* (Oxford: Oxford University Press, 1969), 103, 105.

12. H. S. Harris, "Cows in a Dark Night," *Dialogue* 26 (1987): 627–644, defends the veracity of Hegel's claim and suggests that Reinhold or possibly Steffens was the object. Michael Vater, "Hymns to the Night," ibid., 645–652, opts for Reinhold.

13. Jean W. Sedlar, *India in the Mind of Germany: Schelling, Schopenhauer and Their Times* (Washington, D.C.: University Press of America, 1982), 126n.

14. Dale E. Snow, *Schelling and the End of Idealism* (Albany: State University of New York Press, 1996), 229 n. 8.

15. Schwab, *Oriental Renaissance*, 218.

16. Snow, *Schelling*, 98–118.

17. Ibid.,117.

18. Edward Beach, *The Potencies of God(s)* (Albany: State University of New York Press, 1994), 85.

19. Beach, *Potencies*, 117.

20. Ibid.,116, translation here quoted from F.W.J. Schelling, *Of Human Freedom*, trans. James Gutman (La Salle, Ill.: Open Court, 1954), 30–31.

21. Beach, *Potencies*, 229 n. 8.

22. Friedrich Engels, *Dialectics of Nature* (New York: International Publishers, 1940), 38, 44.

23. John Hendry, "Monopoles before Dirac," *SHPS* 14 (1983): 81–87.

24. P.A.M. Dirac, "Quantitized Singularities in the Electromagnetic Field," *Proceeding of the Royal Society*, Ser. A, 133 (1931): 60–72.

25. Dietrich Thompson, "Monopoles Oughtn't to Be a Monopoly," *Science News* 109 (February 1, 1976): 122–123.

26. Sidney Coleman, "Classical Lumps and Their Quantum Descendants," in *Aspects of Symmetry* (Cambridge: Cambridge University Press, 1985), 215–223, and Lewis H. Ryder, *Quantum Field Theory* (Cambridge: Cambridge University Press), 1985, 413–425.

27. Lawrence M. Krauss, *The Fifth Essence: The Search for Dark Matter in the Universe* (New York: Basic Books, 1989), 247.

28. P. Buford Price, Edward K. Shirk, Weymar Zack Osborne, and Lawrence S. Pinsky,

Physical Review Letters, August 25, 1975; see "Apparent Discovery of Long-Sought Monopole: 'Controlled Excitement,'" *Science News* 108 (August 23 and 30, 1975): 118–120.

29. "Monopole Claim: Storm of Scrutiny," *Science News* 108 (September 13, 1975): 104; Kendrick Frazier, "High Stakes in the Monopole Claim Game; Alvarez: 'Too Bad It Wasn't Right,'" *Science News* 108 (October 4, 1975): 222–223.

30. Hy Cohen and Lester (Hank) Talkington, *Science and Nature*, No. 1 (1978): 55.

31. Helge Kragh, "The Concept of the Monopole," *SHPS* 12 (1981): 168–170. An interesting sidelight is that one physicist, A. S. Goldhaber, who raised an objection to monopoles, did so on the basis that the theory of monopoles is the only classical physical theory not derivable from an action principle. So perhaps a "teleological" physicist, who emphasizes action principles as the ultimate criteria for legitimacy of theories, would also be close to the Schellingian camp.

CHAPTER 15   *Coleridge: Poet of Nature*

1. See E. S. Shaffer, *"Kublai Khan" and "The Fall of Jerusalem": The Mythological School in Biblical Criticism and Secular Literature, 1770–1880* (Cambridge: Cambridge University Press, 1975).

2. E. S. Shaffer, "Coleridge and Natural Philosophy," *HS* 12 (1974): 293.

3. Robert Schofield, "Joseph Priestley, Eighteenth-Century British Neoplatonism and S. T. Coleridge," in *Transformation and Tradition in the Sciences*, ed. Everett Mendelsohn (Cambridge: Cambridge University Press, 1984), 237–254. Schofield in some respects follows up on some of the suggestions of Shaffer but, given that Schofield had already done the major study of the Lunar Society in Birmingham, it is disappointing that he does not specifically pursue the German sources of Beddoes and the communication to Priestley, Davy, and Coleridge by Beddoes.

4. See, for instance, W. Schrickx, "Coleridge and the Cambridge Platonists," *Review of English Literature* 7 (1966): 71–91.

5. René Wellek, *Immanuel Kant in England* (Princeton, N.J.: Princeton University Press, 1931), 44.

6. Norman Fruman, *Coleridge: The Damaged Archangel* (N.Y.: Georges Braziller, 1971), 132–133. Thomas McFarland, *Coleridge and the Pantheist Tradition* (Oxford: Oxford University Press, 1969), 249. R. Schofield, "Southey, Coleridge and Company," *Contemporary Review*, no. 230 (1977): 148–150 .

7. Trevor Levere, *Poetry Realized in Nature: Samuel Taylor Coleridge and Early Nineteenth Century Science* (Cambridge: Cambridge University Press, 1981).

8. Jan Golinski, *Science as Public Culture* (Cambridge: Cambridge University Press, 1992), chap. 6, "'Dr. Beddoes's Breath': Nitrous Oxide and the Culmination of Enlightenment Chemistry," esp. last section, "The End of Enlightenment Science?" 176–187.

9. Gian N. G. Orsini, *Coleridge and German Idealism* (Carbondale: Southern Illinois University Press, 1969), 43.

10. Urbach, *Bacon*; Yehuda Elkana, "Scientific and Metaphysical Problems: Euler and Kant," *Methodological and Historical Essays in the Social and Natural Sciences*, BSPS, vol. 14 (1974), 277–305.

11. Fruman, *Archangel*.

12. Shaffer, "Coleridge," 289.

13. H. J. Jackson, "Coleridge on the King's Evil," *SR* 16 (1967): 337–347.

14. Ibid.
15. Postscript to S. T. Coleridge, *Hints towards the Formation of a More Comprehensive Theory of Life,*" ed. Seth B. Watson (London: John Churchill, 1848), 95.
16. Preface by Seth B. Watson to ibid., 8–9.
17. Coleridge, *Life,* 44.
18. Ibid., 30.
19. Ibid., 56.
20. Fruman, *Archangel*, 123–133, using Henri Nidecker's dissertation, "Praeliminariien zur Neuausgabe der Abhandlung ueber die Lebenstheorie (*Theory of Life*) von Samuel Taylor Coleridge," University of Basel, 1926, via Joseph Warren Beach, "Coleridge's Borrowings from the Germans," *Journal of English Literary History* 9 (1942): 36–58.
21. Coleridge, *Life*, 68.
22. "I must reject fluids and ethers of all kinds, magnetical, electrical, and universal, to whatever quintessential thinness they may be treble distilled and (as it were) super-substantiated" (ibid., 34).
23. Ibid., 45.
24. Ibid., 55.
25. L. Pearce Williams, *Michael Faraday* (1965; reprint, New York: Da Capo, n.d.), 32.

CHAPTER 16   ***Ørsted: Romanticism and Nature Philosophy***

1. Hans Christian Oersted, "Effect of a Current of Electricity on the Magnetic Needle," *Annals of Philosophy* 16 (1820): 273–276, reprinted in Bern Dibner, *Oersted and the Discovery of Electromagnetism* (New York: Blaisdell, 1962), 71–76.
2. Now known for the law of Dulong and Petit, which states that the product of the specific heat and the atomic weight (for many solid elements at moderate temperatures) is constant.
3. H.A.M. Snelders, "Oersted's Discovery of Electromagnetism," in *Romanticism and the Sciences*, ed. Andrew Cunningham and Nicholas Jardine (Cambridge: Cambridge University Press, 1990), 236.
4. Dibner, *Oersted*, 32–33.
5. Timothy Shanahan, "Kant, *Naturphilosophie*, and Oersted's Discovery of Electromagnetism: A Reassessment," *SHPS* 20 (1989): 287–288.
6. *Recherches sur l'identité des forces chimiques et électriques* (Paris, 1813) discussed in Pearce Williams, "Oersted, Hans Christian," in *DSB*, 184–185; and in Barry Gower, "Speculation in Physics: The Theory and Practice of *Naturphilosophie*," *SHPS* 3 (1973): 345–346. Williams emphasizes the anticipation while Gower emphasizes the difference between the early speculations and later experiment.
7. Oersted, "Thermoelectricity," in *Edinburgh Enclyclopedia*, ed. David Brewster (1830), 573–589, cited in Dibner, *Oersted*, 31.
8. Marx, *Capital* 1: 29.
9. Justus von Liebig, "Ueber das Studium der Naturwissenschaft und ueber der Zustand der Chemie in Preussen," in *Redun unde Abhanlungen* (Leipzig and Heidelberg, 1874), 24. I owe this citation to Michael Heidelberger.
10. Maria Trumpler, "Questioning Nature: Experimental Investigations of Animal Electricity in Germany, 1791–1810" (Ph.D. diss., Yale University, 1992), 159–160, 189–190.
11. Johann Ritter, "Die Entdeckung der Ultravioletten Strahlen (Februar 1801)," in *Die

Begründung der Elektrochemie, ed. Armin Hermann (Frankfurt: Akademische Verlagsgesellschaft, 1968), 57–73.

12. Joseph Esposito, *Schelling's Idealism and Philosophy of Nature* (Lewisburg, Penna.: Bucknell University Press, 1977), 148. Esposito devotes a short paragraph to Ritter, mentioning only his "work on galvanism" and table of electrochemical affinities, but not his discovery of ultraviolet light, work on bioelectricity, or other work. Esposito writes several times as much on each of several fringe figures without contributions to natural science who justify the worst stereotypes of the nature philosopher.

13. Snelders, "Oersted's Discovery," 233.

14. Stuart Strickland, "Circumscribing Science: Johann Wilhelm Ritter and the Physics of Sidereal Man," (Ph.D. diss., Harvard University, 1992), chap. 2.

15. Louis Claude de Saint-Martin, *Man: His True Nature and Ministry*, trans. Edward Burton Penny (1865; reprint, Kila: Mont.: Kessinger, n.d.).

16. Paul Gottfried, *Conservative Millenarians: The Romantic Experience in Bavaria* (New York: Fordham University Press, 1979).

17. Strickland, "Circumscribing Science," 138–139.

18. Ibid., 182–183.

19. Keld Nielsen, "Another Kind of Light, Part 1," *Historical Studies in the Physical and Biological Sciences* 20 (1989): 133–134; Strickland, "Circumscribing Science," 191–192.

20. Simon Schaffer, "Self Evidence," in *Questions of Evidence: Proof, Practice and Persuasion Across Disciplines*, ed. James Chandler, Arnold I. Davidson, and Harry Harootunian (Chicago: University of Chicago Press, 1994), 90 n. 75, citing Nielsen, "Another Kind of Light, Part 1," 133.

21. Hegel to Schelling, February 3, 1807, in Clark Butler and Christiane Seiler, eds., *Hegel: The Letters* (Bloomington: Indiana University Press, 1984), 76–77.

22. Johann Wolfgang Goethe, *Elective Affinities* (South Bend, Ind.: Gateway, 1963), 246–248.

23. Stuart Strickland, "Galvanic Disciplines," *HS* 33 (1995): 461–462.

24. See Christa Jungnickel and Russell McCormmach, *Intellectual Mastery of Nature*, 2 vols. (Chicago: University of Chicago Press, 1986), 1: 9, 40, 61. See index for numerous other references to the role of physics in preparing pharmacy students.

25. Nielsen, "Another Kind of Light, Part 1," 134.

26. Trumpler, "Questioning Nature," 189.

27. Robert M. Maniquis, "The Puzzling Mimosa: Sensitivity and Plant Symbols in Romanticism," *SR* 8 (1969): 129–155.

28. Walter D. Wetzels, "J. W. Ritter and Romantic Physics," in Cunningham and Jardine, *Romanticism and the Sciences*, 199–211.

29. Dan C. Christensen, "The Oersted-Ritter Partnership," *AS* 52 (1995): 153–185.

30. Ibid. 164–165.

31. Ibid. 166–167.

32. Simon Schaffer, "Self Evidence," 90 and n 76, citing Walter Wetzels, *Johann Wilhelm Ritter: Physik im Wirkungsfeld der deutschen Romantik* (Berlin: de Gruter, 1973), 52–53.

33. Christensen, "Oersted-Ritter," 169.

34. Quoted in Snelders, "Oersted's Discovery," 232.

35. Ibid., 234; Pearce Williams, *Faraday*, 183.

36. Gower, "Speculation," 347.

37. Oersted, "Effect of a Current," 71, 75.

CHAPTER 17 **Davy, Faraday, and Field Theory**

1. Nancy Nersessian, *Faraday to Einstein: Constructing Meaning in Scientific Theories* (The Hague: Martinus Nijhoff, 1984), 34.
2. Geoffrey Cantor, *Michael Faraday: Sandemanian and Scientist* (New York: St. Martin's, 1991).
3. Fruman, *Archangel*, 422–424, 431.
4. Ibid., 266
5. Cantor, *Faraday*, 229–233.
6. Ibid., 367–368.
7. Rumford devised numerous schemes for workers' well-being, including the low-cost diet that Marx criticizes (Marx, *Capital*, 1: 653).
8. John Meurig Thomas, *Michael Faraday and the Royal Institution: The Genius of the Man and the Place* (Bristol: Adam Hilger, 1991).
9. Jan Golinski, *Science as Public Culture: Chemistry and Enlightenment in Britain, 1760–1820* (Cambridge: Cambridge University Press, 1992), depicts Davy's place in society, his ambition, and the place of the public lectures in his quest for social standing.
10. Humphry Davy, *Elements of Agricultural Chemistry, in a Course of Lectures for the Board of Agriculture* (London: W. Bulmer for Longman, Hurst, Rees Orme and Brown, 1813).
11. Levere, *Poetry Realized*, 202–203.
12. Golinski, *Public Culture*, chap. 6, "'Dr. Beddoes's Breath.'"
13. Ibid., 29.
14. Ibid., 175.
15. Ibid., 173–187.
16. Humphry Davy, *Salmonia; or, Days of Fly-Fishing, in a Series of Conversations with Some Account of the Habit of Fishes Belonging to the Genus Salmo* (London: John Murray, 1828).
17. Humphry Davy, *Consolations in Travel; or, The Last Days of a Philosopher*, 5th ed. (London: John Murray, 1851).
18. Hartley, *Humphry Davy*, 108.
19. Ibid., 147.
20. J. Z. Fulmer, "The Poetry of Humphry Davy," *Chymia* 6 (1960): 102–120.
21. Trevor Levere, "Coleridge, Chemistry, and the Philosophy of Nature," *SR* 16 (1977): 354.
22. Coleridge, *Collected Letters*, 1: 557.
23. Robert C. Stauffer, "Persistent Errors Regarding Oersted's Discovery of Electromagnetism," *Isis* 44 (1953): 307–310. Idem, "Speculation and Experiment in the Background of Oersted's Discovery of Electromagnetism," *Isis* 48 (1957): 33–50.
24. Thomas Kuhn, "Energy Conservation as an Example of Simultaneous Discovery," in *Critical Problems in the History of Science*, ed. Marshall Claggett (Madison: University of Wisconsin Press, 1955), 321–356.
25. Kenneth L. Caneva, *Robert Mayer and the Conservation of Energy* (Princeton: Princeton University Press, 1993), concludes with a successful rebuttal of Kuhn's suggestion that nature philosophy influenced Mayer's formulation of energy conservation.
26. Yehuda Elkana, *The Discovery of the Conservation of Energy* (Cambridge: Harvard University Press, 1974).
27. Timothy Shanahan, "Kant's *Naturphilosophie* and Oersted's Discovery of Electromagnetism: A Reassessment," *SHPS* 20 (1989): 287–305.

28. Trumpler, "Questioning Nature."
29. Joseph Agassi, *Faraday as Natural Philosopher* (Chicago: University of Chicago Press, 1971).
30. Davy *Consolations in Travel*. David Kneightley, "The Scientist as Sage," *SR* 6 (1967): 64–87. Idem, *Humphry Davy* (Oxford: Blackwell, 1992).
31. Michael Faraday, "On the Various Forces of Nature," and "The Chemical History of the Candle," in *Scientific Papers*, in vol. 30 of *The Harvard Classics* (New York: Collier, 1938).
32. Agassi, *Faraday*. William K. Berkson, *Fields of Force* (London: Routledge and Kegan Paul, 1974).
33. Clifford Truesdell, *Essays in the History of Rational Mechanics* (New York: Springer Verlag, 1968), 180.
34. Michael Crowe, *A History of Vector Analysis* (Notre Dame, Ind.: Notre Dame University Press, 1967).
35. Jean Dieudonné, "Cartan, Elie," *DSB* 3: 95.
36. J. B. Spencer, "Boscovich's Theory and Its Relation to Faraday's Researches: An Analytical Approach," *AHES* 4 (1967): 184–202.
37. P. M. Heimann, "Faraday's Theories of Matter and Electricity," *BJHS* 5 (1971): 235–257.
38. Ibid., 237–238.
38. Cantor, *Faraday*, 178.
40. Ibid., 247. P. M. Heimann and J. E. McGuire, "Newtonian Forces and Lockean Powers: Concepts of Matter in Eighteenth-Century Thought," *HSPS* 3 (1971): 233–306.
41. Cantor, *Faraday*, 184.
42. Ibid., 249–252. Joseph Priestley, *Disquisitions Relating to Matter and Spirit* (1777; reprint, Kila, Mont.: Kessingers Publishing, n.d.)
43. Cantor, *Faraday*, 253, n. 119.
44. T. H. Levere, "Faraday, Matter and Natural Theology," *BJHS* 4 (1968): 100.
45. Ibid., 96–97.
46. Arthur Lovejoy, "Kant and the English Platonists," in *Essays, Philosophical and Psychological, in Honor of William James, Professor at Harvard, by his Colleagues at Columbia University* (New York: Longmans, Green, 1908), 265–302.
47. Ibid., 101–104.
48. Cantor, *Faraday*, 175.
49. Ibid., 176.
50. Ibid., 177.
51. Ibid.
52. Ibid., 182.
53. L. Pearce Williams, *The Origins of Field Theory* (Washington, D.C., University Press of America, 1980). Idem, *Faraday*. My account largely follows Williams.
54. Letter to Auguste de La Rive, in Nersessian, *Faraday to Einstein*, 38.

CHAPTER 18   *Maxwell, Field Theory, and Mechanical Models*

1. Lewis Campbell and William Garnett, *The Life of James Clerk Maxwell* (1882; reprint, New York: Johnson, 1969),178.
2. James Clerk Maxwell, "Theory of Perception of Colors," in *Sources of Color Science*, ed. David L. MacAdam (Cambridge: MIT Press, 1970), 62–66.
3. Einstein, *Ideas and Opinions*, 266, 268.

4. John Maxwell, cited in David K. C. MacDonald, *Faraday, Maxwell, and Kelvin* (Garden City, N.Y.: Doubleday Anchor, 1964), 3. See also Campbell and Garnett, *Life*, 27.

5. James Clerk Maxwell, "Thomson and Tait's Natural Philosophy,"(1879) in *Collected Papers* (1890; reprint, N.Y.: Dover, 1965), 783.

6. James Clerk Maxwell, "Analogies, February 1856: Are There Real Analogies in Nature?," chap. 8, "Essays at Cambridge," in Campbell and Garnett, *Life*, 235–245.

7. Richard Olson, *Scottish Philosophy and British Physics* (Princeton: Princeton University Press, 1975), 51–52, 114–118.

8. Nersessian, *Faraday to Einstein*, 72.

9. Bruce Clark, "Allegories of Victorian Thermodynamics," *Configurations: A Journal of Literature and Science* 4, no. 1 (Winter 1996): 70.

10. Olson, *Scottish*, 48–53, 252–270, discusses both Herschel and the role of analogy in the Scottish school but not the role of analogy in Herschel's method.

11. Michael Ruse, *Darwinism Defended* (Reading, Mass.: Addison Wellesley, 1979), 48; idem, "Darwin's Debt to Philosophy," in *The Darwinian Paradigm: Essays on Its History, Philosophy and Religious Implications* (New York: Routledge, 1989), 24–25.

12. Mary Hesse, "Logic of Discovery in Maxwell's Electromagnetic Theory," in *Scientific Method in the Nineteenth Century*, eds. Ronald Giere and Richard S. Westfall (Bloomington: University of Indiana Press, 1973).

13. Hesse, "Discovery," 90; A. F. Chalmers, "Maxwell's Methodology and His Application of It to Electromagnetism," *SHPS* 4 (1973): 111.

14. Chalmers, "Maxwell's Methodology," 107–164.

15. Daniel M. Siegel, *Innovation in Maxwell's Electromagnetic Theory: Molecular Vortices, Displacement Current, and Light* (Cambridge: Cambridge University Press, 1991), 85–143.

16. Ibid., 83.

17. P. M. Heimann, "Maxwell and the Modes of Consistent Representation," *AHES* 6 (1970): 183, citing Maxwell's remarks that "molecules have laws of their own" and that disturbing forces are products of our mind and do not exist in nature.

18. Heimann, "Modes," 183, 190, citing James C. Maxwell, "Physical Lines of Force," in *Collected Papers*, 2 vols. (1890; reprint, New York: Dover, 1965), 1: 160, 468.

19. Siegel, *Innovation*, 157.

20. Barbara Gusti Doran, "Origins and Consolidation of Field Theory in Nineteenth Century Britain: From the Mechanical to the Electromagnetic View of Nature," *HSPS* 6 (1975): 133–260; Russell McCormach, "H. A. Lorentz and the Electromagnetic View of Nature," *Isis* 51 (1970): 459–497.

21. John Hendry, *James Clerk Marxwell and the Theory of the Electromagnetic Field* (Bristol: Adam Hilgar, 1986), 272, citing D. M. Siegel, "Thomson, Maxwell and the Universal Ether in Victorian Physics," in *Conceptions of Ether*, ed. G. N. Cantor and M. J. S. Hodge (Cambridge: Cambridge University Press, 1981), 259.

22. Hendry, *Maxwell*, 4–25.

23. Siegel, *Innovation*.

24. Mario Bunge, "Lagrangian Formulation and Mechanical Interpretation," *American Journal of Physics* 25 (1957): 216.

25. Heimann, "Modes," 212. Campbell and Garnett, *Life*, 439.

26. Martin J. Klein, "Maxwell, His Demon, and the Second Law of Thermodynamics," *American Scientist* 58 (1970): 89–90; but see Norton Wise, "The Maxwell Literature and British Dynamical Theory, " *HSPS* 13 (1982): 199.

27. Ibid., 201.

CHAPTER 19   *Conclusion: Metaphysics of Science—Creation and Justification*

1. Heinrich Hertz, *Electric Waves* (1892; reprint, New York: Dover, 1962), 21.
2. Steven Weinberg, reply in "Sokal's Hoax: An Exchange," *New York Review of Books* 43, no. 15 (October 3, 1996): 56.
3. Hendry, *Maxwell*, 260.
4. This point is made in Imre Lakatos, *Proofs and Refutations: The Logic of Mathematical Discovery* (Cambridge: Cambridge University Press, 1976); and Michael Polanyi, *Personal Knowledge: Towards a Post-Critical Philosophy* (Chicago: University of Chicago Press, 1958).
5. Laudan, *Between Positivism and Relativism*, 82–83 ("pursuit").
6. Karl Popper, *The Logic of Scientific Discovery* (1934; reprint, New York: Basic Books, 1959).
7. Wesley Salmon, *Foundations of Scientific Inference* (Pittsburgh: University of Pittsburgh Press, 1967), suggests that Popper's prior improbability could be incorporated into the "prior probabilities" of Bayesian testing along with other considerations of the "logic of discovery."
8. Norwood Russell Hanson, "The Logic of Discovery," *Journal of Philosophy* 55, no. 25 (December 4, 1958), reprinted in Baruch Brody, ed., *Science: Men, Methods, Goals* (New York: W. A. Benjamin, 1968), 150–162, mention of symmetry on 156. See also "Is There a Logic of Scientific Discovery?" in *Current Issues in the Philosophy of Science*, ed. Herbert Feigl and Grover Maxwell (New York: Holt, Rinehart, and Winston, 1961), reprinted in Baruch Brody and Richard E. Grandy, eds., *Reading in the Philosophy of Science*, 2nd ed. (Englewood Cliffs, N.J.: Prentice Hall, 1989), 398–409.
9. Gilbert Harman, "Inference to the Best Explanation," *Philosophical Review* 74 (1965): 22–35, reprinted in Brody and Grandy, *Reading*, 323–328.
10. Duhem, *Aim and Structure*, 180–190.
11. Imre Lakatos, "Falsification and the Methodology of Scientific Research Programmes," in *Philosophical Papers*, 1: 8–101.
12. Ibid., 48–49.
13. Laudan, *Between Positivism and Relativism*, 50–51.
14. Larry Laudan, *Progress and Its Problems* (Berkeley: University of California Press, 1977), 61–64.
15. Bas van Fraassen, *Formal Semantics and Logic* (New York: Macmillan, 1971), 154–159.
16. Marx Wartofsky, "Metaphysics as Heuristic for Science," BSPS, vol. 3, ed. Robert S. Cohen and Marx Wartofsky (Dordrecht: D. Reidel, 1967), 123–172.
17. Gilbert Ryle, *Dilemmas* (Cambridge: Cambridge University Press, 1954), and idem, *The Concept of Mind* (London: Hutchinson, 1967), 121–123. P. F. Strawson and H. P. Grice, "Science and Metaphysics," reprinted in *The Linguistic Turn*, ed. Richard Rorty (Chicago: University of Chicago Press, 1964).
18. The godfather of logical positivism, Ernst Mach, wrote about the sometimes helpful psychological role of metaphysics and theology in physics. Ernst Mach, *The Science of Mechanics* (La Salle, Ill.: Open Court, 1960), chap. 3, pt 2.
19. Duhem, *Aim and Structure*, 274.
20. Lakatos, *Philosophical Papers*; and idem, "The Role of Crucial Experiments in Science," SHPS 4 (1974): 309–325. Mario Bunge, *Scientific Research* (New York: Springer, 1967), 2: 254.
21. Israel Scheffler, *The Anatomy of Inquiry* (Indianapolis: Bobbs-Merrill, 1963), 137–150.

22. Emile Meyerson, *Identity and Reality* (1930; reprint, New York: Dover, 1962); Duhem, *Aim and Structure*; idem, *To Save the Phenomena*; Burtt, *Metaphysical Foundation*; Ernst Cassirer, *Das Erkenntnisproblem*, 3 vols. (Berlin: Bruno Cassirer, 1906–1920); Alexandre Koyré, *Etudes Galiléennes* (Paris: Hermann, 1966).

23. Stephan Koerner, "Philosophical Arguments in Physics," in *Observation and Interpretation* (New York: Dover, 1957), 97–101. Wartofsky, "Metaphysics as Heuristic."

24. Margaret Masterman, "On the Nature of a Paradigm," in Lakatos and Musgrave, *Criticism*, 72.

25. Joseph Agassi, "The Nature of Scientific Problems and Their Roots in Metaphysics," in *The Critical Approach*, ed. Mario Bunge (Glencoe, Ill.: Free Press, 1964), 198, 193, 204–205.

26. Wartofsky, "Metaphysics as Heuristic," 168–169. Ruth Anna Putnam, "Comments," BSPS, vol. 3 (Dordrecht: D. Reidel, 1967), 176,

27. Lakatos, "Research Programmes," 184.

28. Van Fraassen, *Formal Semantics*, 154–159.

29. P. F. Strawson, *Introduction to Logical Theory* (London: Methuen, 1952), 18.

30. Agassi, "Scientific Problems," 198–200.

31. Wartofsky, "Metaphysics as Heuristic," 145–146.

32. Karl Popper, *Postscript to the Logic of Scientific Discovery*, vol. 3, *Quantum Theory and the Schism in Physics* (Totowa, N.J.: Rowan and Littlefield, 1982), 30–34, 160–162, quoted in Lakatos, "Research Programmes," 183, n. 3; Agassi, "Scientific Problems"; Wartofsky, "Metaphysics as Heuristic"; Koerner, "Philosophical Arguments."

33. Lakatos, "Research Programmes," 184.

34. Laudan, *Progress and Its Problems*, 76.

35. Lakatos, "Research Programmes."

36. Harold I. Brown, *The New Image of Science* (Chicago: University of Chicago Press, 1979), 104–106.

37. Carl R. Kordig, *The Justification of Scientific Change* (Dordrecht: D. Reidel, 1971), 26–27.

38. J. L. Heilbron, "The Electrical Field before Faraday," in Cantor and Hodge, *Conceptions of Ether*, 189.

39. Gerald L. Geison, "Review of Robert E. Kohler, *Lords of the Fly*," *Isis* 87 (1996): 331.

40. Karl Marx, "Critique of Hegel's Philosophy of Right: Introduction," in Karl Marx, *Early Writings*, trans. Rodney Livingstone and Gregor Benton (New York: Penguin, 1992), 251.

SELECTED BIBLIOGRAPHY

Ackroyd, Peter. *Blake: A Biography*. New York: Knopf, 1996.

Agassi, Joseph. *Faraday as Natural Philosopher*. Chicago: University of Chicago Press, 1971.

———. "Leibniz's Place in the History of Physics." *Journal of the History of Ideas* 30 (1969): 331–344.

Agrippa, Henry Cornelius. *Declamation on the Nobility and Preeminence of the Female Sex*. Trans., ed. Albert Rabil, Jr. Chicago: University of Chicago Press, 1996.

———. *Three Books of Occult Philosophy or Magic*. Trans. J. F. 1651. Reprint. Kila, Mont: Kessinger, n.d.

Aiton, Eric J., and Eikoh Shima. "Gorai Kinzo's Study of Leibniz and the *I Ching* Hexagrams." *Annals of Science* 36 (1981): 71–92.

———. "Johannes Kepler in the Light of Recent Research." *History of Science* 14 (1976): 77–100.

———. *Leibniz: A Biography*. Bristol, England: Adam Hilgar, Ltd, 1985.

———. *The Vortex Theory of Planetary Motions*. New York: American Elsevier, 1972.

Alexander, H. G., ed. *The Leibniz-Clarke Correspondence*. 1956. Reprint. New York: Barnes and Noble, 1984.

Allinson, Robert E., ed. *Understanding the Chinese Mind: The Philosophical Roots*. Oxford: Oxford University Press, 1989.

Amrine, Frederick, Francis J. Zucker, and Harvey Wheeler, eds. *Goethe and the Sciences: A Reappraisal*. Boston Studies in the Philosophy of Science, vol. 91. Dordrecht: D. Reidel, 1987.

Aris, Reinhold. *History of Political Thought in Germany: From 1789 to 1815*. Foreword by G. P. Gooch. London: Frank Cass; New York: Russell and Russell, 1965.

Ault, Donald, *Visionary Physics: Blake's Response to Newton*. Chicago: University of Chicago Press, 1974.

Bahro, Rudolf. *Die Alternative: Zur Kritik des real existierenden Sozialismus*. Reinbek bei Hamburg: Rowohlt Taschenbuch Verlag, 1977. Trans. as *The Alternative in Eastern Europe*. London: New Left Books, [1980].

Bailey, Anne M., and Josep R. Llobera, eds. *The Asiatic Mode of Production*. London: Routledge and Kegan Paul, 1981.

Bakhtin, Mikhail. *Rabelais and His World*. 1968. Reprint. Bloomington: Indiana University Press, 1984.

Balazs, Etienne. *Chinese Civilization and Bureaucracy*. Trans. H. M. Wright; ed. Arthur F. Wright. New Haven: Yale University Press, 1964.

Baldwin, M. R. "Magnetism and the Anti-Copernican Polemic." *Journal of the History of Astronomy* 16 (1985): 155–160.

Baldwin, Martha. "Alchemy and the Society of Jesus in the Seventeenth Century: Strange Bedfellows?" *Ambix* 40, no. 2 (July 1993): 41–54.

Barnes, Jonathan. *Early Greek Philosophy*. London: Penguin, 1987.

Barstow, Anne Llewellyn. *Witchcraze: A New History of the European Witch Hunts*. San Francisco: HarperCollins, Pandora, 1994.

Bauer, Wolfgang, *China and the Search for Happiness: Recurring Themes in Four Thousand Years of Chinese Cultural History*. Trans. Michael Shaw. New York: Continuum, 1976.

Beauvoir, Simone de. *The Second Sex*. Trans. H. M. Parshley. New York: Bantam, 1952.

Beiser, Frederick. *Enlightenment, Revolution, Romanticism*. Cambridge: Harvard University Press, 1992.

———. *The Fate of Reason: German Philosophy from Kant to Fichte*. Cambridge: Harvard University Press, 1987.

Bendix, Reinhard. *Max Weber: An Intellectual Portrait*. Garden City, N.Y.: Doubleday, 1960.

Bennett, J. A. "Cosmology and the Magnetical Philosophy, 1640–1680," *Journal of the History of Astronomy* 12 (1981): 165–177.

———. "The Mechanics' Philosophy and the Mechanical Philosophy." *History of Science* 24 (1986): 1–28.

Bennett, Jonathan. *A Study of Spinoza's Ethics*. Indianapolis: Hackett, 1984.

Bergman, Torben. *Dissertation on Elective Affinities*. Trans., intro. J. A. Schufle. New York: Johnson, 1968.

Berkson, William K. *Fields of Force*. London: Routledge and Kegan Paul, 1974.

Bernal, Martin. *Black Athena*. 2 vols. New Brunswick, N.J.: Rutgers University Press, 1987.

Blackwell, Jeanine, and Susanne Zantop, eds. *Bitter Healing: German Women Writers, 1700–1830*. Lincoln: University of Nebraska Press, 1990.

Bloom, Alfred. *The Linguistic Shaping of Thought: A Study in the Impact of Language on Thinking in China and the West*. Hillsdale, N.J.: Lawrence Erlbaum, 1981.

Boas, Marie. *The Scientific Renaissance, 1450–1630*. New York: Harper and Row, 1962.

Bodde, Derk. *Chinese Thought, Society, and Science: The Intellectual and Social Background of Science and Technology in Pre-Modern China*. Honolulu: University of Hawaii Press, 1991.

Bonelli, M. L. Righini, and William R. Shea, eds. *Reason, Experiment and Mysticism in the Scientific Revolution*. New York: Science History Publications, 1975.

Borkenau, Franz. "The Sociology of the Mechanistic World-Picture." Trans. Richard W. Hadden. *Science in Context* 1, no. 1 (March 1987): 109–127.

Boscovich, Roger Joseph. *A Theory of Natural Philosophy*. Trans. J. M. Child. Cambridge: MIT Press, 1966.

Bowie, Andrew. *Schelling and Modern European Philosophy*. London: Routledge, 1993.

Brahe, Tycho. *Learned Tycho Brahe His Astronomical Conjectur of the New and Much Admired * Which Appeared in the Year 1572*. Trans. V.V.S. 1632. Facsimile reprint, New York: Da Capo, 1969.

———. "Tycho Brahe's System of the World." Trans. Marie Boas and A. R. Hall. *Royal Astronomical Society Occasional Notes* 3, no. 21 (1959): 253–263.

Bruford, W. H. *Culture and Society in Classical Weimar, 1775–1806*. Cambridge: Cambridge University Press, 1962.

Bruno, Giordano. *The Ash Wednesday Supper: La Cena de le Ceneri.* Trans., intro., notes Stanley Jaki. The Hague: Mouton, 1975.

———. *Cause, Principle and Unity.* Trans., intro. Jack Lindsay. New York: International Publishers, 1962.

Bukharin, Nikolai I., et al. *Science at the Crossroads.* 2nd ed., 1931. Reprint with new prefaces by Gary Werskey and Joseph Needham. London: Cass, 1971.

Bunge, Mario. "Lagrangian Formalism and the Mechanical Model." *American Journal of Physics* 25, no. 4 (April 1957): 211–218.

Burkert, Wilhelm. *Lore and Science in Ancient Pythagoreanism.* Cambridge, Mass.: Harvard University Press, 1972.

Burtt, Edwin Arthur. *The Metaphysical Foundations of Modern Physical Science.* Rev. ed. Garden City, N.Y.: Doubleday Anchor, 1954.

Bynum, W. F., E. J. Browne, and Roy Porter, eds. *Dictionary of the History of Science.* Princeton, N.J.: Princeton University Press, 1981.

Campbell, Lewis, and William Garnett, *The Life of James Clerk Maxwell.* 1st ed. 1882. Reprinted with letters from 2nd ed. New York: Johnson, 1969.

Caneva, Kenneth L. "Ampère, the Etherians, and the Oersted Connection." *British Journal for the History of Science* 13, no. 4 (1980): 122–138.

———. "From Galvanism to Electrodynamics: The Transformation of German Physics and Its Social Context." *Historical Studies in the Physical Sciences* 9 (1978): 63–159.

Cantor, Geoffrey. *Michael Faraday: Sandemanian and Scientist.* New York: St. Martin's, 1991.

———, and M.J.S. Hodge, eds., *Conceptions of Ether.* Cambridge: Cambridge University Press, 1981.

Caspar, Max. *Kepler.* Trans., ed. C. Doris Hellman. New York: Dover, 1993.

Cassirer, Ernst. "Giovanni Pico della Mirandola: A Study in the History of Renaissance Ideas." *Journal of the History of Ideas* 3 (1942): 123–144, 319–346.

———. *Rousseau, Kant, Goethe.* Trans. James Guttmann, Paul Oscar Kristeller, and John Hermann Randall, Jr. 1945. Reprint. New York: Harper and Row, 1963.

Chalmers, A. F. "Maxwell's Methodology and His Application of It to Electromagnetism." *Studies in the History and Philosophy of Science* 4 (1973): 107–164.

Chu Hsi. *Learning to Be a Sage.* Intro. Daniel K. Gardner. Berkeley: University of California Press, 1990.

Christensen, Dan C. "The Oersted-Ritter Partnership and the Birth of Romantic Natural Philosophy." *Annals of Science* 52 (1995): 153–185.

Chuang Tzu. *Chuang Tzu: Basic Writings.* Trans. Burton Watson. New York: Columbia University Press, 1964.

———. *The Sayings of Chuang Chou.* Trans. James R. Ware. New York: Mentor, 1963.

Clark, Bruce. "Allegories of Victorian Thermodynamics." *Configurations: A Journal of Literature and Science* 4, no. 1 (Winter 1996): 67–90.

Clulee, Nicholas H. *John Dee's Natural Philosophy: Between Science and Religion.* London: Routledge, 1988.

Coe, Michael D. *Breaking the Maya Code.* London: Thames and Hudson, 1992.

Cohen, I. Bernard. *Introduction to Newton's "Principia."* Cambridge, Mass.: Harvard University Press, 1978.

———. ed. *Puritanism and the Rise of Modern Science: The Merton Thesis.* New Brunswick, N.J.: Rutgers University Press, 1990.

Coleridge, Samuel Taylor. *Biographia Literaria.* Princeton: Princeton University Press, 1983.

————. *Hints Towards the Formation of a More Comprehensive Theory of Life*. Ed. Seth B. Watson. Reprint. Farnborough, Eng.: Gregg, 1970.

Conway, Anne. *Principles of the Most Ancient and Modern Philosophy*. Trans. Allison Coudert and Taylor Corse. Cambridge: Cambridge University Press, 1996.

Cook, Daniel J., and Henry Rosemont, Jr. "The Pre-Established Harmony between Leibniz and Chinese Thought." *Journal of the History of Ideas* 42 (1981): 253–267.

Copenhaver, Brian P. "Jewish Theologies of Space in the Scientific Revolution: Henry More, Joseph Raphson, Isaac Newton and Their Predecessors." *Annals of Science* 37 (1980): 489–548.

————. ed., trans., and intro. *Hermetica: The Greek "Corpus Hermeticum" and the Latin "Asclepius" in a New English Translation, with Notes and Introduction*. Cambridge: Cambridge University Press, 1992.

————. and Charles B. Schmitt. *Renaissance Philosophy*. Oxford: Oxford University Press, 1992.

Coudert, Allison P. *Leibniz and the Kabbalah*. Boston: Kluwer, 1995.

Crombie, A. C. *Medieval and Early Modern Science*. 2 vols., Garden City, N.Y.: Doubleday Anchor, 1959.

Crowe, Michael. *A History of Vector Analysis*. 1967. Reprint. New York: Dover, 1985.

Cullen, Christopher. "Some Further Points on the *Shih*." *Early China* 6 (1980–1981): 31–46.

Cunningham, Andrew, and Nicholas Jardine, eds. *Romanticism and the Sciences*. Cambridge: Cambridge University Press, 1990.

Darrigol, Oliver. "The Origin of Quantitized Matter Waves." *Historical Studies in the Physical Sciences* 16 (1976): 197–253.

Davy, Humphry. *Consolations in Travel; or, The Last Days of a Philosopher*. 5th ed. London: John Murray, 1851.

De Bary, William Theodore. *Learning for One's Self*. New York: Columbia University Press, 1991.

————, Wing-Tsit Chan, and Burton Watson, eds. *Sources of Chinese Tradition*. Vol. 1. New York: Columbia University Press, 1960.

Debus, Allen G. *The English Paracelsians*. London: Oldbourne, 1965.

————. *Man and Nature in the Renaissance*. Cambridge: Cambridge University Press, 1978.

Dee, John. *General and Rare Memorials Pertayning to the Perfect Arte of Navigation*. 1577. Reprint. Amsterdam: Da Capo, 1968.

————. *The Private Diary of John Dee and a Catalogue of Manuscripts*. Ed. James Orchard Halliwell, Esq. F.R.S. 1842. Reprint. Kila, Mont.: Kessinger, n.d.

DeFrancis, John. *The Chinese Language: Fact and Fantasy*. Honolulu: University of Hawaii Press, 1984.

Derrida, Jacques. *Dissemination*. Trans. Barbara Johnson. Chicago: University of Chicago Press, 1981.

————. *Of Grammatology*. Trans. Gayatri Chakravorty Spivak. Baltimore: Johns Hopkins University Press, 1976.

Desaguliers, J. T. *The Newtonian System of the World as the Best Model for Government: An Allegorical Poem*. London: A. Campbell for J. Roberts, 1728.

Dibner, Bern. *Oersted and the Discovery of Electromagnetism*. New York: Blaisdell, 1962.

Dijksterhuis, E. J. *The Mechanization of the World Picture*. Trans. C. Dikshoorn. Oxford: Oxford University Press, 1961.

Dobbs, Betty Jo Teeter. *The Foundations of Newton's Alchemy: The Hunting of the Greene Lyon*. Cambridge: Cambridge University Press, 1975

————. *The Janus Faces of Genius: The Role of Alchemy in Newton's Thought.* Cambridge: Cambridge University Press, 1991.

————, and Margaret C. Jacob. *Newton and the Culture of Newtonianism.* Atlantic Highlands, N.J.: Humanities Press, 1995.

Doran, Barbara Gusti. "Origins and Consolidation of Field Theory in Nineteenth Century Britain: From the Mechanical to the Electromagnetic View of Nature." *Historical Studies in the Physical Sciences* 6 (1975): 133–260.

Dreyer, J.L.E. *A History of Astronomy from Thales to Kepler.* [Formerly titled *History of Planetary Systems from Thales to Kepler.* Cambridge: Cambridge University Press, 1905.] 2nd ed., rev. with foreword by W. H. Stahl. New York: Dover, 1953.

————. *Tycho Brahe: A Picture of Scientific Life and Work in the Sixteenth Century.* 1890. Reprint. New York: Dover, 1963.

Duhem, Pierre. *The Aim and Structure of Physical Theory.* Trans. Philip P. Wiener; foreword by Prince Louis de Broglie. Princeton, N.J.: Princeton University Press, 1962.

————. *To Save the Phenomena.* Trans. Edmund Dolland and Chaninah Maschler; intro. Stanley Jaki. Chicago: University of Chicago Press, 1969.

Durkheim, Emile. *The Elementary Forms of the Religious Life.* New York: Free Press, 1965.

————. and Marcel Mauss. *Primitive Classification.* Chicago: University of Chicago Press, 1963.

Eamon, William. *Science and the Secrets of Nature: Books of Secrets in Medieval and Early Modern Culture.* Princeton: Princeton University Press, 1994.

Easlea, Brian. *Witch Hunting, Magic, and the New Philosophy: An Introduction to Debates of the Scientific Revolution, 1450–1750.* Atlantic Highlands, N.J.: Humanities Press, 1980.

Edwards, Paul, ed. *The Encyclopedia of Philosophy.* 8 vols. New York: Macmillan and Free Press, 1967.

Ehrenreich, Barbara, and Deirdre English. *Witches, Midwives and Nurses: A History of Women Healers.* Old Westbury, N.Y.: Feminist Press, 1973.

Einstein, Albert. *Ideas and Opinions.* Trans. Sonja Bargmann. New York: Crown, 1954.

————. *Sidelights on Relativity.* Trans. G. B. Jeffery and W. Perrett. 1922, Reprint. New York: Dover, 1983.

————, and Leopold Infeld. *The Evolution of Physics.* New York: Simon and Schuster, 1961.

Eisele, Carolyn, ed. *Historical Perspectives on Peirce's Logic of Science: A History of Science.* 2 vols. Berlin: Walter de Gruyter, 1985.

Eliade, Mircea. *The Forge and the Crucible.* New York: Harper and Row, 1971.

————, gen. ed. *Encyclopedia of Religion.* New York: Macmillan, 1987.

Engels, Friedrich. *Dialectics of Nature.* New York: International Publishers, 1940.

————. *The Origin of the Family, Private Property and the State*, ed., intro. Eleanor Burke Leacock. New York: International Publishers, 1972.

Esposito, Joseph L. *Schelling's Idealism and Philosophy of Nature.* Lewisburg, Pa.: Bucknell University Press, 1977.

Evans, R.J.W. *Rudolf II and His World.* Oxford: Oxford University Press, 1973

Faraday, Michael. "On the Various Forces of Nature," and "The Chemical History of the Candle." In *Scientific Papers.* Vol. 30 of *The Harvard Classics.* New York: Collier, 1938.

Farrar, Cynthia. *The Origins of Democratic Thinking: The Invention of Politics in Classical Athens.* Cambridge: Cambridge University Press, 1988.

Fauvel, John, Raymond Flood, Michael Shortland, and Robin Wilson, eds. *Let Newton Be! A New Perspective on His Life and Works.* Oxford: Oxford University Press, 1988.

Ferber, Michael. *The Poetry of William Blake.* London: Penguin, 1991.

Ferro, Marc, and the Staff of *Annales*, eds. *Social Historians in Contemporary France: Essays from "Annales."* New York: Harper and Row, 1972.

Feuer, Lewis Samuel. *Spinoza and the Rise of Liberalism.* Boston: Beacon, 1958.

Field, J. V. *Kepler's Geometrical Cosmology.* Chicago: University of Chicago Press, 1988.

———. "Kepler's Star Polyhedra." *Vistas in Astronomy* 23 (1980): 109–141.

———. "A Lutheran Astrologer, Johannes Kepler." *Archive for the History of the Exact Sciences* 31 (1984): 191–225.

Figala, Karin. "Newton as Alchemist." *History of Science* 15 (1977): 102–137.

Findlen, Paula. *Possessing Nature: Museums, Collecting, and Scientific Culture in Early Modern Italy.* Berkeley: University of California Press, 1994.

Forster, Georg. *A Voyage round the World in His Britannic Majesty's Sloop, "Resolution" Commanded by Captain James Cook, during the Years 1772, 3, 4 and 5.* 2 vols. London: B. White, J. Robson, P. Elmsly, and G. Robinson, 1777.

Foucault, Michel. *Madness and Civilization: A History of Madness in the Age of Reason.* New York: New American Library of World Literature, Anchor, 1964.

———. *The Order of Things: A Translation of "Les Mots et les Choses."* 1968. Reprint. New York: Vintage, 1973.

Fowden, Garth. *The Egyptian Hermes: A Historical Approach to the Late Pagan Mind.* 1986, Reprint, with new preface. Princeton: Princeton University Press, 1993.

Fraassen, Bas van, *Formal Semantics and Logic*, New York: Macmillan, 1971.

Fraser, J. T., ed. *The Voices of Time: A Cooperative Survey of Man's Views of Time as Expressed by the Sciences and by the Humanities.* 2nd ed. Amherst: University of Massachusetts Press, 1981.

French, Peter. *John Dee.* New York: Dorset Press, 1989.

Freudenthal, Gad. *Atom and Individual in the Age of Newton.* Dordrecht: D. Reidel, 1986.

Fruman, Norman. *Coleridge: The Damaged Archangel.* New York: George Braziller, 1971.

Fung Yu-lan. *A History of Chinese Philosophy.* 2 vols. Princeton: Princeton University Press, 1983.

———. *A Short History of Chinese Philosophy.* Trans. Derk Bodde. New York: Macmillan, 1948.

Galilei, Galileo. *Dialogue Concerning the Two Chief World Systems.* Trans. Stillman Drake. 1953. Reprint. Berkeley: University of California Press, 1960.

Garrett, Don, ed. *The Cambridge Companion to Spinoza.* Cambridge: Cambridge University Press, 1996.

Gavroglu, Kostas, Jean Christianidis, and Efthymios Nicolaidis, eds. *Trends in the Historiography of Science.* Boston Studies in the Philosophy of Science, vol. 151. Dordrecht: Kluwer, 1994.

Gilbert, William. *De Magnete.* Trans. P. Fleury Mottelay. New York: Dover, 1958.

Gillispie, Charles Coulston. *The Edge of Objectivity: An Essay in the History of Scientific Ideas.* Princeton: Princeton University Press, 1960.

———, gen. ed. *A Dictionary of Scientific Biography.* 18 vols. New York: Scribner, 1970–1990.

Gingerich, Owen, ed. *The Eye of Heaven: Ptolemy, Copernicus, Kepler.* New York: American Institute of Physics, 1993.

Ginzburg, Carlo. *The Cheese and the Worms.* Trans. John Tedeschi and Anne Tedeschi. Harmondsworth, Eng.: Penguin, 1982.

———. *Ecstacies: Deciphering the Witches' Sabbath.* 1991. Reprint. New York: Penguin, 1992.

Girardot, Norman J. *Myth and Meaning in Early Taoism: The Theme of Chaos (Hun-tun)*. Berkeley: University of California Press, 1983.

Gjertsen, Derek. *The Newton Handbook*. London: Routledge and Kegan Paul, 1986.

Gleich, James. *Genius: The Life and Science of Richard Feynman*. New York: Pantheon, 1992.

Godwin, Joscelyn. *Athanasius Kircher*. London: Thames and Hudson, 1979.

———. *Robert Fludd: Hermetic Philosopher and Surveyor of Two Worlds*. Boulder, Colo.: Shambhala, 1979.

Goethe, Johann Wolfgang von. *Elective Affinities*. Trans. James Anthony Froude and R. Dillon Boylan; intro. Frederick Ungar. New York: Frederick Ungar, 1962.

———. *Goethe's Botanical Writings*. Trans. Bertha Mueller. Woodbridge, Conn.: Ox Bow, 1989.

———. *Scientific Studies*. Ed. and Trans. Douglas Miller. Vol. 12 of *Goethe Edition*. New York: Suhrkamp Verlag, 1988.

———. *Theory of Colors*. Trans., notes Charles Lock Eastlake; intro. Deane B. Judd. Cambridge: MIT Press, 1970.

Golinski, Jan. *Science as Public Culture: Chemistry and Enlightenment in Britain, 1760–1820*. Cambridge: Cambridge University Press, 1992.

Gooch, G. P. *Germany and the French Revolution*. 1920. Reprint. London: Russell and Russell, Inc., 1966.

Gooding, David. "Conceptual and Experimental Bases of Faraday's Denial of Action at a Distance." *Studies in the History and Philosophy of Science* 9 (1978): 117–149.

———. *Experiment and the Making of Meaning*. Dordrecht: Kluwer, 1990.

———. "Metaphysics vs. Measurement: The Conservation of Force in Faraday's Physics." *Annals of Science* 37 (1980): 1–29.

———, and F. James, eds. *Faraday Rediscovered*. London: Macmillan, 1985.

Goodman, Katherine R., and Edith Waldstein, eds. *In the Shadow of Olympus: German Women Writers around 1800*. Albany: State University of New York Press, 1992.

Gower, Barry. "Speculation in Physics: The History and Practice of *Naturphilosophie*." *Studies in the History and Philosophy of Science* 3 (1973): 301–356.

Graham, A. C. *Disputers of the Tao*. La Salle, Ill.: Open Court, 1989.

———. *Later Mohist Logic, Ethics and Science*. Hong Kong: Chinese University Press, 1978.

———. *Studies in Chinese Philosophy and Philosophical Literature*. Albany: State University of New York Press, 1990.

———. *Unreason within Reason: Essays on the Outskirts of Rationality*. La Salle, Ill.: Open Court, 1992.

Gramsci, Antonio. *Selections from the Prison Notebooks of Antonio Gramsci*. New York: International Publishers, 1977.

Granet, Marcel. *Chinese Civilization*. London: Kegan Paul, Trench, Trubner, 1930.

———. *Festivals and Songs of Ancient China*. New York: E. P. Dutton, 1932.

———. *La Pensée Chinoise*. 1934. Reprint. Paris: Albin, Michel, 1968.

———. *The Religion of the Chinese People*. Trans. Maurice Freedman. New York: Harper and Row, 1977.

Greene, John C. *The Death of Adam*. New York: New American Library of World Literature, 1961.

Grene, Marjorie, and Debra Nails, eds. *Spinoza and the Sciences*. Dordrecht: D. Reidel, 1986.

Hall, A. Rupert. *Science and Society: Historical Essays on the Relations of Science, Technology and Medicine*. Brookfield, Vt.: Ashgate, 1994.

Hall, David L., and R. T. Ames. *Thinking Through Confucius*. Albany: State University of New York Press, 1987.

Hallyn, Fernand. *The Poetic Structure of the World: Copernicus and Kepler*. New York: Zone Books, 1990.

Hammer, Christy. "A Sociological Study of the Manifestation of Multi-Culturalism in Public Schools." Ph.D. diss., University of New Hampshire, 1994.

Hankins, Thomas. *Jean d'Alembert*. Oxford: Oxford University Press, 1970.

————. *William Rowan Hamilton*. Baltimore: Johns Hopkins University Press, 1979.

Hansen, Chad. *A Daoist Theory of Chinese Thought*. Oxford: Oxford University Press, 1992.

————. *Language and Logic in Ancient China*. Ann Arbor: University of Michigan Press, 1983.

Hanson, Norwood Russell. "Contra-Equivalence: A Defense of the Originality of Copernicus." *Isis* 55 (1964): 308–325.

————. "The Copernican Disturbance and the Keplerian Revolution." *Journal of the History of Ideas* 22 (1961): 169–184.

————. *Patterns of Discovery*. Cambridge: Cambridge University Press, 1958.

Hardin, James, and Christoph E. Schweitzer. *Dictionary of Literary Biography*. Vol. 90, *German Writers in the Age of Goethe , 1789–1832*. Detroit: Gale Research, 1989.

Harper, Donald J. "The Han Cosmic Board: A Response to Christopher Cullen." *Early China* 6 (1980–1981): 47–56.

————. "The Han Cosmic Board (*Shih*)." *Early China* 4 (1978–1979): 1–10.

Harré, Rom. *The Philosophies of Science*. Oxford: Oxford University Press, 1972, 1984.

————. *Principles of Scientific Thinking*. Chicago: University of Chicago Press, 1970.

Harrington, Anne. *Reenchanted Science: Holism in German Culture from Wilhelm II to Hitler*. Princeton: Princeton University Press, 1996.

Harris, Marvin. *Cultural Materialism*. New York: Random House, 1979.

————. *The Rise of Anthropological Theory: A History of Theories of Culture*. New York: Columbia University Press, 1968.

Hartley, Harold. *Humphry Davy*. Wakefield, Eng.: EP Publishing, 1972.

Hartwell, Robert M. "Historical Analogism, Public Policy and Social Science in Eleventh- and Twelfth-Century China." *American Historical Review* 76 (1971): 690–727.

Heer, Friedrich. *The Intellectual History of Europe*. Trans. Jonathan Steinberg. London: Weidenfeld and Nicholson, 1966.

————. *The Medieval World: Europe 1100–1350*. Trans. Janet Sondheimer. London: Weidenfeld and Nicholson, 1962; New York: New American Library of World Literature, 1963.

Heimann, P. M. "Faraday's Theories of Matter and Electricity." *British Journal for the History of Science* 5, no. 19 (1971): 235–257.

————. "Maxwell and the Modes of Consistent Representation." *Archive for the History of the Exact Sciences* 6 (1970): 171–213.

————, and J. E. McGuire. "Newtonian Forces and Lockean Powers: Concepts of Matter in Eighteenth-Century Thought." *Historical Studies in the Physical Sciences* 3 (1971): 233–306.

Heisenberg, Werner. *Philosophical Problems of Quantum Physics*. [Originally *Philosophical Problems of Nuclear Science*. New York: Pantheon Books, 1952.] Reprint. New Haven, Conn.: Ox Bow, 1979.

————. *Physics and Philosophy: The Revolution in Modern Science.* Intro. F.S.C. Northrop. 1958. Reprint. New York: Harper Torchbooks, 1962.

Henderson, John B. *The Development and Decline of Chinese Cosmology.* New York: Columbia University Press, 1984.

Hendry, John. "Discussion: Monopoles before Dirac." *Studies in the History and Philosophy of Science* 14 (1983): 81–87.

————. *James Clerk Maxwell and the Theory of the Electromagnetic Field.* Bristol, Eng.: Adam Hilgar, 1986.

Herder, Johann Gottfried von. *God: Some Conversations.* Trans. Frederick H. Burkhardt. Indianapolis: Bobbs-Merrill, 1940.

Hertz, Deborah. *Jewish High Society in Old Regime Berlin.* New Haven, Conn.: Yale University Press, 1988.

Hertz, Heinrich. *Electric Waves.* Trans. D. E. Jones; preface by Lord Kelvin. 1893. Reprint. New York: Dover, 1962.

————. *The Principles of Mechanics Presented in a New Form.* Preface by Hermann von Helmholtz; trans. D. E. Jones and J. T. Walley. 1899. Reprint with new intro. Robert S. Cohen. New York: Dover, 1956.

Hesse, Mary. *Forces and Fields: The Concept of Action at a Distance in the History of Physics.* Totowa, N.J.: Littlefield, Adams, 1965.

————. "Gilbert and the Historians." *British Journal for the Philosophy of Science* 11 (1960): 1–10, 130–142.

————. *Models and Analogies in Science.* Notre Dame, Ind.: University of Notre Dame Press, 1966.

————. "Logic of Discovery in Maxwell's Electromagnetic Theory." In *Scientific Method in the Nineteenth Century,* ed. Ronald Giere and Richard S. Westfall. Bloomington: University of Indiana Press, 1973.

Hill, Christopher. *The Century of Revolution, 1630–1714.* New York: Norton, 1961.

————. *Intellectual Origins of the English Revolution.* London: Panther Books, 1972.

————. *Some Intellectual Consequences of the English Revolution.* Madison: University of Wisconsin Press, 1980.

————. *The World Turned Upside Down.* New York: Penguin, 1991.

Hill, Frances. *The Delusion of Satan: The Full Story of the Salem Witch Trials.* New York: Doubleday, 1995.

Holton, Gerald. *Thematic Origins of Scientific Thought: Kepler to Einstein.* Cambridge, Mass.: Harvard University Press, 1973.

Holzman, Donald. *Poetry and Politics: The Life and Art of Juan Chi, A.D. 210–263.* Cambridge: Cambridge University Press, 1976.

Home, R. W. *Electricity and Experimental Physics in Eighteenth-Century Europe.* Brookfield, Vt.: Ashgate Variorum, 1992.

Hsi Kang. *Philosophy and Argumentation in Third-Century China: The Essays of Hsi Kang.* Transl., intro. Robert G. Henricks. Princeton, N.J.: Princeton University Press, 1983.

Hunt, Bruce J. *The Maxwellians.* Ithaca, N.Y.: Cornell University Press, 1991.

Hutchinson, Keith. "What Happened to Occult Qualities in the Scientific Revolution?" *Isis* 73 (1982): 233–253.

I Ching; or, The Book of Changes. Trans. Richard Wilhelm, and Cary F. Baynes. Princeton, N.J.: Princeton University Press, 1967.

Iverson, Erik. *The Myth of Egypt and Its Heiroglyphs in European Tradition.* 1961. Reprint with new preface. Princeton, N.J.: Princeton University Press, 1993.

Jackson, Miles. "A Spectrum of Belief: Goethe's 'Republic' versus Newtonian Despotism." *Social Studies of Science* 24 (1994): 673–701.

Jacob, Margaret C. *The Cultural Meaning of the Scientific Revolution.* Philadelphia: Temple University Press, 1988.

———. *Living the Enlightenment: Freemasonry and Politics in Eighteenth Century Europe.* Oxford: Oxford University Press, 1991.

———. *The Newtonians and the English Revolution, 1689–1720.* Ithaca, N.Y.: Cornell University Press, 1976.

Jammer, Max. *Concepts of Space.* Foreword by Albert Einstein. New York: Harper, 1960.

Jolley, Nicholas, ed. *The Cambridge Companion to Leibniz.* Cambridge: Cambridge University Press, 1995.

Jung, Carl Gustav. *Psychology and Alchemy.* Trans. R.F.C. Hull. 2nd ed. Princeton: Princeton University Press, Bollingen Series, 1968.

Kaltenmark, Max. *Lao Tzu and Taoism.* Trans. Roger Greaves. Stanford, Calif.: Stanford University Press, 1969.

Kant, Immanuel. *Critique of Judgement.* Trans. Werner Pluhar; foreword by Mary Gregor. Indianapolis, Ind.: Hackett, 1987.

———. *Immanuel Kant's Critique of Pure Reason.* Trans. Norman Kemp Smith. London: Macmillan, 1929.

———. *Kant's Latin Writings.* Ed. Lewis White Beck. New York: Peter Lang, 1986.

———. *Opus Posthumum.* Ed. Eckhart Föster; trans. Eckhart Föster and Michael Rosen. Cambridge: Cambridge University Press, 1993.

———. *Philosophical Correspondence, 1759–99.* Ed. and Trans. Arnult Zweig. Chicago: University of Chicago Press, 1967.

———. *Philosophy of Material Nature.* Trans. James Ellington. Indianapolis: Hackett, 1985.

Kargon, Robert Hugh. *Atomism in England from Hariot to Newton.* Oxford: Clarendon, 1966.

———. "W. R. Hamilton and Boscovichean Atomism." *Journal of the History of Ideas* 26 (1965): 137–141.

———. "William Rowan Hamilton, Michael Faraday, and the Revival of Boscovichean Atomism." *American Journal of Physics* 32 (1964): 792–795.

Karlsen, Carol F. *The Devil in the Shape of a Woman.* New York: Vintage, 1987.

Kearney, Hugh. *Science and Change, 1500–1700.* New York: McGraw-Hill, 1971.

Keller, Evelyn Fox, *Reflections on Gender and Science.* New Haven, Conn.: Yale University Press, 1985.

Kepler, Johannes. *Epitome of Copernican Astronomy, Books 4 and 5,* and *The Harmonies of the World, Book 5,* Trans. Charles Glenn Wallis. Vol. 16 of *The Great Books of the Western World.* ed. Robert Maynard Hutchins. Chicago: Encyclopaedia Britannica and William Benton, 1952.

———. "Johannes Kepler on the More Certain Fundamentals of Astrology: Prague 1601." Foreword, Notes, and Analytical Outline by J. Bruce Brackenridge; trans. Mary Ann Rossi. *Proceedings of the American Philosophical Society* 123 (1979): 85–116.

———. *Kepler's "Somnium: The Dream; or, Posthumous Work on Lunar Astronomy."* Trans. Edward Rosen. Madison: University of Wisconsin Press, 1967.

———. *Mysterium Cosmographicum: The Secret of the Universe.* Trans. A. M. Duncan. New York: Abaris, 1981.

———. *New Astronomy.* Trans. William H. Donahue. Cambridge: Cambridge University Press, 1992.

Keynes, John Maynard. "Newton the Man." In *Essays in Biography*, ed. Geoffrey Keynes. New York: Norton, 1963.

Kircher, Athanasius. *China Illustrated*. Trans. Charles D. Van Tuyl from the 1677 original Latin ed. Muskogee, Okla.: Indian University Press, 1987.

———. *Magnes: Sive de Arte Magnetica*. 3 parts. Coloniae Agrippinae apud Iodocum Kalcoven, 1643.

———. *Mundus Subterraneus*. 3rd ed. Amstelodami: Joannem Jansnssonium a Waesberg & Filios, 1678.

Kirk, G. S., and J. E. Raven. *The Presocratic Philosophers*. Cambridge: Cambridge University Press, 1957.

Kneightly, David N. "The Origin of the Ancient Chinese City: A Comment." *Early China* 1 (1975): 3–5.

Knight, David. " German Science in the Romantic Period." In *The Emergence of Science in Western Europe*, ed. Maurice Crosland. Cambridge: Cambridge University Press, 1975.

———. *Humphry Davy*. Oxford: Blackwell, 1992.

———. "The Scientist as Sage." *Studies in Romanticism* 6 (1967): 64–87.

———. *The Transcendental Part of Chemistry*. Folkestone, Eng.: William Dawson, 1978.

———. *The Origins of Chinese Civilization*. Berkeley: University of California Press, 1983.

Koerner, Lisbet. "Goethe's Botany: Lessons of a Feminine Science." *Isis* 84 (1993): 470–495.

Koerner, Stephan, ed. *Observation and Interpretation*. New York: Dover, 1957

Koestler, Arthur. *The Roots of Coincidence*. New York: Vintage, 1973.

———. *The Watershed*. Garden City, N.Y.: Doubleday, 1960.

Kohn, Livia Knaul. *Early Chinese Mysticism: Philosophy and Soteriology in the Taoist Tradition*. Princeton: Princeton University Press, 1991.

———. *Laughing at the Tao Debates among Buddhists and Taoists in Medieval China*. Princeton: Princeton University Press, 1995.

———. "Lost Chuang-Tzu Passages." *Journal of Chinese Religions* 10 (fall 1982): 53–79.

Koyré, Alexandre. *From the Closed World to the Infinite Universe*. Baltimore: Johns Hopkins University Press, 1957.

———. *Newtonian Studies*. Chicago: University of Chicago Press, 1968.

Kozhamthadam, Job. *The Discovery of Kepler's Laws: The Interaction of Science, Philosophy and Religion*. Notre Dame, Ind.: University of Notre Dame Press, 1994.

Kragh, Helge. " The Concept of the Monopole: A Historical and Analytical Case Study." *Studies in the History and Philosophy of Science* 12 (1981): 141–172.

Kuhn, Thomas S. *The Copernican Revolution: Planetary Astronomy in the Development of Western Thought*. 1957. Corrected reprint. New York: Vintage, 1959.

———. *The Structure of Scientific Revolutions*. 1962. 2nd ed. Chicago: University of Chicago Press, 1970.

Lach, Donald F. "Leibniz and China." *Journal of the History of Ideas* 6 (1945): 436–455.

———, ed. and trans. *The Preface to Leibniz' "Novissima Sinica."* Honolulu: University of Hawaii Press, 1957.

———, and Edwin J. Van Kley (in vol. 3). *Asia in the Making of Europe*. 3 vols. in 9. Chicago: University of Chicago Press, 1965–1993.

Lakatos, Imre. *Philosophical Papers*. 2 vols. Ed. John Worrall and Gregory Currie. Cambridge: Cambridge University Press, 1978.

Lao Tzu. *Tao Te Ching*. Trans. Victor H. Mair. New York: Bantam, 1990.

————. *Te-Tao Ching.* Trans. Robert G. Hendricks. New York: Ballantine, 1989.

Laudan, Larry. *Between Positivism and Relativism: Theory, Method and Evidence.* Boulder, Colo.: Westview, 1996.

————. *Progress and Its Problems.* Berkeley: University of California Press, 1977.

Le Blanc, Charles. *Huai-nan-tzu: Philosophical Synthesis of Early Han Thought—The Idea of Resonance (Kan-Ying), with a Translation and Analysis of Chapter Six.* Hong Kong: Hong Kong University Press, 1985.

Lefkowitz, Mary. *Not Out of Africa: How Afrocentrism Became an Excuse to Teach Myth as History.* New York: Basic Books, 1996.

————, and Guy MacLean Rogers, eds. *Black Athena Revisited.* Chapel Hill: University of North Carolina Press, 1996.

Leibniz, Gottfried Wilhelm. *Discourse on the Natural Theology of the Chinese.* Trans. Henry Rosemont, Jr., and Daniel J. Cook. Honolulu: University of Hawaii Press, 1977.

————. *Philosophical Essays.* Trans. Roger Ariew and Daniel Garber. Indianapolis: Hackett, 1989.

————. *Writings on China.* Trans., ed., intro. Daniel J. Cook and Henry Rosemont, Jr. Chicago: Open Court, 1994.

Lessing, Gotthold Ephraim, Friedrich Heinrich Jacobi, and Moses Mendelssohn. *The Spinoza Conversations between Lessing and Jacobi: Text with Excerpts from the Ensuing Controversy.* Trans. Gérard Vallée, J. B. Lawson, and C. G. Chapple. Lanham, Md.: University Press of America, 1988.

Levenson, Joseph R. *Modern China and Its Confucian Past: The Problem of Intellectual Continuity.* [First published as *Confucian China and Its Modern Fate.* Berkeley: University of California Press, 1958.] Garden City, N.Y.: Doubleday, Anchor, 1964.

Levere, Trevor H. *Affinity and Matter.* Oxford: Clarendon, 1971.

————. *Chemists and Chemistry in Nature and Society, 1770–1878.* Brookfield, Vt: Ashgate Variorum, 1994.

————. "Coleridge, Chemistry, and the Philosophy of Nature." *Studies in Romanticism* 16 (1977): 349–380.

————. "Faraday, Matter and Natural Theology." *British Journal for the History of Science* 4 (1968): 95–107.

————. *Poetry Realized in Nature.* Cambridge: Cambridge University Press, 1981.

Lévi-Strauss, Claude. *The Elementary Structure of Kinship.* Trans. J. S. Bell, J. H. von Sturmer, and Rodney Needham. Boston: Beacon, 1968.

————. *Structural Anthropology.* Trans. Claire Jacobson. New York: Basic Books, 1963.

Lichtheim, George. *The Concept of Ideology.* New York: Vintage, 1967.

Lieh Tzu, The *Book of Lieh Tzu: A Classic of the Tao.* Trans. A. C. Graham. 1960. Reprint. New York: Columbia University Press, 1990.

Lindberg, David C., and Robert S. Westman, eds. *Reappraisals of the Scientific Revolution.* Cambridge: Cambridge University Press, 1990.

Liu I-ch'ing. "*Shih-shuo Hsin-yü*": *A New Account of Tales of the World.* Commentary by Liu Chün; trans. Richard B. Mather. Minneapolis: University of Minnesota Press, 1976.

Lloyd, G.E.R. *Adversaries and Authorities: Investigations into Ancient Greek and Chinese Science.* Cambridge: Cambridge University Press, 1996.

————. *Demystifying Mentalities.* Cambridge: Cambridge University Press, 1990.

————. *Polarity and Analogy.* 1966. Reprint. Indianapolis: Hackett, 1992.

Loemker, Leroy. "A Note on the Origin and Problem of Leibniz's Discourse of 1686." *Journal of the History of Ideas* 8 (1947): 449–466.

————. *The Struggle for Synthesis: The Seventeenth Century Background of Leibniz's Synthesis of Order and Freedom*. Cambridge, Mass.: Harvard University Press, 1972.

Loewe, Michael. *Chinese Ideas of Life and Death: Faith, Myth and Reason in the Han Period (202 B.C. to A.D. 220)*. London: George Allen and Unwin, 1982.

————. *Ways to Paradise: The Chinese Quest for Immortality*. London: George Allen and Unwin, 1979.

MacAdam, David L., ed. *Sources of Color Science*. Cambridge, Mass.: MIT Press, 1970.

McCormach, Russell. "H. A. Lorentz and the Electromagnetic View of Nature." *Isis* 51 (1970): 459–497.

McGuire, J. E. "Force, Active Principles, and Newton's Invisible Realm." *Ambix* 15 (1968): 154–208.

————, and Martin Tammy. *Certain Philosophical Questions: Newton's Trinity Notebook*. Cambridge: Cambridge University Press, 1983.

————, and P. M. Rattansi. "Newton and the 'Pipes of Pan.'" *Notes and Records of the Royal Society* 21 (1966): 108–142.

Mach, Ernst. *The Science of Mechanics*. La Salle, Ill.: Open Court, 1960.

Machamer, Peter, and Robert G. Turnbull, eds. *Motion and Time, Space and Matter*. Columbus: Ohio State University Press, 1976.

McMullin, Ernan. *Newton on Matter and Activity*. Notre Dame, Ind.: University of Notre Dame Press, 1978.

————, ed. *Galileo, Man of Science*. Princeton Junction, N.J.: Scholar's Bookshelf, 1988.

Major, John S. *Heaven and Earth in Early Han Thought: Chapters Three, Four and Five of the Huainanzi*. Albany: State University of New York Press, 1993.

Malebranche, Nicholas. *Dialogue between a Christian Philosopher and a Chinese Philosopher on the Existence and Nature of God*. Trans. Dominick A. Iorio. Washington, D.C.: University Press of America, 1980.

Mandrou, Robert. *From Humanism to Science, 1480–1700*. Trans. Brian Pearce. New York: Penguin, 1978.

Mannheim, Karl. *Ideology and Utopia*. Trans. Louis Wirth and Edward Shils. 1936. Reprint. N.p.: Harcourt Brace and World, n.d.

Manuel, Frank E. *A Portrait of Isaac Newton*. Cambridge, Mass.: Harvard University Press, 1968.

Marshack, Alexander. *The Roots of Civilization*. Rev. ed. Mount Kisco, N.Y.: Moyer Bell, 1991.

Marx, Karl. *Capital*. Vol. 1. Trans. Samuel Moore and Edward Aveling. New York: International Publishers, 1967.

————. *The Economic and Philosophic Manuscripts of 1844*. Ed. Dirk J. Struik. New York: International Publishers, 1964.

————. *The Eighteenth Brumaire of Louis Bonaparte*. New York: International Publishers, 1963.

————. *The Grundrisse*. Ed. Martin Niklaus. Harmondsworth, Eng.: Penguin, 1973.

————. *Precapitalist Economic Formations*. Ed. E. J. Hobsbawm. London: Lawrence and Wisehart, 1964.

Maverick, Lewis A. *China: A Model for Europe*. San Antonio, Tex.: Paul Anderson, 1946.

Maxwell, James Clerk. "Analogies, February 1856: Are There Real Analogies in Nature?" Chap. 8, "Essays at Cambridge." In *The Life of James Clerk Maxwell*, by L. Campbell and W. Garnett. 1882. Reprint. New York: Johnson, 1969.

————. *Collected Papers*, 1890. Reprint. New York: Dover, 1965.

————. *Matter and Motion*. 1877. Reprint. London: Sheldon Press, 1920; New York: Dover [1951].

————. *A Treatise on Electricity and Magnetism*. 3rd ed. 2 vols. 1891. Reprint. New York: Dover, 1954.

Meinecke, Friedrich. *The Age of German Liberation, 1795–1815*. Trans. Peter Paret and Helmuth Fischer. Berkeley: University of California Press, 1977.

Mencius. *Mencius*. Trans. D. C. Lau. London: Penguin, 1970.

Merchant, Carolyn. *The Death of Nature: Women, Ecology and the Scientific Revolution*. San Francisco: Harper and Row, 1980.

————. "The Vitalism of Frances Mercury von Helmont: Its Influence on Leibniz." *Ambix* 26 (1979): 170–183.

Merkel, Ingrid, and Allen G. Debus. *Hermeticism and the Renaissance: Intellectual History and the Occult in Early Modern Europe*. Washington, D.C.: Folger Books, 1988.

Merton, Robert K. *Science, Technology, and Society in Seventeenth-Century England*. New York: Harper and Row, 1970.

————. *Social Theory and Social Structure*. Rev. ed. Glencoe, Ill.: Free Press, 1957.

Meyer, R. W. *Leibnitz and the Seventeenth Century Revolution*. Trans. J. P. Stern. Cambridge, Eng.: Bowes and Bowes, 1952.

Meyerson, Emile. *Explanation in the Sciences*. Trans. Mary-Alice Sipfle and David Sipfle. Boston Studies in the Philosophy of Science, vol. 128. Boston: Kluwer, 1991.

————. *Identity and Reality*. Trans. Kate Loewenberg. 1930. Reprint. New York: Dover, 1962.

————. *The Relativistic Deduction: Epistemological Implications of the Theory of Relativity*. Trans. Mary-Alice Sipfle and David Sipfle. Boston Studies in the Philosophy of Science, vol. 83, Dordrecht: D. Reidel, 1985.

Miller, Arthur I. *Imagery in Scientific Thought: Creating Twentieth-Century Physics*. 2nd ed. Cambridge: MIT Press, 1986.

Mills, C. Wright. *Power, Politics and People*. New York: Ballantine, 1963.

Misner, Charles W., Kip S. Thorne, and John Archibald Wheeler. *Gravitation*. San Francisco: W. H. Freeman, 1973.

Moss, Jean Dietz. *Novelties in the Heavens: Rhetoric and Science in the Copernican Controversy*. Chicago: University of Chicago Press, 1993.

Mungello, David E. *Leibniz and Confucianism: The Search for Accord*. Honolulu: Hawaii: University of Hawaii Press, 1977.

————. "Malebranche and Chinese Philosophy." *Journal of the History of Ideas* 41 (1980): 551–578.

Murray, Margaret A. *The Witch-Cult in Western Europe*. 1921. Reprint. New York: Barnes and Noble, 1996.

Murtfeld, Rudolf. *Caroline Schlegel-Schelling: Moderne Frau in revolutionärer Zeit*. Bonn: Bouvier Verlag Herbert Grundman, 1973.

Nakayama, Shigeru. *Academic and Scientific Traditions in China, Japan, and the West*. Trans. John Dusenbury. Tokyo: University of Tokyo Press, 1984.

Nauert, Charles G., Jr. *Agrippa and the Crisis of Renaissance Thought*. Urbana: University of Illinois Press, 1965.

Needham, Joseph. *Clerks and Craftsmen in China and the West*. Cambridge: Cambridge University Press, 1970.

————. *The Grand Titration*. Toronto: University of Toronto Press, 1969.

————. *Science and Civilization in China*. 7 vols. Cambridge: Cambridge University Press, 1954–.

————. *Science in Traditional China.* Cambridge: Harvard University Press, 1981.

Nersessian, Nancy. *Faraday to Einstein: Constructing Meaning in Scientific Theories.* The Hague: Martinus Nijhoff, 1984.

Newton, Isaac. *Isaac Newton: Papers and Letters on Natural Philosophy.* New York: Hafner, 1953.

————. *Mathematical Principles of Natural Philosophy and His System of the World.* Trans. Andrew Motte; rev., ed. Florian Cajori. Berkeley: University of California Press, 1966.

————. *Opticks.* With foreword by Albert Einstein; introduction by Edmund Whittaker; preface by I. Bernard Cohen. New York: Dover, 1952.

Nicolson, Marjorie Hope. *The Breaking of the Circle: Studies in the Effect of the "New Science" on Seventeenth Century Poetry.* New York: Columbia University Press, 1960.

Nielsen, Keld. "Another Kind of Light: The Work of T. J. Seebeck and His Collaboration with Goethe." *Historical Studies in the Physical and Biological Sciences* 20 (1989): 107–182; 21 (1990): 317–397.

Noble, David F. *A World without Women: The Christian Clerical Culture of Western Science.* New York: Knopf, 1992.

Oersted, Hans Christian. *The Soul in Nature.* Trans. Leonora Horner and Joanna B. Horner. Reprint. London: Dawsons of Pall Mall, 1966.

Olson, Richard. "The Reception of Boscovich's Ideas in Scotland." *Isis* 60 (1969): 91–103.

————. *Scottish Philosophy and British Physics.* Princeton: Princeton University Press, 1975.

O'Meara, Dominic, J. *Pythagoras Revived: Mathematics and Philosophy in Late Antiquity.* Oxford: Clarendon, 1989.

Orsini, Gian N. G. *Coleridge and German Idealism.* Carbondale: Southern Illinois University Press, 1969.

Pachter, Henry M. *Magic into Science: The Story of Paracelsus.* New York: Henry Schuman, 1951.

Pagel, Walter. *Joan Baptista van Helmont.* Cambridge: Cambridge University Press, 1982.

————. *Paracelsus: An Introduction to Philosophical Medicine in the Era of the Renaissance.* 2nd ed. Basel: Karger, 1982.

————. *The Smiling Spleen: Paracelsianism in Storm and Stress.* Basel: Karger, 1984.

Pagels, Elaine. *The Gnostic Gospels.* Reprint. New York: Vintage, 1989.

Palter, Robert M. "The Earliest Measurements of Magnetic Force." *Isis* 63 (1972): 544–558.

————. "Philosophic Principles and Scientific Theory." *Philosophy of Science* 23 (1956): 111–135.

————. "Saving Newton's Text: Documents, Readers, and the Ways of the World." *Studies in the History and Philosophy of Science* 18 (1987): 385–439.

————, ed. *The "Annus Mirabilis" of Sir Isaac Newton.* Cambridge: MIT Press, 1970.

Paracelsus, Theophrastus Bombastus. *The Hermetic and Alchemical Writings of Paracelsus the Great.* Ed., trans. Arthur Edward Waite. 2 vols. in 1. 1894. Reprint. Edmonds, Wash.: Alchemical Press, 1992.

————. *Paracelsus: Essential Writings.* Sel., trans. Nicholas Goodrick-Clarke. Wellingborough, Eng.: Aquarian Press, 1990.

————. *Paracelsus: Selected Writings.* Ed. Jolande Jacobi. 1952. Reprint. Princeton, N.J.: Princeton University Press, 1988.

————. *Vom Licht der Natur und des Geistes.* Ed. Kurt Goldhammer. Stuttgart: Philip Reclam Jun., 1979.

Parry, R. H., ed. *The English Civil War and After, 1642–1658.* Berkeley: University of California Press, 1970.

Patai, Raphael. *The Jewish Alchemists: A History and Source Book*. Princeton, N.J.: Princeton University Press, 1994.

Pauli, Wolfgang. *Essays: Writings on Physics and Philosophy*. Ed. Charles P. Enz and Karl von Meyenn; trans. Robert Schlapp. Berlin: Springer Verlag, 1994.

Peerenboom, R. P. *Law and Morality in Ancient China: The Silk Manuscripts of Huang-Lo*. Albany: State University of New York Press, 1993.

Peirce, Charles S. *Collected Papers*. 6 vols. Ed. Charles Hartshorne and Paul Weiss. Cambridge, Mass.: Harvard University Press, Belknap Press, 1931–1935.

Pinker, Steve. *The Language Instinct: How the Mind Creates Language*. New York: William Morrow, 1994.

Poggi, Stefano, and Maurizio Bossi. *Romanticism in Science: Science in Europe, 1790–1840*. Dordrecht: Kluwer, 1994.

Poincaré, Henri. *Science and Hypothesis*. Trans. W. J. Greenstreet. 1905. Reprint. New York: Dover, 1952.

———. *Science and Method*. Trans. Francis Maitland; preface by Bertrand Russell. 1914. Reprint. New York: Dover, n.d.

———. *The Value of Science*. Trans. G. Bruce Halstead. 1907. Reprint. New York: Dover, n.d.

Polanyi, Michael. *Personal Knowledge: Towards a Post-Critical Philosophy*. Chicago: University of Chicago Press, 1958.

Popper, Karl. *The Logic of Scientific Discovery*. [1934]. Reprint. New York: Basic Books, 1959.

———. *Postscript to the Logic of Scientific Discovery*. Vol. 3, *Quantum Theory and the Schism in Physics*. Totowa: N.J.: Rowan and Littlefield, 1982.

Porta, Giambattista della. *Natural Magick*. 1558; trans. 1658. Reprint. New York: Basic Books, 1957.

Power, Henry. *Experimental Philosophy: In Three Books Containing New Experiments, Microscopical, Mercurial. Magnetical*. 1664. Reprint. with intro. Marie Boas Hall. New York: Johnson, 1966.

Pumfrey, Stephen. "Mechanizing Magnetism in Restoration England: The Decline of Magnetic Philosophy." *Annals of Science* 44 (1987): 1–22.

———. "O Tempora, O Magnes! A Sociological Analysis of the Discovery of Secular Magnetic Variation in 1634." *British Journal of the History of Science* 22 (1989): 181–214.

———. "William Gilbert's Magnetic Philosophy, 1580–1684: The Creation and Dissolution of a Discipline." Ph.D. diss., University of London. 1987.

Randall, John Herman. *The Career of Philosophy*. 2 vols. New York: Columbia University Press, 1962, 1965.

Rattansi, Piyo M. "The Intellectual Origins of the Royal Society." *Notes and Records of the Royal Society of London* 23 (1968): 129–143.

———. "Paracelsus and the Puritan Revolution." *Ambix* 11 (1963): 24–32.

Ravetz, Jerome. *Scientific Knowledge and Its Social Problems*. Oxford: Oxford University Press, 1971.

Reiss, H. S., ed. *The Political Thought of the German Romantics, 1793–1815*. Oxford, U.K.: Basil Blackwell, 1955.

Revel, Jacques, and Lynn Hunt, eds. *Histories: French Constructions of the Past*. Trans. Arthur Goldhammer. New York: New Press, 1995.

Ricci, Matteo. *China in the Sixteenth Century: The Journals of Matthew Ricci, 1583–1610*. Trans. Louis J. Gallagher; ed. Nicholas Trigault. New York: Random House, 1953.

Ricochet, W. Allyn, ed. and trans. *Guanzi: Political. Economic and Philosophical Essays from Early China*. 3 vols. Princeton: Princeton University Press, 1985–.

Ringer, Fritz. *The Decline of the German Mandarins*. 1969. Reprint. Hanover, N.H.: University Press of New England, 1990.

Ritchie, Gisella Fuhrmann. *Caroline Schlegel Schelling in Wahrheit und Dichtung*. Bonn: Bouvier, 1968.

Ritter, Johann Wilhelm. *Die Begründung der Elektrochemie*. Ed. Armin Hermann. Ostwald's Klassiker der Exakten Wissenschaft, new series, vol. 2. Frankfurt am Main: Akademische Gesellschaft, 1968.

———. *Entdeckungen zur Elektrochemie. Bioelektrochemie und Photochemie*. Ed., intro. Hermann Berg and Klaus Richter. Ostwald's Klassiker der Exakten Wissenschaft, vol. 272. Leipzig: Akademische Gesellschaft Geest und Portig, 1986.

Robinet, Isabelle. *Taoist Meditation*. Trans. Julian F. Pas and Norman G. Girardot. Albany: State University of New York Press, 1993.

Ronan, Colin, ed. *The Shorter Science and Civilization in China*. 5 vols. Cambridge: Cambridge University Press, 1978–.

Rosemont, Henry, Jr. "Gathering Evidence for Linguistic Innateness." *Synthese* 38 (1978): 127–148.

———. "On Representing Abstractions in Archaic Chinese." *Philosophy East and West* 24 (1974): 71–88.

———, ed. *Chinese Texts and Philosophical Contexts*. La Salle, Ill.: Open Court, 1991.

Rosen, Edward. *Three Imperial Mathematicians: Kepler Trapped between Tycho Brahe and Ursus*. New York: Abaris, 1986.

Rossi, Paolo. *Francis Bacon: From Magic to Science*. London: Routledge and Kegan Paul, 1968.

———. *Philosophy, Technology and the Arts in the Early Modern Era*. New York: Harper and Row, 1970.

———. "The Twisted Roots of Leibniz' Characteristic." In *The Leibniz Renaissance: International Workshop, Firenze. 2–5 giugno 1986*. Florence: Leo S. Olschki Editore, 1989.

Russell, Bertrand. *A Critical Exposition of the Philosophy of Leibniz*. London: George Allen and Unwin, 1900.

———. *An Essay on the Foundations of Geometry*. 1897. Reprint with new foreword by Morris Kline. New York: Dover, 1956.

———. *A History of Western Philosophy and Its Connection with Political and Social Circumstances from Earliest Times to the Present Day*. New York: Simon and Schuster, 1945.

———. *Logic and Knowledge*, ed. R. C. Marsh. 1956. Reprint. New York: Capricorn Books, 1971.

Ryle, Gilbert. *The Concept of Mind*. 1949. Reprint. New York: Barnes and Noble, 1968.

———. *Dilemmas*. Cambridge: Cambridge University Press, 1954.

Sachs, Mendel. "Maimonides, Spinoza, and the Field Concept in Physics." *Journal of the History of Ideas* 37 (1976): 125–131.

Sahlins, Marshall. *Culture and Practical Reason*. Chicago: University of Chicago Press, 1976.

Saine, Thomas P. *Georg Forster*. Boston: Twayne, 1972.

Sambursky, Shmuel. *Physics of the Stoics*. 1959. Reprint. London: Hutchinson, 1971.

Sapir, Edward. *Language: An Introduction to the Study of Speech*. New York: Harcourt, Brace, 1949.

Saunders, Simon, and Harvey R. Brown, eds. *Philosophy of the Vacuum*. Oxford: Oxford University Press, 1991.

Sayre, Robert. "The Young Coleridge: Romantic Utopianism and the French Revolution." *Studies in Romanticism* 28 (1989): 397–415.

Schafer, Edward H. *Ancient China*. New York: Time-Life Books, 1967.

———. *The Golden Peaches of Samarkand*. Berkeley: University of California Press, 1962.

———. *Pacing the Void: T'ang Approaches to the Stars*. Berkeley: University of California Press, 1977.

Schaffer, Simon. "Self Evidence." In *Questions of Evidence: Proof, Practice and Persuasion across Disciplines*, by James Chandler, Arnold I. Davidson, and Harry Harootunian. Chicago: University of Chicago Press, 1994.

Schelling, Friedrich Wilhelm Joseph von. *The Ages of the World*. Trans. Frederick de Wolfe Boman, Jr. New York: Columbia University Press, 1942.

———. *Bruno; or, On the Divine Principle of Things*. Trans. Michael Vater. Albany: State University of New York Press, 1984.

———. *Idealism and the Endgame of Theory: Three Essays by F.W.J. Schelling*. Trans., ed. Thomas Pfau. Albany: State University of New York Press, 1994.

———. *Ideas for a Philosophy of Nature as Introduction to the Study of This Science, 1797*. Trans. Errol E. Harris and Peter Heath. Cambridge: Cambridge University Press, 1988.

———. *On University Studies*. Trans. E. S. Morgan; ed. Norbert Guterman. Athens: Ohio University Press, 1966.

———. *Schelling on Human Freedom*. Trans. James Gutman. Chicago: Open Court, 1936.

———. *System of Transcendental Idealism (1800)*. Trans. Peter Heath. Charlottesville: University Press of Virginia, 1978.

———. *The Unconditional in Human Knowledge: Four Early Essays, 1794–1796*. Trans. Fritz Marti. Lewisburg, Pa.: Bucknell University Press, 1980.

———. ed. *Zeitschrift fuer Spekulative Physik*. 1800. Reprint. Hildesheim: Georg Olms Verlagsbuchhandlung, 1969.

———. [Caroline Schlegel-Schelling?] *Die Nachtwachen des Bonaventura*. Ed., trans. Gerald Gillespie. Austin: University of Texas Press, 1971.

Schilpp, Paul A., ed. *Albert Einstein: Philosopher Scientist*. 1949. Reprint in 2 vols. New York: Harper and Row, 1959.

Schipper. Kristofer. *The Taoist Body*. Trans. Karen C. Duval; foreword by Norman Girardot. Berkeley: University of California Press, 1993.

Schlegel-Schelling, Caroline. *Caroline: Briefe aus der Frühromantik*. Ed. Georg Waitz; aug. Erich Schmidt. 2 vols. 1913. Reprint. Bern: Herbert Lang, 1970.

Schrecker, Ellen W. *No Ivory Tower: McCarthyism and the Universities*. Oxford: Oxford University Press, 1986.

Schrickx, W. "Coleridge and the Cambridge Platonists." *Review of English Literature* 7 (1966): 71–91.

Schrödinger, Erwin. *Nature and the Greeks*. 1954. Reprinted in *Nature and the Greeks and Science and Humanism*, foreword by Roger Penrose. Cambridge: Cambridge University Press, 1996.

Schwab, Raymond. *The Oriental Renaissance*. Trans. Gene Patterson-Black and Victor Reinking; foreword by Edward W. Said. New York: Columbia University Press, 1984.

Schwartz, Benjamin. *China and Other Matters*. Cambridge: Harvard University Press, 1996.

———. *The World of Thought in Ancient China*. Cambridge: Harvard University Press, 1985.

Schweber, Silvan S. *QED and the Men Who Made It: Dyson, Feynman, Schwinger and Tomonaga*. Princeton: Princeton University Press, 1994.

Sedlar, Jean. *India in the Mind of Germany*. Washington, D.C.: University Press of America, 1982.

Seidel, Anna. "The Image of the Perfect Ruler in Early Taoist Messianism: Lao Tzu and Li Hung." *History of Religions* 9 (1969–1970): 161–180.

————. "Taoist Messianism." *Numen* 31 (1984): 161–174.

————, and Marcel Strickmann. "Taoism," and "Taoist Literature." In *Encyclopaedia Britannica*, 15th ed. Chicago: Encyclopaedia Britannica, 1992.

Sepper, Dennis L. *Goethe contra Newton: Polemics and the Project for a New Science of Color*. Cambridge: Cambridge University Press, 1988.

Shaffer, Elinor S. "Coleridge and Natural Philosophy: A Review of Recent Literary and Scientific Research." *History of Science* 12 (1974): 284–295.

Shanahan, Timothy. "Kant's *Naturphilosophie* and Oersted's Discovery of Electromagnetism: A Reassessment." *Studies in the History and Philosophy of Science* 20 (1989): 287–305.

Shchutskii, Iulien K. *Researches on the "I Ching."* Trans. William L. MacDonald, Tsuyoshi Hasegawa, and Hellmut Wilhelm. Princeton: Princeton University Press, 1979.

Shea, William R. *The Magic of Numbers and Motion: The Scientific Career of René Descartes*. Canton, Mass.: Science History Publications, 1991.

Siegel, Daniel M. *Innovation in Maxwell's Electromagnetic Theory: Molecular Vortices, Displacement Current, and Light*. Cambridge: Cambridge University Press, 1991.

Sivin, Nathan. *Chinese Alchemy: Preliminary Studies*. Cambridge. Mass.: Harvard University Press, 1968.

————. "Discovery of Spagyrical Invention." *Harvard Journal of Asiatic Studies* 41 (1981): 219–235.

————. *Medicine. Philosophy and Religion in Ancient China*. Aldershot, Eng.: Ashgate Variorum, 1995.

————. "On the Word 'Taoist' as a Source of Perplexity, with Special Reference to the Relations of Science and Religion in Traditional China." *History of Religions* 17 (1978): 303–329.

————. "Ruminations on the Tao and Its Disputers." *Philosophy East and West* 42 (1992): 21–29.

————. *Science in Ancient China*. Aldershot. Eng.: Ashgate Variorum, 1995.

————. "Shen Kua." In *Dictionary of Scientific Biography*. 18 vols. Ed. Charles Coulston Gillispie. New York: Scribner, 1970–1990.

————. *Traditional Medicine in Contemporary China: A Partial Translation of "Revised Outline of Chinese Medicine (1972)" with an Introductory Study on Change in Present Day and Early Medicine*. Ann Arbor: University of Michigan Center for Chinese Studies, 1987.

Smith, Richard J. *Fortune-Tellers and Philosophers: Divination in Traditional Chinese Society*. Boulder, Colo.: Westview, 1991.

Smuts, Jan Christian. *Holism and Evolution*. New York: Viking, 1961.

Snelders, H.A.M. "Romanticism and *Naturephilosophie* in the Inorganic Natural Sciences, 1797–1840: An Introductory Survey." *Studies in Romanticism* 9 (1970): 193–215.

Soble, Alan. "In Defense of Bacon." *Philosophy of the Social Sciences* 25 (1995): 192–215.

Soothill, William. *The Hall of Light: A Study of Early Chinese Kingship*. New York: Philosophical Library, 1952.

Spence, Jonathan. *Emperor of China: Self-Portrait of K'ang-hsi*. New York: Vintage, 1975.

————. *The Memory Palace of Matteo Ricci*. New York: Viking Penguin, 1984.

Stauffer, Robert C. "Persistent Errors Regarding Oersted's Discovery of Electromagnetism." *Isis* 44 (1953): 307–310.

―――. "Speculation and Experiment in the Background of Oersted's Discovery of Electromagnetism." *Isis* 48 (1957): 33–50.

Stephenson, Bruce. *Kepler's Physical Astronomy*. 1987. Reprint. Princeton, N.J.: Princeton University Press, 1994.

―――. *The Music of the Heavens: Kepler's Harmonic Astronomy*. Princeton, N.J.: Princeton University Press, 1994.

Steuwer, Roger H., ed. *Minnesota Studies in the Philosophy of Science*. Vol. 5, *Historical and Philosophical Perspectives of Science*. 1970. Reprint. New York: Gordon and Breach, 1989.

Stiefel, E. L., and G. Scheifele. *Linear and Regular Celestial Mechanics*. New York: Springer Verlag, 1971.

Strawson, P. F. *Introduction to Logical Theory*. London: Methuen, 1952.

Strickland, Stuart. "Circumscribing Science: Johann Wilhelm Ritter and the Physics of Sidereal Man." Ph.D. diss. Harvard University, 1992.

―――. "Galvanic Disciplines: The Boundaries, Objects, and Identities of Experimental Science in the Era of Romanticism." *History of Science* 33 (1995): 450–467.

Strickmann, Marcel. "History, Anthropology and Chinese Religion." *Harvard Journal of Asiatic Studies* 40 (1981): 201–248.

―――. "The Longest Taoist Scripture." *History of Religions* 17 (1978): 331–353.

―――. "The Mao-shan Revelations: Taoism and the Aristocracy." *T'oung Pao* 63 (1978): 1–64.

Strong, Edward W. *Procedures and Metaphysics: A Study in the Philosophy of Mathematical-Physical Science in the Sixteenth and Seventeenth Centuries*. 1936. Reprint. Hildesheim: Georg Olms Verlagsbuchhandlung, 1966.

Tambiah, Stanley Jeyaraja. *Magic, Science, Religion, and the Scope of Rationality*. New York: Cambridge University Press, 1990.

Taton, René, and Curtis Wilson, eds. *Planetary Astronomy from the Renaissance to the Rise of Astrophysics: Tycho Brahe to Newton*. Vol. 2, pt. A of *The General History of Astronomy*, ed. Michael Hoskin. Cambridge: Cambridge University Press, 1989.

Taub, Liba Chaia. *Ptolemy's Universe: The Natural Philosophical and Ethical Foundations of Ptolemy's Philosophy*. La Salle, Ill.: Open Court, 1993.

Teich, Mikuláš, and Robert Young, eds. *Changing Perspectives in the History of Science*. London: Heineman Educational Books, 1973.

Temple, Robert. *The Genius of China*. New York: Simon and Schuster, 1986.

Thomas, John Meurig. *Michael Faraday and the Royal Institution: The Genius of the Man and the Place*. Bristol: Adam Hilger, 1991.

Thomas, Keith. *Religion and the Decline of Magic*. New York: Charles Scribner's Sons, 1971.

Thompson, Laurence G. "Taoism: Classic and Canon." In *The Holy Book in Comparative Perspective*, by Frederick M. Denny and Rodney L. Taylor. Columbia: University of South Carolina Press, 1985.

Thoren, Victor E. *The Lord of Uraniborg: A Biography of Tycho Brahe*. Cambridge: Cambridge University Press, 1991.

Thorndike, Lynn. *A History of Magic and Experimental Science*. Vols. 1–8. New York: Macmillan, 1923.

Toulmin, Stephen. *Cosmopolis: The Hidden Agenda of Modernity*. New York: Free Press, 1990.

Treneer, Anne. *The Mercurial Chemist: A Life of Sir Humphry Davy*. London: Methuen, 1963.

Trevor-Roper, H. R. *The European Witch-Craze of the Sixteenth and Seventeenth Centuries and Other Essays*. New York: Harper and Row, 1967.

————. *Renaissance Essays*. Chicago: University of Chicago Press, 1985.

Truesdell, Clifford. *Essays on the History of Rational Mechanics*. New York: Springer Verlag, 1968.

Trumpler, Maria. "Questioning Nature: Experimental Investigation of Animal Electricity in Germany, 1791–1810." Ph.D. diss. Yale University, 1992.

Twitchett, Denis, and Michael Loewe, eds. *The Cambridge History of China*. Vol. 1, *The Ch'in and Han Empires, 221 B.C.–A.D. 220*. Cambridge: Cambridge University Press, 1986.

Ulman, Gregory. *Applied Grammatology: Post(e) Pedagogy from Jacques Derrida to Jacob Beuys*. Baltimore: Johns Hopkins University Press, 1985.

Urbach, Peter. *Francis Bacon's Philosophy of Science: An Account and a Reappraisal*. La Salle, Ill.: Open Court, 1987.

Vernant, Jean-Pierre. *Myth and Society in Ancient Greece*. London: Methuen, 1982..

————. *The Origins of Greek Thought*. Ithaca: Cornell University Press, 1982

Verschuur, Gerrit L. *Hidden Attraction: The Mystery and History of Magnetism*. New York: Oxford University Press, 1993.

Vickers, Brian. "Frances Yates and the Writing of History." *Journal of Modern History* 51 (1979): 287–316.

————. *Occult and Scientific Mentalities in the Renaissance*. Cambridge: Cambridge University Press, 1984.

Wagner, Rudolph G. "Interlocking Parallel Style: Laozi and Wang Bi." *Asiatische Studien/ Etudes Asiatiques* 34 (1980): 18–58.

Waithe, Mary Ellen, ed. *A History of Women Philosophers*. Vol. 1, *Ancient Women Philosophers, 600 B.C.–500 A.D.* Dordrecht: D. Reidel, 1987.

Walker, D. P. *Spiritual and Demonic Magic: From Ficino to Campanella*. Notre Dame, Ind.: University of Notre Dame Press, 1975.

Wang Bi. "Commentary on the *Book of Changes*." In *The Classic of Changes: A New Translation of the "I Ching" as Interpreted by Wang Bi*, trans. Richard Lynn John. New York: Columbia University Press, 1994.

————. [Wang Pi.] *Commentary on the "Lao Tzu."* Trans. by Ariane Rump. Honolulu: University of Hawaii Press, 1979.

Wang Ch'ung. *"Lun-Heng": Philosophical Essays of Wang Chhung*. 2 vols. Ed. A. Forke. 1907–1911. Reprint. New York: Paragon, 1962.

Ware, James R., trans. *Alchemy, Medicine, Religion in the China of A.D. 320: The "Nei P'ien" of Ko Hung ("Pao-p'u tzu")*. Cambridge: MIT Press, 1966.

Wartofsky, Marx. "Is Science Rational?" In *Science, Technology, and Freedom*, ed. Willis H. Truitt and T. W. Graham Solomons. Boston: Houghton Mifflin, 1974.

————. "Metaphysics as Heuristic for Science." *Boston Studies in the Philosophy of Science*. Vol. 3. Ed. Robert S. Cohen and Marx Wartofsky. Dordrecht: D. Reidel, 1967.

————. *Models: Representation and the Scientific Understanding*. Boston Studies in the Philosophy of Science, vol. 48. Dordrecht: D. Reidel, 1979.

Weber, Max. *The City*. Trans. Don Martindale and Gertrud Neuwirth. New York: Collier, 1962.

————. *The Religion of China*. New York: Free Press, 1951.

Webster, Charles, ed. *From Paracelsus to Newton: Magic and the Making of Modern Science*. Cambridge: Cambridge University Press, 1982.

―――. *The Great Instauration*. London: Duckworth, 1975.

―――. *The Intellectual Revolution of the Seventeenth Century*. London: Routledge, 1974.

Wechsler, Judith, ed. *On Aesthetics in Science*. Cambridge: MIT Press, 1978.

Wedgewood, Cicely Veronica. *The King's Peace*. New York: Book-of-the-Month Club, 1995.

―――. *The King's War*. New York: Book-of-the-Month Club, 1995.

―――. *The Thirty Years War*. New York: Book-of-the-Month Club, 1995.

Weiland, Werner. *Der junge Friedrich Schlegel: Oder die Revolution in der Frühromantik*. Stuttgart: W. Kohlhammer Verlag, 1968.

Weinberg, Steven. *Dreams of a Final Theory*. New York: Vintage, 1994.

―――. "Sokal's Hoax." *New York Review of Books* 43, no. 13 (1996): 11–15.

―――. "Sokal's Hoax: An Exchange." *New York Review of Books* 43, no. 15 (1996): 55–56.

Welch, Holmes. "The Bellagio Conference on Taoist Studies." *History of Religions* 9 (1969–February 1970): 107–136.

―――, and Anna Seidel, eds. *Facets of Taoism: Essays in Chinese Religion*. New Haven: Yale University Press, 1979.

Wellek. René. *Immanuel Kant in England, 1793–1838*. Princeton, N.J.: Princeton University Press, 1931.

Werskey, Gary. *The Visible College*. New York: Holt, Rinehart and Winston, 1978.

Westfall, Richard S. *Never at Rest: A Biography of Isaac Newton*. Cambridge: Cambridge University Press, 1980.

Westman, Robert S., ed. *The Copernican Achievement*. Berkeley: University of California Press, 1975.

―――, and John E. McGuire, eds. *Hermeticism and the Scientific Revolution: Papers Read at a Clark Library Seminar. March 9, 1974*. Los Angeles: Clark Memorial Library and University of California Press, 1977.

Wetzels, Walter D. "Aspects of Science in German Romanticism." *Studies in Romanticism* 10 (1971): 44–59.

Weyer, Johann. *Witches, Devils, and Doctors in the Renaissance: Johann Weyer, de Praestigiis Daemonum*. Binghamton, N.Y.: Medieval and Renaissance Texts and Studies, 1991.

Weyl, Hermann. *Philosophy of Mathematics and Natural Science*. Trans. Olaf Helmer; rev., and aug. English ed. Princeton: Princeton University Press, 1949.

Wheatley, Paul. *The Pivot of the Four Quarters: A Preliminary Inquiry into the Origins and Character of the Ancient Chinese City*. Chicago: Aldine, 1971.

Wheelwright, Philip, ed. *The Presocratics*. New York: Macmillan, 1966

White, Lynn, Jr. *Medieval Technology and Social Change*. Oxford: Oxford University Press, 1966.

Whorf, Benjamin Lee. *Language, Thought, and Reality*. 1956. Reprint. Cambridge, Mass.: MIT Press, 1964.

Whyte, Lancelot Law. *The Unconscious before Freud*. New York: Doubleday Anchor, 1960.

―――, ed. *Roger Joseph Boscovich*. London: George Allen and Unwin, 1961.

Widmaier, Rita. *Die Rolle der chinesischen Schrift in Leibniz' Zeichentheorie. Studia Leibnitiana Supplementa*. Vol. 24. Wiesbaden: Franz Steiner Verlag, 1983.

Wiener, Philip P., and Aaron Noland, eds. *The Roots of Scientific Thought: A Cultural Perspective*. New York: Basic Books, 1957.

Wilhelm, Hellmut. *Change: Eight Lectures on the "I Ching."* Princeton, N.J.: Princeton University Press, 1973.

―――. *Heaven, Earth, and Man in the "Book of Changes."* Seattle: University of Washington Press, 1977.

Wilhelm, Richard. *Lectures on the "I Ching": Constancy and Change*. Trans. Irene Eber. Princeton: Princeton University Press, 1979.

Wilson, Curtis. "Kepler's Derivation of the Elliptical Path." *Isis* 59 (1968): 1–25.

Wilson, Katharina M., ed. *An Encyclopedia of Continental Women Writers*. New York: Garland, 1991.

Wilson, Margaret. *Leibniz's Metaphysics: A Historical and Comparative Study*. Princeton, N.J.: Princeton University Press, 1989.

Williams, L. Pearce. *Michael Faraday*. New York: Basic Books, 1965.

———. *The Origins of Field Theory*. Washington, D.C.: University Press of America, 1980.

Williams, Selma R., and Pamela Williams Adelman. *Riding the Nightmare: Women and Witchcraft from the Old World to Colonial Salem*. New York: HarperCollins, 1978.

Wing-Tsit Chan, ed. *Chu Hsi and Neo-Confucicanism*. Honolulu: University of Hawaii Press, 1986.

———. *Chu Hsi: Life and Thought*. Hong Kong: Chinese University Press, 1987.

———. *Chu Hsi: New Studies*. Honolulu: University of Hawaii Press, 1989.

———, ed. *A Source Book in Chinese Philosophy*. Princeton, N.J.: Princeton University Press, 1963.

Wise, B. Norton. "The Maxwell Literature and British Dynamical Theory." *Historical Studies in the Physical Sciences* 13 (1982): 175–205.

———. "Pascual Jordan: Quantum Mechanics, Psychology, National Socialism." In *Science, Technology and National Socialism*, ed. Monika Renneberg and Mark Walker. Cambridge: Cambridge University Press, 1994.

Wittfogel, Karl August. *Oriental Despotism*. New York: Random House, 1981.

Woodward, William, and Robert S. Cohen, eds. *World Views and Scientific Discipline Formation: Science Studies in the German Democratic Republic*. Boston Studies for the Philosophy of Science, vol. 134. Dordrecht: Kluwer, 1991.

Yates, Frances A. *The Art of Memory*. London: Routledge, 1966.

———. *Giordano Bruno and the Hermetic Tradition*. New York: Random House, 1964.

———. "The Hermetic Tradition in Renaissance Science." In *Art, Science and History in the Renaissance*, ed. Charles Singleton. Baltimore: Johns Hopkins University Press, 1968.

———. *The Occult Philosophy in the Elizabethan Age*. London: Routledge and Kegan Paul, 1979.

———. *The Rosicrucian Enlightenment*. Boulder, Colo.: Shambhala, 1972.

———. *The Theater of the World*. Chicago: University of Chicago Press, 1969.

Yovel, Yirmiyahu. *Spinoza and Other Heretics*. Vol. 1, *The Marrano of Reason*. Princeton, N.J.: Princeton University Press, 1989.

Zilsel, Edgar. "The Sociological Roots of Science." *American Journal of Sociology* 47 (1941–1942): 544–562.

———. *Die sozialen Ursprünge der neuzeitlichen Wissenschaft*. Ed., trans. Wolfgang Krohn. Frankfurt: Suhrkamp Verlag, 1976.

SUBJECT INDEX

Abdera, Greece, 5, 18
aborigines, Australian, 80
Absolute (F.W.J. Schelling), 213–214, 218, 237
abduction. *See* logic of discovery.
accommodation. *See* accord.
"accord" (of China and Catholicism), 195, 197, 200
action at a distance, 143, 146, 151, 188–189, 269, 76
active powers. *See* active principles.
active principles, 128, 178, 184, 186, 190
affinity, 237, 270
Africa, 16, 108–112, 191,
Afrocentrism, 108, 110–112
alchemy: and Francis Bacon 145, 150; Chinese 62–65; Egyptian 110; European 95, 99; 106–107, of Carl G. Jung 162; moral model 114–117; Isaac Newton and 178–190; romantics 214, 215, 261
allegory, 276
Anabaptists, 100, 209
analogy, 11, 276–281, 343n. 10
ancient wisdom, 107, 111, 131
animism, 124, 141, 142, 144, 145
Annales historians, 6
anthropology, ecological, 73, 74, 76
Annus Mirabilis, 128, 317n. 23
apothecaries. *See* pharmacists.
Arabic science. *See* Islamic science.
archetypal form (Goethe), 219–220
Aristotelianism. *See also* scholasticism. 20,

68, 70, 53, 68–70, 100, 103–104, 127, 158, 186, 196, 246
artisans. *See* also craftspeople. 113, 122–124, 147, 259, 266, 273
Asiatic Society, 9, 73–77, 307n. 9
astral body, 104
astrology: Francis Bacon on 145; and Tycho Brahe 103, 105–106; Johannes Kepler on 151, 154–157; J. W. Ritter 254; Edward Rosen on 322n. 13; and tides 165 astronomy: mathematical versus physical 164, 324n. 64; Johannes Kepler's 152–154, 157–174
atheism 211–212
atomism: Boscovichian 261, 266, 267, 279; and *ch'i* 310n. 91, Daltonian 266–267, Democritus 4, 25, India 77, and individualism 18, 296n. 61, and Indo-European language 91 logical 19; James C. Maxwell 277, 280; and nuclear family 78; psychological 19, 245, 274; Thomas Hobbes 190; René Descartes 222; and Max Weber 223
attraction, versus coition 143–144, as metaphor for political rule 193
auxiliary hypotheses 283, 285

Baconianism 247, 264
Bayesianism 344n. 7
Bell's theorem 22
benevolent despot. *See* enlightened despotism.
Berlin: romantic circle 228; salons, 231

NAME INDEX

Abernathy, John, 249
Abraham, Max, 278
Adelard of Bath, 93
Agassi, Joseph, 287–290
Agrippa, Henry Cornelius, 10, 116, 124, 131, 133, 183
Alberti, Leon Battista, 9, 123
Ames, R. T., 88
Ampère, André-M., 268, 270, 271, 276, 277, 281
Anaxagoras, 4, 23, 261
Anaximander, 72
Anaximenes, 72, 204
Andreä, Johann Valentin, 137–139
Anhalt, Christian, 138
Aquinas, Thomas, 68, 142, 164, 165, 203
Arago, François, 251
Archimedes, 271
Aristotle: Afrocentric claim 111; antiatomism 19; and Giordano Bruno 53; Samuel T. Coleridge versus 246; elements 103, 156; and William Gilbert 143; and manual labor 123; organic model 20; and Neo-Confucian scholasticism 68, 70; Parcelsus and 103, 104; Jesuits in China and196; 201; sexism of 53; universe 163–165
Ashmole, Elias, 215
Aubrey, John, 134
Ault, Donald, 216, 217
Averroës, 165, 167
'Awfi, Muhammad al-, 93

Baader, Franz von, 215, 237, 252, 255, 257
Bacon, Francis: William Blake and 217; and Chinese inventions 29; on ideographs 200; and S. T. Coleridge 247; and Humphry Davy 264; on magnet 144–146; hermeticism and 113; nature as female 10, 129, 185; and Rosicrucianism 138
Bacon, Roger 94–95
Bahro, Rudolf, 74, 76
Bailak al-Qabajaqi, 93
Bakhtin, Mikhail, 7, 132
Bauer, Wolfgang, 66
Beauvoir, Simone de, 6
Beddoes, Thomas, 244, 245, 260, 261
Becquerel, Antoine Cesar, 271
Bell, John S,. 17, 22
Bentham, Jeremy, 222, 246
Bentley, Richard, 133, 191–192
Bergman, Torbern, 222–223
Bergson, Henri, 85
Bernal, John Desmond, 21
Bernal, Martin, 108–111
Bertalanffy, Ludwig von, 21
Berzelius, Jons, 251, 260
Biot, Jean Baptiste, 184
Blake, William, 11, 20, 120, 216–217, 247
Bloomfield, Leonard, 88–89
Boas, Marie, 143, 179
Bodde, Derk, 82
Bodin, Jean, 131, 135
Bohm, David, 17, 22

Schelling, Friedrich (*continued*)
  and Ritter 254–255; and transcendental
  philosophy 113, 213–214
Schiller, Friedrich 232
Schipper, Kristofer, 61
Schlegel, August, 227, 230, 232–233, 238,
  252, 257, 263
Schlegel, Dorothea, 229–233
Schlegel, Friedrich, 226, 228–233, 236,
  246, 252, 256–257, 263
Schlegel-Schelling, Caroline, 11, 226, 228–
  230, 233, 238–239, 254–255
Schleiermacher, Friedrich, 230–231, 257
Schlick, Morris, 121
Schopenhauer, Arthur, 232, 252
Schopenhauer, Johanna, 230, 232
Schrecker, John, 48
Schrödinger, Erwin, 54, 264
Schwartz, Benjamin, 48, 52
Schwinger, Julian, 4
Seebeck, Thomas, 221, 256
Shanahan, Timothy, 262–264
Shao Young, 69–70
Shelley, Mary, 11, 261
Shelley, Percy Bysshe, 11, 226, 246
Shen Kua, 34, 37, 41, 66, 68
Siegel, Daniel, 277, 279
Simplicius, 144
Sivin, Nathan, 49, 55, 64–66
Smith, Adam, 74, 193, 226
Smuts, Jan, 16, 21
Socrates, 40, 88, 110, 246
Soemmerring, Samuel Thomas von, 137,
  215, 255
Sosigenes, 163
Southey, Robert, 245–246, 260–261
Spinoza, Baruch: and Giordano Bruno 105,
  192; and Confucianism 332n. 44; and
  Denis Diderot 105, 212; and Albert
  Einstein 214; and God 211; and
  Wolfgang Goethe 213, 218; and
  Marranos 210–211; and Marxism 105,
  212; and monism 210, 211; and
  pantheism 209–210; Pantheism Dispute
  209–214, 218; and freethinkers 182,
  192; and Mennonites 209–210; and
  Quakers 126, 209; and Friedrich
  Schelling 234–235, 237, 239
Sprat, Thomas, 129, 190

Staël, Madame de, 11, 231, 232
Stallin, Joseph 32, 75
Stauffer, Robert, 262
Steffens, Henrik, 232, 250, 252, 257
Stephenson, Bruce, 175
Stewart, Dugald, 276
Strickmann, Michel, 55–56, 61, 64–65
Strong, Edward, 175
Su Shih, 69
Sylvester, James, 266

Thackray, Arnold, 14
Thales, 123, 143
Theodoric of Freiburg, 94
Theosebia, 110
Thomas, Keith, 133, 135
Thompson, Benjamin, Count Rumford 260
Thomson, Joseph J., 281
Thomson, William, Lord Kevin, 272, 276,
  278–279
Thoth, 107, 109–110, 180
Tieck, Ludwig, 238
Toland, John, 192
Tolman, E. C., 19
Trevor-Roper, H. R., 130, 135
Truesdell, Clifford, 265
Trumpler, Maria, 256, 263
Ts'ao Ts'ao, 61
Tsou Yen, 49–50
Tung Chung-shu, 42
Tycho Brahe, 103–105, 138, 157, 162, 166,
  172–173
Tyndall, John, 272

Urbach, Peter, 146, 247

Van den Ende, Francis, 211
Veit, Dorothea. *See* Schlegel, Dorothea;
  Mendelssohn, Dorothea
Verschuur, Gerrit L., 141
Vickers, Brian, 138–139
Vinci, Leonardo da, 9, 123
Vogt, Karl, 252
Volta, Alessandro, 254–257
Voltaire, 74, 212, 217, 252

Wackenfels, Wacker von, 158
Waddington, C. H., 21, 120
Wagstaffe, John, 134

ABOUT THE AUTHOR

Val Dusek is an associate professor of philosophy at the University of New Hampshire. He studied chemistry, zoology, mathematics, and philosophy at Yale and the University of Texas at Austin. His field of scholarship is the philosophy, history, and social theory of science. He is a former member of the Sociobiology Study Group of Science for the People and has written articles on the sociobiology debate and the science wars.

1-MONTH